European Consortium for
Mathematics in Industry 7

Heiliö (Ed.)

Proceedings of the Fifth European Conference
on Mathematics in Industry

European Consortium for Mathematics in Industry

Edited by
Michiel Hazewinkel, Amsterdam
Helmut Neunzert, Kaiserslautern
Alan Tayler, Oxford
Hansjörg Wacker, Linz

ECMI Vol. 7

Within Europe a number of academic groups have accepted their responsibility towards European industry and have proposed to found a European Consortium for Mathematics in Industry (ECMI) as an expression of this responsibility.

One of the activities of ECMI is the publication of books, which reflect its general philosophy; the texts of the series will help in promoting the use of mathematics in industry and in educating mathematicians for industry. They will consider different fields of applications, present casestudies, introduce new mathematical concepts in their relation to practical applications. They shall also represent the variety of the European mathematical traditions, for example practical asymptotics and differential equations in Britain, sophisticated numerical analysis from France, powerful computation in Germany, novel discrete mathematics in Holland, elegant real analysis from Italy. They will demonstrate that all these branches of mathematics are applicable to real problems, and industry and universities in any country can clearly benefit from the skills of the complete range of European applied mathematics.

Proceedings of the

Fifth European Conference on Mathematics in Industry

June 6–9, 1990 Lahti

Edited by

Matti Heiliö

Rolf Nevanlinna Institute
University of Helsinki
Helsinki, Finland

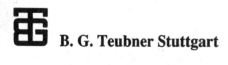 B. G. Teubner Stuttgart

KLUWER ACADEMIC PUBLISHERS
DORDRECHT / BOSTON / LONDON

Library of Congress Cataloging-in-Publication Data

European Conference on Mathematics in Industry (5th : 1990 : Lahti,
Finland)
    Proceedings of the Fifth European Conference on Mathematics in
Industry : June 6-9, 1990, Lahti / edited by Matti Heiliö.
      p.   cm. -- (European Consortium for Mathematics in Industry ;
ECMI vol. 7)
    ISBN 0-7923-1317-8 (HB : acid free paper)
    1. Engineering mathematics--Congresses.   I. Heiliö, Matti.
II. Title. III. Series: European Consortium for Mathematics in
Industry (Series) ; vol. 7.
TA329.E96 1990
620'.00151--dc20   91-19546

ISBN 0-7923-1317-8 (Kluwer)
CIP-Titelaufnahme der Deutschen Bibliothek
CIP-data available from publisher (Teubner)

ISBN 3-519-02176-5 (Teubner)

Sold and distributed in Continental Europe (excluding U.K.)
by B. G. Teubner GmbH, P.O. Box 801069, D-7000 Stuttgart-80

Sold and distributed in the U.S.A. and Canada
by Kluwer Academic Publishers,
101 Philip Drive, Norwell, MA 02061, U.S.A.

Kluwer Academic Publishers incorporates
the publishing programmes of
D. Reidel, Martinus Nijhoff, Dr W. Junk and MTP Press.

In all other countries (including U.K.), sold and distributed
by Kluwer Academic Publishers Group,
P.O. Box 322, 3300 AH Dordrecht, The Netherlands.

Printed on acid-free paper

Printed in the Netherlands

PREFACE

The Fifth European Conference on Industrial Mathematics (ECMI 90) took place at Lahti, Finland on June 6-9, 1990. The conference was organised by the Rolf Nevanlinna Institute together with the Lahti Research and Training Centre of the University of Helsinki. Like its predecessors the Lahti meeting was devoted to the exchange of experience, ideas and methods from various fields of industrial mathematics. The series of ECMI conferences have clearly established an important forum of interaction between the advancing front of technology and one of its crucial development resources, modern applications-oriented mathematics.

The precise title of the conferences has been the subject of some discussion and it has been argued that there is no such area which can be labelled as "industrial mathematics". This is certainly true if one thinks only in terms of the range of ideas, theorems, methods and algorithms constituting mathematics all of which may be applied. However with another viewpoint industrial mathematics is not a collection of topics but refers to the interactive process in which mathematics, the science, meets the real world of applications. Ideally this interaction involves both good mathematics and technological advance. The computer revolution has created a new era in technology with the increased computational capability making it possible to simulate complex industrial processes, devices, and other technological systems. This simulation depends on mathematical modelling and analysis and these techniques, sometimes ingenious but often quite routine, have provided a powerful tool for industrial scientists and creative research management.

The series of ECMI conferences, which began in Amsterdam in 1985, is intended to offer a state-of-the-art survey on the developments in this vital area of technology transfer. The Lahti conference confirmed both the relevance of the subject matter and the growth of the field of application. Despite the remote geographical location, and hence unduly high air fares, about 175 participants from 24 countries attended the meeting. The programme was composed of 11 invited lectures and 90 contributed papers. The range of the topics was broad reflecting the variety of applications that mathematical methods have in industry.

From the application areas displayed in the programme some examples could be mentioned: the supply and distribution of energy, electromagnetic field computations, fluid- and gas dynamics, welding and casting phenomena, phase and shape transitions, process and device simulation, chemical reactors, physical measurements and signals, picture processing, systems analysis and control, robotics and education in industrial mathematics. The invited lectures covered topics like chemical engineering, shell problems, sedimentation and aggregation phenomena, stochastic systems, radar technology, shape design and nondestructive testing, VLSI industry, random functions in mechanical systems, wavelets and modelling the ocean surface.

These proceedings consist of 7 invited lectures and 64 contributed papers. The first section of this volume contains the invited lectures. The second section consists of the papers presented at the special session on energy production, a minisymposium organised by Hansjörg Wacker. Because of the wide range of topics no specific grouping has been applied to the rest of the papers. Within each section the papers appear in alphabetical order according to the first author.

We would like to thank the authors for their cooperation in editing this volume. Also the valuable work of the referees is gratefully acknowledged. We would also like to thank the conference secretary Mrs. Sinikka Vaskelainen and her staff at the Lahti Research and Training Centre for successful running of the local arrangements. Finally I express my sincere thanks to my secretary Ms. Tarja Nieminen at the Rolf Nevanlinna Institute for her enormous help in preparing this volume. Without her reliable and efficient work this book would still be a long way from print.

Lappeenranta, April 1991   Matti Heiliö

TABLE OF CONTENTS

INVITED LECTURES

Mini-symposium:
OPTIMAL SUPPLY AND DISTRIBUTION OF ENERGY

CONTRIBUTED PAPERS

INVITED LECTURES

SOME PDE AND STATISTICAL PROBLEMS
FROM THE VLSI INDUSTRY

by

Ellis Cumberbatch

My talk consisted of descriptions of problems and results for three projects submitted to the Claremont Mathematics Clinic by local industry. They were (i) temperature estimation for Joule heating in current-carrying metal lines; (ii) an experimental design problem for the optimum choice of the width and spacing of metal lines; (iii) an asymptotic result for the resistance of current flowing into a small contact in a MOSFET source/drain region. I also described a number of results obtained in W. Fang's thesis at CGS concerning the inverse problem of the estimation of interfacial contact resistance and of the location/size/shape of the contact region from boundary measurements. Due to limitations on space, and the fact that the latter two topics will be published elsewhere, I shall write about (i) and (ii) only.

Heat Transfer in Transistor Lines

Transistors are getting smaller and more of them are being packed on a chip. The lines which carry current between devices are also getting narrower, thereby increasing current densities and the heat generated by current. A consequence of this is an increase in electromigration, the migration of ions of polycrystalline materials, [1], [2]. When there is differential migration

3

M. Heiliö (ed.), Proceedings of the Fifth European Conference on Mathematics in Industry, 3–11.
© 1991 B. G. Teubner Stuttgart and Kluwer Academic Publishers. Printed in the Netherlands.

sufficient to create a void across a wire, current flow is interrupted. Electromigration is found to be dependent on temperature gradients, and the
goal of the Clinic project, [3], was to obtain temperature profiles in regions
of large gradients, say where the line cross-sectional area changes rapidly,
giving large current density changes.

Typical cross-sectional and plan-view geometries are shown in Figures
1,2. The metal line is sheathed in an electrical insulator (SiO_2) and is above
an insulating pad on a silicon base.

The steady-state heat conduction equation applies, with a source term
modeling the Joule heating effect. The latter is proportional to the product
of the square of the current density and the material resistivity, which is
taken linear in the temperature, [4]. Hence the field equations are

$$(1) \quad \Delta u_1 \;=\; -au_1 + b \text{ in the metal region,}$$

$$(2) \quad \Delta u_2 \;=\; 0 \text{ in the insulator,}$$

$$\text{where} \quad a \;=\; \frac{I^2 r_1}{K_{AL}\delta^2\omega^2}, \quad b = \frac{I^2 r_2}{K_{AL}\delta^2\omega^2}.$$

I is the current input, K_{AL} the thermal conductivity of aluminum, δ, ω the
thickness and width of the rectangular line, and $r_1, r_2 > 0$ two resistivity
constants. Typical dimensions are $\omega = 2$ microns, and δ and the insulator
thicknesses fractions of microns.

Boundary conditions are taken as

$$(3) \qquad u = u_0 \quad \text{on } \Gamma_1, \Gamma_3,$$

$$(4) \quad K_i \frac{\partial u}{\partial n} = -h(u - u_0) \quad \text{on } \Gamma_2,$$

where u_0 is the ambient air temperature, Γ_1, Γ_3 represent upstream and down-stream locations, and Γ_2 is the insulator-air interface where a linear convective model is taken with coefficient h, see [4]. At the $A\ell - SiO_2$ interface the temperature and heat flux are continuous, giving

$$(5) \qquad u_1 = u_2 \quad , \quad K_{AL} \frac{\partial u_1}{\partial n} = K_i \frac{\partial u_2}{\partial n} \quad \text{on } \Gamma_4.$$

It is clear that the complicated geometry inhibits analytic approaches, so a numerical investigation was begun. There are powerful 3-D elliptic solvers, but they are restrictive in the geometries and boundary conditions they can handle. ELLPACK provides a sub-routine for modifications necessary for a jump discontinuity in normal derivative, present on Γ_4 since $K_{A\ell} \neq K_i$, but the numerical precision was found poor, giving false output near corners of the domains. These difficulties forced a reduction to less complicated geometries. Three 2-D problems were solved: (I) in the cross-section plane, (II) in the plan-view plane where the effect of varying cross-section could be modeled, with convective boundary conditions taken on the sides of the region, and (III) a hybrid model with the same geometry as in (II) but a constant temperature sink taken on one side of the region to gauge the effect of the base. These solutions were obtained using PLTMG, a finite-element 2-D elliptic solver of good precision.

Interesting results for problems (II) and (III) were obtained for a wedge-shaped region of unit slope narrowing from a width of 103 to a rectangular region of width 3, see Figure 2. Of particular note are the appearance at high current densities of local maxima followed by an extremely fast decrease in temperature towards a local minimum inside the narrow rectangular region. The temperature then increases again. The minimum has a negative temperature which indicates meaningless results for the model. The threshold for the minimum is for currents around 0.4 amps for II, 1.0 amps for III (larger than operating currents). Speculation on the cause of this focused on the possibility of the appearance of multiple solutions. However no definite results were obtained in the short time available.

Experimental Design

The determination of the width of electrical lines is required for quality control in the VLSI industry. One method is to measure line resistance, the inverse of which is proportional to width. A test apparatus consists of a number of lines laid parallel to one another; the width, W, of each line, and the spacing, S, change from one section of the line to another. Current is passed along the lines and resistances of sections of the line measured. The model adopted to relate the difference between W_α, the design linewidth, and its measured value $W_{\alpha\beta}$, when it is located at a spacing S_β to its neighbor, is

$$(6) \qquad W_{\alpha\beta} - W_\alpha = \Delta W + \frac{a}{W_\alpha} + \frac{b}{S_\beta} + \frac{c}{W_\alpha S_\beta} + \xi_\alpha$$

where ΔW is a constant called the over/under etch; a, b, c are process effect coefficients and ξ represents an error term. There are upper and lower bounds on W and S. The aim of this investigation is to determine optimum choices for W, S within these bounds for a set of tests used to obtain the constants $\Delta W, a, b, c$ by regression analysis.

Most Mathematics Clinics are year projects. Due to exceptional circumstances this investigation involved three different teams over three semesters. Their reports are [5], [6], [7]. This material is taken from [7].

In the standard form of regression analysis, (6) is written

$$(7) \qquad y_i = \beta_0 + \frac{\beta_1}{x_{1i}} + \frac{\beta_2}{x_{2i}} + \frac{\beta_3}{x_{1i}x_{2i}} + \epsilon_i$$

where $i = 1, 2, \cdots, n$ indexes the number of experiments. Models are called linear if they are linear in the parameters $\beta_0, \cdots, \beta_3$. The experimental region is a rectangle in the $x_1 - x_2$ plane, and the design problem is to specify n-points in this rectangle in order that the measurements y_i yield estimates of the parameters under some optimality criteria.

The experimental design problem is usually referred to the linear model

$$(8) \qquad y_i = \beta_0 + \beta_1 x_{1i} + \beta_2 x_{2i} + \cdots + \beta_r x_{ri} + \epsilon_i$$

or, in vector form,

(9) $$Y = X\beta + \epsilon$$

where X is an $n \times (r+1)$ design matrix, and ϵ is the error vector assumed to be of zero mean, uncorrelated and to have uniform variance and normal distribution.

Using the least squares method, the maximum likelihood method, or searching for the best unbiased linear estimater, provides the same estimator for β, viz

(10) $$\hat{\beta} = [X^T X]^{-1} X^T Y$$

so that

(11) $$\hat{Y} = HY \text{ where } H = X[X^T X]^{-1} X^T$$

Given a design region R, a design X is thought of as a probability distribution over R. If X gives weight $\frac{1}{n}$ to each of n not necessarily distinct points in R then X is called an exact design, since it can be realized in practice exactly.

There is considerable literature on the general experimental design problem, see [8] for a recent review. Most optimality criteria are based on minimizing some form of "variance" associated with $\hat{\beta}$ or $\hat{Y}$. The designs are said to be A, D, E, G or V optimal depending on the criterion chosen. The Clinic concentrated on:

1. D optimality, which maximizes $\det(X^T X)$, the inverse of the generalized variance of $\hat{\beta}$, and thus reduces the confidence region for the true value of the parameters, and showed that observations should be placed at the vertices of R. The value of $\det(X^T X)$ may be increased by increasing the number of experiments, so a penalty for this was included via

(12) $$D_n = n^{1-r} \det(X^T X)$$

From simulations it was seen that for a square region in R_2, D_n took its maximum value when $n = 4m$, and m observations were made at the 4 vertices; addition of extra observations reduced D_n until symmetry returned.

2. G optimality, which minimizes the maximum of the variances of the predicted response over R. This implies finding a design that equalizes the leverage of the measurements, and it was shown that G optimality is achieved by placing the observations at the vertices of a regular polygon in R.

Reference [7] contains a review of the literature, its relation and application to the problem posed by the experimental design problem for (6), extensions of the theory and various simulations.

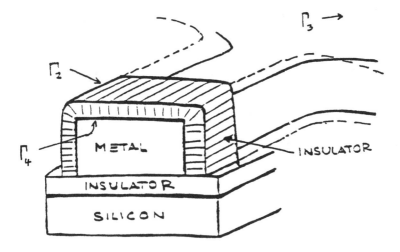

Figure 1: Sketch of the general conductor.

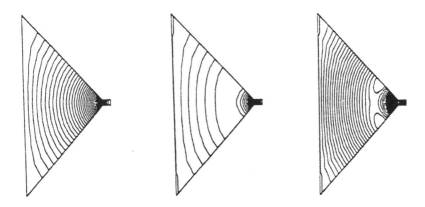

Figure 2: Series of different runs for different input current for problem II. From right to left the values of I are 0.25, 1.0, and 2.0 A respectively. Notice the appearance of a relative maximum at the right end of the triangular region and of a relative minimum at the left extreme of the rectangular region.

References

[1] Schreiber, H. U., *Activation Energies for the Different Electromigration Mechanisms in Aluminum.* Solid State Electronics 24: 583-589, 1981.

[2] Wu, C. J. and J. McNutt, *Effects of Substrate Thermal Characteristics on the Electromigration Behavior of A1 Thin Film Conductors.* IEEE Proceedings, IRPS, 1983.

[3] Velasco, J., D. Babai, H. Jung, *Heat Transfer in Transistors.* Claremont Math Clinic, May 1990.

[4] Holman, J. P., Heat Transfer. 2nd edition. MacGraw Hill, New York, 1981.

[5] Herring, S., G. Andersson, *Experimental Design for Linear and Non-Linear Parameter Estimation.* Claremont Math Clinic, August 1989.

[6] Tsui, D., K. Lee, Y.M. Lee, P. Tran, P. Good, *Robust Experimental Design.* Claremont Math Clinic, December 1989.

[7] Helfgott, E.S., A. Romberger, *Optimality Criteria and Exact Experimental Designs for Regression Models.* Claremont Math Clinic, May 1990.

[8] Atkinson, A.C., *Recent Developments in the Methods of Optimum and Related Experimental Designs.* International Statistical Review, Vol. 56, No. 2, pp. 99-115, 1988.

The Claremont Graduate School

Department of Mathematics

Claremont, California 91711-3988, U.S.A.

REFLECTOR DESIGN AS AN INVERSE PROBLEM[1]

Heinz W.Engl and Andreas Neubauer

Abstract: We report about an industrial project which concerns the computer–aided design of surfaces in $I\!R^3$ that can be used as reflectors where the light coming from a point source is reflected in such a way that the intensity distribution of the outgoing light in a given plane ("near field problem") or sphere ("far field problem") can be prescribed. We present a mathematical model and investigate stability properties of this inverse problem. Finally, we report about our numerical approach and results.

1. The Problem

We report about a project that has been done for and in coorporation with an Austrian company which designs and manufactures lightings for industrial illumination purposes. The lightings consist of a lamp and a reflector. Our task was to develop a mathematical model and an algorithm (resulting in computer software) for the problem of designing the shape of the reflector in such a way that the illumination distribution on a given plane or sphere behaves in a prescribed way and that various additional constraints are fulfilled. In this algorithm, the reflector should be described in a way that could be directly used for CAM purposes in that company.

This problem is a very old one (see [6]), several patents are connected with it (e.g. [2],[10]). The "state of the art" seems to be that one can handle 2–dimensional problems in a practically satisfactory way, i.e., cases where the light distribution of the lamp, the desired illumination distribution, and the reflector have either rotational or translational symmetry ([12],[13]). For a general background about problems in lighting design see e.g. [7],[11].

Our aim is to handle the fully 3–dimensional problem, where the reflector is not required to have any symmetry. Since we deal with an inverse problem (cf., e.g., [3],[4],[8]), stability questions arise. These will be treated in Section 2.

[1]Supported in part by the Austrian Fonds zur Förderung der wissenschaftlichen Forschung (Project S32/03).

M. Heiliö (ed.), Proceedings of the Fifth European Conference on Mathematics in Industry, 13–24.
© *1991 B. G. Teubner Stuttgart and Kluwer Academic Publishers. Printed in the Netherlands.*

When the illumination distribution is prescribed on a plane, we call our problem the
"near field problem"; the problem will be referred to as **"far field problem"**, if
the illumination distribution is prescribed on a sphere, i.e., as a function of the (solid)
angle. The latter problem can be interpretted as the limiting case of the former when
the plane moves out to infinity, which explains the nomenclature. See also [9] for a
discussion of these problems.

2. Stability

For illustration and simplicity, we study the stability question for the 2–dimensional
case. This gives at least some guidance for the 3–dimensional case. We derive a simple
model for the 2–dimensional far field problem:

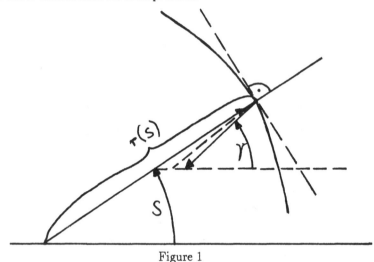

Figure 1

A normal vector at $(s, r(s))$ is given by $\left(\dot{r}(s) \sin s + r(s) \cos s, r(s) \sin s - \dot{r}(s) \cos s \right)$.
Since we base our model on geometric optics, the reflection condition says that the tan-
gent of the angle between this normal vector and the x–axis has to equal $\text{tg}\left(\frac{s+\gamma(s)}{2} \right)$,
i.e.,

$$r(s) \sin s - \dot{r}(s) \cos s = \text{tg}\left(\frac{s + \gamma(s)}{2} \right)\left[\dot{r}(s) \sin s + r(s) \cos s \right]. \qquad (2.1)$$

If we denote by $I(\gamma)$ the illumination distribution in dependence of the direction γ of
the outgoing light rays and by $L(s)$ the light distribution of the lamp in dependence

of the direction s of the incoming light rays, then the condition to be satisfied is

$$I(\gamma(s)).\dot{\gamma}(s) = L(s) \qquad (2.2)$$

for the case of translational invariance and

$$I(\gamma(s)). \sin\gamma(s).\dot{\gamma}(s) = L(s). \sin s \qquad (2.3)$$

for the case of rotational symmetry.

Thus, for the case of translational invariance, we have the system of ordinary differential equations

$$\begin{pmatrix} \dot{r}(s) \\ \dot{\gamma}(s) \end{pmatrix} = \begin{pmatrix} \frac{\cos\gamma(s)-\cos s}{\sin\gamma(s)+\sin s} & 0 \\ 0 & \frac{1}{I(\gamma(s))} \end{pmatrix} \begin{pmatrix} r(s) \\ L(s) \end{pmatrix} \qquad (2.4)$$

together with initial conditions .

$I(\gamma)$ is given only up to a multiplicative constant; this constant has to be determined from a light–balance relation, which says that the total amount of light going into the reflector has to come out again. This relation can and has to be used to control the illuminated angular sector.

There are two problems connected with this model:

The direct problem: Given L and the reflector $r = r(s)$, compute $(\gamma = \gamma(s))$ and $I = I(t)$.

Here, t is thought of as an independent angular variable, I is given as a function of this variable as opposed to being a function of s, the direction of the incoming rays, via $I = I(\gamma(s))$.

From (2.1), we obtain

$$\gamma(s) = 2\mathrm{arctg}\left[\frac{\dot{r}(s)\sin s + r(s)\cos s}{r(s)\sin s - \dot{r}(s)\cos s}\right] - s; \qquad (2.5)$$

here, γ depends on r and $\dot{r}$. In order to compute $I = I(t)$, we have to "solve" $\gamma(s) = t$ for s: $s = \gamma^{-1}(t)$.

From (2.2), we then obtain

$$I(t) = \frac{L(\gamma^{-1}(t))}{\dot{\gamma}(\gamma^{-1}(t))}; \qquad (2.6)$$

hence, I depends on $r, \dot{r}$ and $\ddot{r}$. Because of the convergence properties for cubic spline interpolation, this indicates that cubic spline interpolation of r should give a good approximation for I for a sufficiently fine grid. However, computations show that on a coarse grid, spline interpolation of r changes I dramatically.

The inverse problem: Given L and I, determine r (and γ).

Note that we do not need stability for r and for γ, but we only need that the resulting illumination distribution $I(t)$ is stable: We are not looking for approximations to some "ideal" reflectors, but just for any reflector that approximates the resulting illumination distribution.

Standard continuous dependence results for ordinary differential equations imply that r and γ depend (under computable conditions) locally in a Lipschitz way on the data L and I (considered as the **desired** illumination distribution).

Thus, the problem of determining (r, γ) from data (L, I) is (locally) well–posed in the uniform norm. However, a method that approximates (r, γ) in the uniform norm does not yield a good approximation for I, since the resulting illumination distribution depends on $r, \dot{r}$ and $\ddot{r}$, as we have seen above. Thus, we need to approximate (r, γ) in $C^2 \times C^1$. Differentiation of the second equation in (2.4) and application of the continuous dependence results for ordinary differential equations yields that locally, (r, γ) depend continuously (in this sense) on $L, \dot{L}, I,$ and $\dot{I}$. Hence, a stable approximation of $I = I(\gamma(s))$ requires accurate knowlegde of the data $(L, I = I(t))$ in the C^1–sense.

3. The 3–Dimensional Problem: Near Field

For deriving a mathematical model, we use the following sketch:

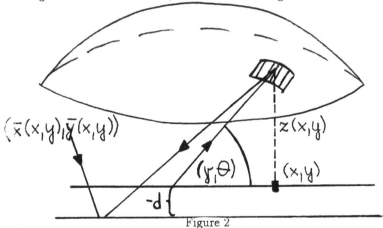

Figure 2

Here, we use the following notation:

$(x, y, z(x, y))$ Cartesian reflector coordinates, $(x, y) \in G$

$(\bar{x}(x, y), \bar{y}(x, y))$ coordinates of the point where the ray reflected in $(x, y, z(x, y))$ meets the plane $z = -d$ (depend on the reflector!)

$I = I(\bar{x}, \bar{y})$ desired illumination distribution in the plane $z = -d$

$L = L(\gamma, \theta)$ known light distribution of the lamp (situated in the origin)

(γ, θ) spherical coordinates of the incoming rays.

The basic requirement is that the amount of light leaving the lamp in any solid angle increment should equal the amount of light reflected into the corresponding area increment in the illuminated plane, i.e.,

$$ \text{``}I dA = L d\Omega\text{''} , \tag{3.1} $$

where dA denotes the area increment in the illuminated plane, and $d\Omega$ is the solid angle increment of incoming rays; the mathematical meaning of the heuristic formula (3.1) is that for all sufficiently regular subregions ΔA and $\Delta \Omega$,

$$ \int_{\Delta A} I dA = \int_{\Delta \Omega} L d\Omega \tag{3.2} $$

has to hold. Via a substitution that transforms the first integral into one with respect to $d\Omega$, this leads (since then both integrands have to coincide) to

$$ I(\bar{x}, \bar{y}).|\bar{x}_x.\bar{y}_y - \bar{x}_y.\bar{y}_x| = L(\gamma, \theta)\frac{|z - x.z_x - y.z_y|}{(x^2 + y^2 + z^2)^{\frac{3}{2}}}. \tag{3.3} $$

For this substitution to be possible, one needs the constraints that $z - xz_x - yz_y$ and $\bar{x}_x\bar{y}_y - \bar{x}_y\bar{y}_x$ do not change sign. Since the reflector has to lie above the plane, we need that $z(x, y) > -d$; since the lamp has to lie between reflector and plane, $z(0, 0) > 0$ has to hold.

Now we have to compute $(\bar{x}, \bar{y})$ from the reflection condition; this leads to

$$ \begin{pmatrix} \bar{x}(x, y) \\ \bar{y}(x, y) \end{pmatrix} = \frac{2(xz_x + yz_y - z) - d(1 + z_x^2 + z_y^2)}{2(xz_x + yz_y - z) + z(1 + z_x^2 + z_y^2)} \cdot \begin{pmatrix} x \\ y \end{pmatrix} + $$
$$ + \frac{2(xz_x + yz_y - z)(z + d)}{2(xz_x + yz_y - z) + z(1 + z_x^2 + z_y^2)} \cdot \begin{pmatrix} z_x \\ z_y \end{pmatrix}. \tag{3.4} $$

Thus, we have the following

Mathematical model:

$$ I(\bar{x}, \bar{y}).|\bar{x}_x\bar{y}_y - \bar{x}_y\bar{y}_x| = L(\gamma, \theta)\frac{|z - xz_x - yz_y|}{(x^2 + y^2 + z^2)^{\frac{3}{2}}} \tag{3.5} $$

with

$$\gamma(x,y) = \arccos\left(\frac{z}{(x^2 + y^2 + z^2)^{\frac{1}{2}}}\right) \qquad \left(z(0,0) \neq 0\right) \tag{3.6}$$

$$\theta(x,y) = \begin{cases} \arccos\left(\frac{x}{\sqrt{x^2+y^2}}\right), & y \geq 0 \\ \\ 2\pi - \arccos\left(\frac{x}{\sqrt{x^2+y^2}}\right), & y < 0 \end{cases}, \quad \left((x,y) \neq (0,0)\right), \tag{3.7}$$

$$\begin{pmatrix} \bar{x} \\ \bar{y} \end{pmatrix} = \frac{A}{C}\begin{pmatrix} x \\ y \end{pmatrix} + \frac{B}{C}\begin{pmatrix} z_x \\ z_y \end{pmatrix} \tag{3.8}$$

$$\begin{pmatrix} \bar{x}_x \\ \bar{y}_y \end{pmatrix} = \frac{A_x C - A C_x}{C^2}\begin{pmatrix} x \\ y \end{pmatrix} + \frac{A}{C}\begin{pmatrix} 1 \\ 0 \end{pmatrix} + \frac{B_x C - B C_x}{C^2}\begin{pmatrix} z_x \\ z_y \end{pmatrix} + \frac{B}{C}\begin{pmatrix} z_{xx} \\ z_{xy} \end{pmatrix}, \tag{3.9}$$

$$\begin{pmatrix} \bar{x}_y \\ \bar{y}_y \end{pmatrix} = \frac{A_y C - A C_y}{C^2}\begin{pmatrix} x \\ y \end{pmatrix} + \frac{A}{C}\begin{pmatrix} 0 \\ 1 \end{pmatrix} + \frac{B_y C - B C_y}{C^2}\begin{pmatrix} z_x \\ z_y \end{pmatrix} + \frac{B}{C}\begin{pmatrix} z_{xy} \\ z_{yy} \end{pmatrix} \tag{3.10}$$

with

$$A = 2(x z_x + y z_y - z) - d(1 + z_x^2 + z_y^2), \tag{3.11}$$

$$B = 2(x z_x + y z_y - z)(z + d), \tag{3.12}$$

$$C = 2(x z_x + y z_y - z) + z(1 + z_x^2 + z_y^2), \tag{3.13}$$

$$A_x = 2(x - d z_x) z_{xx} + 2(y - d z_y) z_{xy}, \tag{3.14}$$

$$A_y = 2(x - d z_x) z_{xy} + 2(y - d z_y) z_{yy}, \tag{3.15}$$

$$B_x = 2(x z_{xx} + y z_{xy})(z + d) + 2(x z_x + y z_y - z) z_x, \tag{3.16}$$

$$B_y = 2(x z_{xy} + y z_{yy})(z + d) + 2(x z_x + y z_y - z) z_y, \tag{3.17}$$

$$C_x = 2(x z_{xx} + y z_{xy}) + z_x(1 + z_x^2 + z_y^2) + 2z(z_x z_{xx} + z_y z_{xy}), \tag{3.18}$$

$$C_y = 2(x z_{xy} + y z_{yy}) + z_y(1 + z_x^2 + z_y^2) + 2z(z_x z_{xy} + z_y z_{yy}), \tag{3.19}$$

and the constraints

$$\text{sign}(z - x z_x - y z_y) = \text{constant}, \tag{3.20}$$

$$\text{sign}(\bar{x}_x \bar{y}_y - \bar{x}_y \bar{y}_x) = \text{constant}, \tag{3.21}$$

$$C < 0, \tag{3.22}$$

$$z > -d, z(0,0) > 0. \tag{3.23}$$

If (3.22) is violated, the reflected ray misses the plane.

This is one (quite complicated) nonlinear partial differential equation of second order with additional constraints. The first task which one will perform is its linearization. This is a lot of work; we checked the results with the symbolic computation package REDUCE.

The linearized equation is elliptic, iff

$$\bar{x}_x \bar{y}_y - \bar{x}_y \bar{y}_x > 0, \tag{3.24}$$

and hyperbolic, iff

$$\bar{x}_x \bar{y}_y - \bar{x}_y \bar{y}_x < 0. \tag{3.25}$$

A change of type, which is forbidden by the assumption (3.21), would mean that points in the plane are met by more than one light ray. Thus, (3.21) is not only an (artificial) mathematical constraint, but has a clear interpretation.

For a rotationally symmetric paraboloid reflector, the linearized equation is usually (depending on where the lamp is) elliptic near the vertex and hyperbolic further downward. Thus, not too large "closed" reflectors seem to fall into the elliptic case. Based on this, it was decided to treat the elliptic case. Since the appropriate side conditions are then boundary conditions, we can treat the practically important case that the bounding curve of the reflector is given.

Our first numerical attempt was to use Newton's method with line search for the nonlinear partial differential equation, which leads to a sequence of linear elliptic boundary value problems. These were solved using cubic finite elements. This approach did not work well, most likely because of the fact that as soon as the nonlinear problem was unsolvable in the finite element space used (which is the generic case), Newton's method was trying to approximate an unsolveable problem. This observation motivated the use of an optimization approach:

In our second attempt, we represented the reflector by cubic tensor product splines, which is appropriate for CAM purposes in the company. Then we performed a minimization of the L^2–defect in (3.5) using a conjugate gradient method ([5]) and taking the constraints into account. This worked quite well on academic test examples. However, in order to converge (in a reasonable time), our algorithm needs an initial guess which is a cubic tensor product spline and elliptic in the above sense. For real design problems, such "starting reflectors" are usually not available. Thus, although we implemented our algorithm in the company, it will only be of real use as soon as suitable starting reflectors will be available. We expect that these will be obtainable from the results of our algorithm for the far field problem to be described in Section 4 as soon as that algorithm works well for realistic examples.

4. The 3–Dimensional Problem: Far Field

The model is derived in a similar way as in Section 4; instead of a plane, a sphere (centered in the lamp) is illuminated by the outgoing rays which are all thought to emanate from the lamp. This makes the model a bit more transparent and makes no difference in the far field anyway.

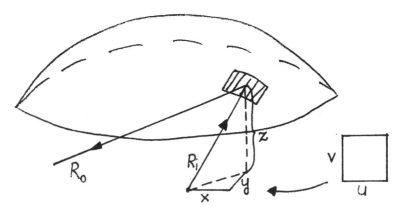

Figure 3

In this figure, R_i denotes the direction of the incoming light ray, normed in such a way that $R_i = \Big(x(u,v), y(u,v), z(u,v)\Big)$ is also the point on the reflector where this ray is reflected; R_o is the reflected ray. The independent variables (u,v) vary in some parameter domain G, over which the (prescribed) projection $(x(u,v), y(u,v))$ of the reflector onto the x–y–plane is parameterized. The unknown variable, which determines the reflector, is then z.

The reflection condition now reads

$$R_o = R_i - \frac{2\langle R_i, n\rangle}{\|n\|^2} n \tag{4.1}$$

with

$$n = \frac{\partial R_i}{\partial u} \times \frac{\partial R_i}{\partial v}. \tag{4.2}$$

The "balance condition" is now

$$\text{``}I(R_o)d\Omega_o = L(R_i)d\Omega_i\text{''}, \tag{4.3}$$

where $d\Omega_o$ and $d\Omega_i$ are solid angle increments for the outgoing and incoming rays.

As in the near field problem, this leads (via a formulation of (4.3) in integral terms and substitution) to

$$I(R_o).|\left\langle R_o, \frac{\partial R_o}{\partial u} \times \frac{\partial R_o}{\partial v} \right\rangle| = L(R_i).|\left\langle R_i, n \right\rangle|, \qquad (4.4)$$

where the expressions in $|\quad|$ are not allowed to change sign.

The unknown function z appears in (4.4) via R_i and via R_o (cf. (4.1), (4.2)) and is differentiated up to two times. Hence, (4.4) is again a nonlinear partial differential equation of second order (with constraints analoguos to those in Section 3). We do not write down this equation explicitly. Analogously to the near field problem, the linearization is complicated and lengthy and was checked using REDUCE. It leads to a linear partial differential equation with sign constraints which is elliptic where

$$\left\langle R_i, n \right\rangle.\left\langle R_o, \frac{\partial R_o}{\partial u} \times \frac{\partial R_o}{\partial v} \right\rangle < 0 \qquad (4.5)$$

and hyperbolic where

$$\left\langle R_i, n \right\rangle.\left\langle R_o, \frac{\partial R_o}{\partial u} \times \frac{\partial R_o}{\partial v} \right\rangle > 0. \qquad (4.6)$$

Elliptic **and** hyperbolic regions are present if different outgoing rays are parallel.

For the same reasons as outlined in Section 3, it was decided to treat the elliptic case, where again the bounding curve of the reflector can be prescribed. Because of our numerical experience with the near field problem, we based our approach from the beginning on defect minimization:

$$\|I(R_o)|\left\langle R_o, \frac{\partial R_o}{\partial u} \times \frac{\partial R_o}{\partial v} \right\rangle| - L(R_i).|\left\langle R_i, n \right\rangle| \, \|_{L^2(G)} \to \quad \min, \qquad (4.7)$$

where the actual variable $z = z(u, v)$ is represented by cubic tensor product splines (using the B–spline representation, see [1]).

Side conditions of the following type are used: The boundary curve of the reflector can either be fixed or be allowed to vary within prescribed bounds; in order to control the total size of the reflector, box constraints on the B–spline coefficients are imposed; finally, there are sign conditions similar to those described in Section 3, e.g. (4.5).

The resulting constrained minimization problem was attacked with a projected conjugate gradient method. Note that the evalution of the objective functional and of its gradient (which is done explicitly) is very expensive, so that only a relatively coarse grid can be used.

The algorithm works well on academic test examples. E.g., on a grid with 5×5 nodes, the illumination distribution generated by a plane reflector was reached from an initial guess perturbed by 300 % compared to the plane after 37 iterations in 13 minutes CPU–time on a MicroVAX 3500, the algorithm found back to the plane. In Figure 4, the starting reflector and the plane are shown (without hidden lines for reasons of transparency).

On "real" problems, the algorithm worked (initially) badly for reasons similar to those explained in Section 3 for the near field problem. The company can provide only rotationally symmetric "starting reflectors", which cannot be directly used since they are in general not cubic tensor product splines. However, as explained in Section 2, their interpolations by such splines on a sufficiently fine grid should be good starting reflectors. But since we can use only coarse grids, their spline interpolations turned out to be much too far off for leading to convergence.

Now, the following question arises: For CAM purposes, the company interpolates by splines. Why do they not see this effect in the reflectors manufactured on this basis? The answer might be that our model is "too sharp": The lamp is not a point, but extends over a small region, which has a smoothing effect compared to our model. This suggests a change in the objective functional in (4.7) to enforce some averaging over small subregions (and thus simulating the effect of an extended lamp). This idea was implemented together with a spatially varying weight in (4.7). Also, instead of interpolation, we now use an **approximation** of the rotationally symmetric starting reflector by splines with respect to the norm $\|z\|_{L^2}^2 + \|\frac{\partial z}{\partial u}\|_{L^2}^2 + \|\frac{\partial z}{\partial v}\|_{L^2}^2 + \|\frac{\partial^2 z}{\partial u \partial v}\|_{L^2}^2$, thus taking into account at least one second derivative, which turned out to be important in Section 2.

The resulting modified algorithm works reasonably well also on first real examples:

Example: A rotationally symmetric "wall–washer", i.e., a reflector generating a uniform illumination distribution on an infinitely distant plane, is restricted to a square domain. The aim is to compute a reflector that generates the same illumination distribution from a lamp whose light distribution is not rotationally symmetric. On a grid with 10×10 nodes, this can be achieved after 48 iterations in 40 minutes of CPU–time at least in a subdomain away from the boundary where the steep spikes of the starting reflector (Figure 5) are located. In Figure 6, the resulting reflector (over this subdomain) is shown together with the incoming and the outgoing rays.

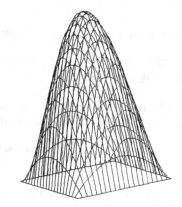

Figure 4

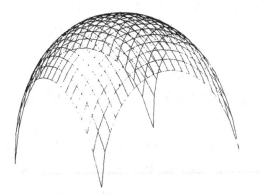

Figure 5

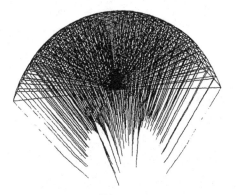

Figure 6

References

[1] de Boor, C. (1978) A Practical Guide to Splines, Springer, New York.

[2] Deutsches Patentamt, 'Spezialreflektor und einen Spezialreflektor aufweisendes Beleuchtungsgerät', Offenlegungsschrift 1 497 303.

[3] Engl, H.W. 'Inverse und inkorrekt gestellte Probleme', to appear in U.Kulisch (ed.), Jahrbuch Überblicke Mathematik, Vieweg, Wiesbaden.

[4] Engl, H.W. and Groetsch, C.W. (eds.) (1987) Inverse and Ill–Posed Problems, Academic Press, Orlando.

[5] Fletcher, R. (1980 & 1981) Practical Methods of Optimization, Vols. 1 & 2, Wiley, Chichester.

[6] Halbertsma, N. (1925) 'Die Vorausbestimmung der Lichtverteilungskurve eines spiegelnden Reflektors', Zeitsch.f.techn.Physik 10, 501–504.

[7] Hentschel, H.-J. (1987) Licht und Beleuchtung – Theorie und Praxis der Lichttechnik, Hüthig, Heidelberg.

[8] Louis, A. (1990) 'Numerik inverser Probleme', GAMM–Mitteilungen 1, 5–27.

[9] Ngai, P.Y. (1987) 'On near–field photometry', Jour.of the Illuminating Engin.Soc., 129–136.

[10] Price, E. and Goodbar, I. 'Lighting Fixtures', US Patent 3 098 612.

[11] Reeb, O. (1962) Grundlagen der Photometrie, Braun, Karlsruhe.

[12] Wolber, W. (1973) 'Optimierung der Lichtlenkung von Leuchten', Dissertation, Universität Karlsruhe.

[13] Wolber, W. (1970) 'Berechung von Reflektoren für beliebige Lichtverteilungen', Lichttechnik 22, 597–598.

Address of both authors:
Institut für Mathematik
Johannes-Kepler-Universität
A-4040 Linz
Austria

Wavelets and applications

Stéphane Jaffard

In the seventies and the eighties, alternative methods to Fourier analysis appeared independently in many fields of science and technology. Let us mention oil detection, analysis of speech, quantum mechanics, image analysis, analysis of turbulent flows, multigrid methods, the theory of interpolation between functional spaces...

Wavelets are a mathematical tool which lies behind these new methods. We have two purposes in this paper. We give a survey of the construction of wavelets and their mathematical properties, and we also show how wavelets unify these methods already existing in different fields, and, in several cases, improve them. For the reader interested in mathematical properties, or in a specific application, we also give at the end a wide bibliography.

1 Mathematical historical background.

The mathematical evolution that led to wavelets can be interpreted as the construction of successive bases of functions with the following aim: the decomposition on these bases yields the sharpest possible information on the time and frequency behavior of the analysed signal.

The first step was obtained by the Fourier series. The two main drawbacks of the Fourier analysis are that it is not local and that it is unstable when dealing with other spaces than L^2 or the Sobolev spaces H^s.

The problem of having a stable decomposition for other spaces than L^2 led Alfred Haar to the so-called Haar system constructed as follows.

Let ϕ be the caracteristic function of $[0, 1/2]$, and $\psi = \phi(x) - \phi(x - 1/2)$. The collection of all the $\psi_{j,k}$ ($j \in Z$, $k \in Z$), defined by

$$\psi_{j,k}(x) = 2^{j/2}\psi(2^j x - k)$$

forms an orthonormal basis of $L^2(R)$, and the decomposition makes sense also in the L^p spaces. The drawback is that the decomposition on this system does not give a sharp frequency information, since the function ψ

M. Heiliö (ed.), Proceedings of the Fifth European Conference on Mathematics in Industry, 25–34.
© 1991 *B. G. Teubner Stuttgart and Kluwer Academic Publishers. Printed in the Netherlands.*

has not a good frequency localization. We shall see that wavelets provide a way to avoid this drawback.

A wavelet basis (in one dimension) is defined as above, except that the function ψ is required to be smoother. Since 1981 several such bases have been constructed, with an arbitrary regularity. For instance, ψ can be compactly supported, or in the Schwartz class, or a spline with exponential decay (see [D2], [M1],[L]). We shall describe in the following part the construction that leads to these wavelets and its link with algorithms used in image analysis. But, before that, let us recall some results on "doubly-localized" orthonormal bases, which are of importance in mathematical physics and signal analysis.

T. Steiger proved (see[B]) that L^2 does not admit a basis of the following form

$$f_j(x) = e^{ib_j x} g_j(x - a_j)$$

where the g_j would be such that sup $\| g_j \|_\epsilon < \infty$, for an $\epsilon > 0$, where

$$\| g_j \|_\epsilon^2 = \int (1 + x^2)^{1+\epsilon} \mid g(x) \mid^2 dx + \int (1 + \xi^2)^{1+\epsilon} \mid \hat{g}(\xi) \mid^2 d\xi.$$

the optimal result was obtained by J.Bourgain who found a basis in the case $\epsilon = 0$ (see[B]).

Actually, if we accept to mix in the same function positive and negative frequencies of the same value, this obstruction no more exists, and there exists an orthonormal basis of $L^2(R)$ of the following form (see[DJJ])

$$
\begin{aligned}
\psi_{0,n}(x) &= \phi(x - n) \\
\psi_{l,n}(x) &= \sqrt{2}\phi(x - \tfrac{n}{2})\cos(2\pi l x) \text{ if } l \neq 0, l + n \in 2Z \\
&= \sqrt{2}\phi(x - \tfrac{n}{2})\sin(2\pi l x) \text{ if } l \neq 0, l + n \in 2Z + 1,
\end{aligned}
$$

where ϕ and $\hat{\phi}$ have exponential decay.

The fact that we do not try to separate positive and negative frequencies of the same amplitude means, in the signal analysis terminology, that we study the real signal, and not the corresponding analytical signal.

2 Construction of wavelets and signal analysis.

2.1 Multiresolution analysis.

We shall first describe a standard way to construct wavelets. We shall stick to the dimension 1 for the sake of simplicity. A multiresolution analysis is an increasing sequence $(V_j)_{j \in Z}$ of closed subspaces of L^2 such that

- $\forall j \ \bigcup_k \overline{S_{j,k}} = \overline{\Omega}$, and $S_{j,k} \cap S_{j,k'}$ is empty if $k \neq k'$.

- $\bigcup V_j = \{0\}$

- $\bigcap V_j$ is dense in L^2

- $f(x) \in V_j \Leftrightarrow f(2x) \in V_{j+1}$

- $f(x) \in V_0 \Leftrightarrow f(x+1) \in V_0$

- There is a function g in V_0 such that the $g(x - k)_{k \in Z}$ form a Riesz basis of V_0.

We also require g to be smooth and well localized.

A simple example of multiresolution analysis is obtained by taking for V_j the space of continuous and piecewise linear functions on the intervals $[k2^{-j}, (k+1)2^{-j}]$, $j, k \in Z$. A possible choice for g is the "hat" function which is the function of V_0 taking the value 1 for $x = 0$ and vanishing at the other integers. It is easy to orthonormalise the set $g(x - k)$ by choosing

$$\hat{\phi}(\xi) = \hat{g}(\xi)(\sum | \hat{g}(\xi + 2k\pi) |^2)^{-1/2}.$$

Then, the $\phi(x - k)$ form an orthonormal basis of V_0.

Define W_j as the orthogonal complement of V_j in V_{j+1}. One immediately checks that the W_j are mutually orthogonal, and their direct sum is equal to L^2. By a similar procedure which led to the construction of ϕ, we can obtain a function ψ such that the $\psi(x - k)$ form an orthonormal basis of W_0. Since the W_j are obtained from each other by dilation, and are mutually orthogonal, the functions $2^{j/2} \psi(2^j x - k)$ form an orthonormal basis of $L^2(R)$.

Let us now come back to the orthogonal decomposition $V_1 = V_0 \oplus W_0$. We have two orthonormal bases of V_1: the first one is the $\sqrt{2}\phi(2x - k)$, $k \in Z$, and the second one is the union of the $\phi(x - k)$ and $\psi(x - k)$. The existence of two bases implies the existence of an isometry mapping the coordinates in the first basis on the coordinates in the second basis. Let

$$\alpha_k = \sqrt{2} \int f(x)\phi(2x - k)dx,$$

$$\beta_{2k} = \int f(x)\phi(x - k)dx, \text{ and}$$

$$\gamma_{2k} = \int f(x)\psi(x - k)dx.$$

The isometry transforming the sequence (α_k) into $(\beta_{2k}, \gamma_{2k})$ can be written $F = (F_0, F_1)$ where F_0 and F_1 are commuting with even translations. One easily checks that

$$Id = F_0^* F_0 + F_1^* F_1.$$

In the terminology of signal analysis, F_0 and F_1 are said to be quadrature mirror filters. This notion has been introduced in 1977 by D.Esteban and C.Galand for improving the quality of digital transmission of sound.

Suppose now that a signal is given by a sequence of discrete values (α_k). We can consider that it is the coefficients of a function of V_0 on the $\phi(x - k)$. Iterating the filters defined above, we obtain the coefficients on the $\psi_{j,k}$ for $j \leq 0$. Each level requires only a discrete convolution. This algorithm constitutes the Fast Wavelet Transform (see [Ma2]).

2.2   Wavelets and signal analysis.

A strong point in favor of orthogonal wavelets is the existence of fast transform algorithms. An important drawback in signal analysis is the non-invariance of the decomposition under translations. There are now two ways to avoid this drawback. The first one is to use the *continuous wavelet transform*, that is, to take the scalar product of the signal with *all* the translations and dilations of the wavelet. Namely, one calculates

$$c(a, b) = \frac{1}{a} \int f(t) \psi(\frac{t - b}{a}) dt.$$

A one dimensional signal will thus be represented by a function of two variables defined in the upper half plane. This representation is commuting with translations and dilations. It is thus especially fitted to study signals which have some scaling invariance properties, and can actually be used to discover these properties just by "looking" at the transform. The continuous wavelet transform is, for instance, becoming an important tool in the study of fractals (see [AA]). The fact that wavelets perform an analysis localized simultaneously in space and frequency is of particular importance in the study of turbulence (see [FR]) since it allows to study the energy repartition of the flow in the spacial domain.

The continuous wavelet transform is computationally much more costly than the the orthogonal decomposition and can be used only when this point can be disregarded. Thus, it could not be used in real-time algorithms or as a tool in large iterative computations (such as in numerical analysis).

The following method was recently introduced by S.Mallat in order to obtain a computationally efficient translation-invariant wavelet transform. We consider the sampling of the continuous wavelet transform at the dyadic frequencies, that is the set of functions

$$g_j(x) = \int f(t)2^{j/2}\psi(2^j t - x)dt,$$

but, instead of sampling these functions at the points $x = k2^{-j}$ (as in the orthonormal case), we keep the position and value of the local extrema. The representation is now translation-invariant. However, the sampling is non linear, and difficult to handle mathematically.

3 Numerical analysis of partial differential equations.

The use of wavelets in numerical analysis is more recent than in signal analysis and the following should rather be considered as a description of "work in progress".

3.1 Wavelet method for elliptic problems.

Consider an elliptic problem, such as the Poisson problem, on a bounded domain. Let us first recall some properties of its resolution by Galerkin methods based on finite elements or of finite differences.

One of the main difficulties in these methods is that, once the problem has been properly discretized, one has to solve a system which is ill-conditioned. Typically, for a second order elliptic problem in two dimensions, one obtains a matrix M such that

$$\kappa = \| M \| \, \| M^{-1} \| = O(1/h^2)$$

where h is the size of the discretization (see [SF]). Such ill-conditioning has two drawbacks; it leads to numerical instabilities and to slow convergence for iterative resolution algorithms. In order to avoid this problem, one usually uses a preconditioning, which amounts to finding an easily invertible matrix D such that $D^{-1}MD^{-1}$ (or $D^{-1}M$, depending on the method used) will have a better condition number κ. For the example we considered, the usual preconditioning methods on general domains (SSOR or DKR on a conjugate gradient method, for instance) make κ become $O(1/h)$. We shall now describe a method based on orthonormal wavelets where this drawback disappears (see[J2]). Namely, the simplest conceivable preconditioning, when D is a diagonal matrix, yields a $\kappa = O(1)$ in any dimension (provided that the domain has a Lipschitz boundary). This result requires the construction of wavelets adapted to the domain Ω (see[JM]); they are no more obtained from each other by translation and dilation, but only keep the same size estimates. Let us first describe this wavelet basis.

Consider the space V_p of functions that are C^{2m-2}, vanish outside Ω, and are polynomials of degree $2m - 1$ in each variable in the cubes

$$k2^{-p} + 2^{-p}[0,1]^n, \text{ with } k \in Z^n$$

(thus, the discretization step is $h = 2^{-p}$). Then there exists an L^2-orthonormal basis of V_p composed of functions $\psi_{j,k}$ $(j \leq p)$ such that

$$\mid \partial^\alpha \psi_{j,k}(x) \mid \leq C2^{j\alpha}2^{nj/2} \exp(-\gamma 2^j \mid x - k2^{-j} \mid)$$

for $\mid \alpha \mid \leq 2m - 2$, and a positive γ.

The wavelets are indexed by j $(0 \leq j \leq p)$ and by the $k \in Z^n$ such that $k2^{-j} + (m + 1)2^{-j}[0,1]^n \subset \overline{\Omega}$. The decay estimates show that $\psi_{j,k}$ and its partial derivatives are essentially centered around $k2^{-j}$ with a width 2^{-j}. In the following, wavelets will be indexed by $\lambda = k2^{-j}$.

Actually, though these wavelets are not the same as in the case $\Omega = R^n$, they are "almost" the same; that is, only the wavelets that are close to the boundary are modified. Thus, we can essentially keep the fast decomposition algorithms, with only small modifications near the boundary.

Let us now describe the method of resolution for the Poisson equation. It is performed by a standard Galerkin method, keeping all the wavelets up to a frequency J_0. If we solve a Laplacian on a domain with Dirichlet boundary conditions, we have to invert a matrix

$$(M_{\lambda,\lambda'}) = (< \nabla\psi_\lambda \mid \nabla\psi_{\lambda'}).$$

We now renormalize the wavelets for the Sobolev H^1 norm, that is we consider that the functions on which the problem is discretized are the

$$\psi'_\lambda = 2^{-j}\psi_\lambda;$$

the condition number of the corresponding matrix is then bounded independently of the size discretization $h = 2^{-J_0}$. Thus, a conjugate gradient method will converge in a bounded number of steps, no matter how precise we require the solution to be. We would like to explain this result in the next part by comparing it with multigrid algorithms. Let us also mention that, if we use smooth wavelets, the order of accuracy of the method is extremely good since it is driven by the *local* regularity of the problem (by opposition to spectral methods, for instance).

3.2   Wavelet and multigrid algorithms.

A conjugate gradient method converges slowly when the condition number is large. Actually, the convergence is rather fast on the subspaces corresponding to the largest eigenvalues, but slow for the small eigenvalues. For an elliptic problem, small eigenvalues are associated to smooth, slowly oscillating functions (i.e. to wavelets indexed by a small j), and large eigenvalues to high frequency functions (i.e. to wavelets indexed by a large j).

Schematically, in a multigrid method, one starts by making a few steps of conjugate gradient, until the high frequency component of the solution is well approximated; the error is then a comparatively low frequency function, which can thus be accurately calculated on a grid with a double-size mesh. The resolution on the larger grid is performed again by the same method, and one iterates this procedure. The part of the solution which has frequencies around 2^j is thus calculated on the grid of size 2^{-j}. This is precisely what is performed on the wavelet method we described. The splitting on functions defined on meshes of different sizes (which is done in multigrid algorithms) is also performed by the wavelet decomposition. The essential difference is the following: it is the decomposition on wavelets and the recomposition which is iterative (it is performed by the fast algorithms described in part 2.2), but the resolution is just performed once by a conjugate gradient. Actually, when the function is written in its wavelet decomposition

$$f = \sum_j \sum_k C_{j,k} \psi_{j,k},$$

each block $\sum_k C_{j,k} \psi_{j,k}$ has its frequencies around 2^j, so that the purpose of the renormalisation that we make (multiply the terms of this block by 2^{-j}) is to bring all the eigenvalues of the matrix M close together so that a conjugate gradient will converge fast. Of course, this renormalization need not be performed in the multigrid algorithm, since, at each step, the problem has been localized in frequency.

3.3   Time-evolution problems.

Though some numerical algorithms using wavelets for time-evolution problems have already been successfully tested (see [GLR] and [LPT]), there is, to our knowledge, no analysis of the properties of such algorithms. However, some remarks can already be made.

First, let us recall that implicit time-evolution algorithms are needed when explicit schemes are either unstable or would require too small time-steps. Such algorithms require at each time-step the resolution of an elliptic problem. Thus, one can use the method described above for this resolution. A drawback of implicit schemes is usually their poor degree of approximation. However, the "locally optimal" approximation property that we already mentioned should improve this point.

Wavelet methods for time-evolution problems also have the advantage of allowing an easy tracking of singularities. First wavelets detect very sharply the position and nature of singularities and, on the other hand, of points of regularity (see[J1]). And second, it is easy to construct adaptative schemes similar to mesh-refinement methods by adding to the set of wavelets we work with some high-frequency wavelets localized in certain regions (they can for instance be added where a singularity has been detected).

4   Bibliography.

[ASH] E.H.ADELSON, E.SIMONCELLI ET R.HINGORANI. *Orthogonal pyramid transforms for image coding.* SPIE Vol. 845 Vision.Comm.and Image Proc. (1987).

[AA]F.ARGOUL, A.ARNÉODO, G.GRASSEAU, Y.GAGNE, E.J.HOPFINGER ET U.FRISCH. *Wavelet analysis of turbulence data reveals the multifractal nature of the Richardson cascade.* Nature Vol. 338, n.6210 (1989).

[Bg] J.BOURGAIN. *A remark on the uncertainty principle of hilbertian basis.* J. of funct. anal. Vol. 79 p136-143 (1988).

[CR] R.R.COIFMAN AND R.ROCHBERG. *Representation theorems for holomorphic and harmonic functions in L^p.* Asterisque $n^o 77$ (1980).

[D1] I.DAUBECHIES. *The wavelet transform, time-frequency localization and signal analysis,* I.E.E.E. on information theory (1989).

[D2] I.DAUBECHIES. *Orthonormal bases of compactly supported wavelets.* Comm. on pure and Appl.Math. Vol. 41 (1988).

[DGM] I.DAUBECHIES, A.GROSSMANN, Y.MEYER. *Painless nonorthogonal expansions.* J.Math.Phys. Vol. 27. p.1271-1283 (1986).

[DGM] I.DAUBECHIES, S.JAFFARD AND J.L.JOURNÉ. *A simple Wilson orthonormal basis with exponential decay.* To appear in S.I.A.M. of Mathematical analysis.

[FR] M.FARGE AND D.RABREAU. *Transformée en ondelettes pour detecter et analyser les structures cohérentes dans les écoulements turbulents bidimensionnels.* C.R.A.S. Vol. 307, Série 2 p1479-1486 (1988).

[GLR] R.GLOWINSKI, W.LAWTON, M.RAVACHOL, *Wavelet solution of linear and nonlinear elliptic, parabolic and hyperbolic problems in one space dimension,* preprint (June 1989).

[J1] S.JAFFARD, *Exposants de Holder en des points donnés et coefficients d'ondelettes* C.R.A.S. Vol308 Série 1 p.79-81 (January 1989).

[J2] S.JAFFARD, *Wavelet methods for fast resolution of elliptic problems,* preprint (1990).

[JM] S.JAFFARD ET Y.MEYER. *Bases d'ondelettes dans des ouverts de R^n.* Jour.Math.pures et appl. Vol.68 p.95-108 (1989).

[LPT] LIANDRAT, V.PERRIER AND P.TCHAMITCHIAN *Numerical resolution of the regularized Burgers equation using the wavelet transform.* Preprint (1990).

[L] P.G.LEMARIÉ. *Ondelettes à localisation exponentielle.* J.Math.Pures et Appl. Vol 67 p.227-236 (1988).

[LM] P.G.LEMARIÉ ET Y.MEYER. *Ondelettes et bases hilbertiennes.* Revista Math.Iberoamericana Vol.1 (1986).

[Ma1] S.MALLAT. *Multiresolution approximations and wavelet orthonor-*

mal bases of $L^2(R)$. Trans.A.M.S. (1989).

[Ma2] S.MALLAT. *A theory for Multiresolution signal decomposition:The wavelet representation.* I.E.E.E. on pattern analysis and machine intelligence, Vol. II,n.7 (dec.89).

[M1] Y.MEYER. *Ondelettes et opérateurs.* Hermann (1990).

[M2] Y.MEYER, *Ondelettes, fonctions splines et analyses graduées* (Cahiers de Mathématiques de la décision, CEREMADE, Université Paris-Dauphine,75" Paris Cedex 16).

[SW] H.C.SCHWEINLER AND E.P.WIGNER, *Orthogonalization methods,* J.Math.Phys. Vol.11 p.1693-1694 (1970).

[SF] G.STRANG AND G.J.FIX, *An analysis of the finite element method,* Prentice hall.

[S] J.O.STROMBERG, *A modified Franklin system and higher order spline systems on* R^n *as unconditionnal bases for Hardy spaces.* Conference in honor of Antoni Zygmund,Vol.2 p.475-493, W.Beckner ed. Wadsworth Math series.

[W] K.G.WILSON . *Generalized Wannier Functions* (preprint 1987) Cornell Theory Center and Laboratory of Nuclear Studies, Cornell U. Ithaca, NY 14853 USA.

Stéphane Jaffard , *Laboratoire d'analyse et de modélisation mathématiqu (C.E.R.M.A.), Ecole Nationale des Ponts et Chaussées, 93167 Noisy le Grand,* **France**

ESTIMATION OF DIRECTIONAL SPECTRA OF OCEAN WAVES

Harald E. Krogstad

"The basic law of the seaway is the apparent lack of any law"

1 Introduction

The above citation, attributed to Lord Rayleigh by Kinsman [10], illustrates the general state of knowledge about wind waves on the ocean surface up to the Second World War. Although of fundamental concern to seamen in all times, the mathematical theory for the confused sea was not really developed until the mid to late fifties, and even today details about the generation of waves by wind are subject to controversy.

The first attempts to consider a random sea were in the mid forties where Barber and Ursell carried out a spectral analysis of data from a bottom mounted pressure recorder by means of a purely mechanical device. The conceptual model of the sea as a random sum of plane waves was developed further by Longuet-Higgins [15], and the evolution of the theory is nicely described in the book by Kinsman [10] from 1965.

The stochastic model of surface waves is a stochastic field solution to the deterministic equations for surface waves. The equations are nonlinear, and no full closed form solutions are possible. However, by scaling the equations for the case of small surface slope, one obtains a solution which together with an assumption that the solution is Gaussian, constitutes what is commonly referred to as Gaussian, linear wave theory. This first order solution is surprisingly accurate, and it is probably fair to say that ocean wind waves represent one of the best real examples of a multivariate Gaussian field. For appropriate scales, the second order stochastic properties of the surface are thus characterised by a spectrum as a function of wavenumber and angular frequency.

It was early realised by observations that the interaction between water waves is weak, but by carrying out an expansion similar to the Stokes perturbation expansion for periodic waves, O. Phillips and K. Hasselmann [24] revealed a mechanism where four waves interact and gradually transfer energy to each other. By carrying the expansion up to the sixth order in the resonant terms leading to energy redistribution, K. Hasselmann was able to derive an equation formally similar to the Boltzmann equation describing the time evolution of the interacting field [8].

However, Gaussian, linear wave theory has remained the basic model for most practical applications of random waves to ocean engineering and naval architecture [20].

Measurements of waves are generally aiming at measuring various forms of the wave spectrum, or parameters related to the spectrum. Wave observations range from visual observations from ships and lighthouses to sophisticated recorders which take simultaneous recordings of a multitude of quantities related to the waves, at possibly different positions. The problem

M. Heiliö (ed.), Proceedings of the Fifth European Conference on Mathematics in Industry, 35–43.

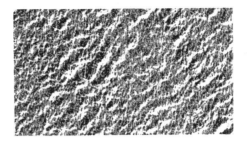

Figure 1: Aerial image of ocean waves illustrating the randomness of the sea (Courtesy Norsk luftfoto og fjernmåling).

of estimating the multidimensional ocean wave spectrum from a set of wave records leads to a nice interplay between multivariate stochastic processes and Fourier analysis. It also represents, because of the dimensionality of the spectrum, a typical inverse problem.

2   The stochastic model of the surface

For scales of the order of 10 km and .5 hour, it is appropriate to model the open ocean surface as a horizontally homogeneous and stationary stochastic field,

$$\eta(x,t) = \int_{k,\omega} e^{i(kx-\omega t)} dB(k,\omega), \tag{1}$$

where x is the horizontal coordinate, t is time, ω is the angular frequency, k is the wavenumber and B is the spectral amplitude. The spectrum, χ, of η is called the (three dimensional) wave spectrum, and defined by $\chi(C) = \mathcal{E}[B(C)B^*(C)]$ where C is a measurable set in the wavenumber-frequency space and $\mathcal{E}$ is the expectation operator.

Assuming inviscid, irrotational and incompressible flow and deep water, the wave induced water velocity may be obtained from a velocity potential function $\Phi(x,z,t)$ satisfying the Laplace equation in the mean square stochastic sense with a spectral representation $\Phi(x,z,t) = \int_{k,\omega} e^{i(kx-\omega t)} e^{kz} dA(k,\omega)$, $k = |k|$ [23]. The potential and the surface are connected by two conditions at the free surface denoted the kinematic and the dynamic boundary conditions, respectively [23]:

$$\eta_t = \Phi_z - (\Phi_x \eta_x + \Phi_y \eta_y) \text{ for } z = \eta(x,t),$$
$$\Phi_t + (\Phi_x^2 + \Phi_y^2)/2 + g\eta = 0 \text{ for } z = \eta(x,t), \tag{2}$$

(g is the acceleration of gravity).

Expanding the boundary conditions, η and Φ in a small parameter, typically the mean slope of the surface, gives a perturbation expansion whose leading order solution is summarised in $\omega^2 = gk$ and $dA = -\frac{i\omega}{k} dB$. By further assuming that η is Gaussian, "since it consists of a random sum of many small waves", this leads to what is commonly referred to as Gaussian Linear Wave Theory (GLWT). In GLWT, B and χ are thus singular measures supported on the dispersion surface, $\omega^2 = gk$. This, and the symmetry of χ, may be used to eliminate ω as an independent variable and express the first order spectrum only in terms of k. This wavenumber spectrum is defined as

$$\Psi(\boldsymbol{k})d\boldsymbol{k} = 2 \int_{\omega > 0} d\chi(\boldsymbol{k}, \omega). \tag{3}$$

For other applications, it is convenient to write Ψ in terms of ω and the direction, θ, of the wavenumber. This is called the directional spectrum and written $E(\omega, \theta) = S(\omega)D(\theta, \omega)$ where S is the frequency spectrum and D the directional distribution having integral with respect to θ normalised to 1. These reduced forms of the wave spectrum are limited to linear wave theory.

When carrying the stochastic perturbation expansion to higher orders, the arithmetic quickly gets very complicated. A readable treatise of higher order spectra are found in [19].

Recently, an interesting alternative approach based on the Hamiltonian of the wave field has been introduced [5]. The Hamiltonian uses surface elevation and the velocity potential at the surface as canonical variables, and the idea is to carry out a canonical transformation of the variables such that the leading order nonlinear term in the perturbation expansion of the Hamiltonian in the new variables vanishes. The linear theory in the new variables is similar to the conventional theory but if this linear solution is transformed back to the original variables, the surface will no longer be Gaussian. In particular, a sine wave in the new set of variables closely approximates the finite amplitude Stokes wave in the original variables.

3 Estimation of wave spectra

Ocean waves are observed by a multitude of instruments as reflected in the report by W.J. Pierson prepared for NASA with the suggestive title *The Theory and Applications of Ocean Wave Measuring Systems at and Below the Sea Surface, on the Land, from Aircraft and from Spacecraft* [22]. Since estimation of the full frequency-wavenumber spectrum of the surface would require an $\mathcal{O}(1$ hour, 1s sampling interval) time series of an $\mathcal{O}(1$ km^2, 5 m sampling distance) area of the sea, this has so far been been out of question. Estimation of the wave spectrum thus represents a typical *inverse problem*: How do we obtain as much information as possible about the spectrum from our limited set of data.

In situ measurements often consist of a set of simultaneously measured time series of *e.g.* surface elevation, slope, velocity etc., taken at fixed, possibly different locations. Within GLWT, the series are sampled realisations $W(m) = Y(t = 2\pi m/\omega_s)$, $m = 0, \cdots, M-1$ of a multivariate stochastic process, Y, with spectral representation

$$Y(t) = \int_{\boldsymbol{k}, \omega} \boldsymbol{T}(\boldsymbol{k}, \omega) e^{i(\boldsymbol{k}\boldsymbol{x}_i - \omega t)} d\boldsymbol{B}(\boldsymbol{k}, \omega) \tag{4}$$

where $\boldsymbol{T}$ are (known) transfer functions and $\{\boldsymbol{x}_i\}$ are the locations of the observations. The cross spectrum for Y, $\boldsymbol{\Sigma}(\omega)$, is thus given by

$$\boldsymbol{\Sigma}(\omega) = S(\omega) \int_{\theta=0}^{2\pi} \boldsymbol{T}(\boldsymbol{k}, \omega) \boldsymbol{T}^H(\boldsymbol{k}, \omega)) D(\theta, \omega) d\theta. \tag{5}$$

In practice, the sampling frequency is usually high enough to avoid excessive aliasing, and the recording interval is long compared to the correlation time of the series. This implies that the series are *well-recorded* in the sense that the asymptotic properties of the finite Fourier transform, $\widehat{W}$, of W apply for all practical purposes. This means, in particular, that for

$\omega_m = m\omega_s/M, |m| < M/2, \widehat{W(\omega_m)}$ are zero mean, multivariate complex Gaussian variables with a covariance matrix essentially defined by Eqn. (5). The coefficients are also virtually independent for different values of m [2]. The standard estimate of the cross spectrum (or the covariance matrix of $\widehat{W}$) is now [2]

$$\widehat{\Sigma(\omega)} = \frac{2}{\nu} \sum_{\beta=1}^{\nu/2} \widehat{W}(\omega_\beta) \widehat{W}^H(\omega_\beta), \tag{6}$$

where ω_β runs over ω_m–s surrounding ω (Note the use of a "hat" both for the Fourier transform and the estimator). With the above property of the Fourier coefficients, the standard estimate is a maximum likelihood (ML) estimate of the spectrum if no further constraints are imposed on the matrix ([17], Sec. 4.2).

A popular way of measuring directional spectra is by utilising so-called *single point triplets* of which the heave/pitch/roll buoy [14] is the most common. After a proper rotation of the data and elimination of tracking errors, the buoy records η, $\partial\eta/\partial x$, and $\partial\eta/\partial y$ at a fixed point. Due to the general validity of the corresponding transfer functions, data from a heave/pitch/roll buoy may also be analyzed without assuming GLWT [1]. Assuming GLWT, the cross spectral matrix of the three time series is

$$\Sigma = S \begin{pmatrix} 1 & -ika_1 & -ikb_1 \\ ika_1 & k^2(1+a_2)/2 & k^2b_2/2 \\ ikb_1 & k^2b_2/2 & k^2(1-a_2)/2 \end{pmatrix}, \tag{7}$$

where the Fourier coefficients of D have been introduced and the dependence on ω has been omitted for clarity.

If one let the wavenumber be an independent parameter, *i.e.* not given by the dispersion relation, the matrix is similar to a real symmetric matrix with six independent parameters. This essentially reduces the ML estimation problem to the unconstrained real matrix case, and proves that the commonly used set of estimators for S, k, a_1, b_1, a_2 and b_2 [13] in fact are ML estimators in this case. By restricting the wavenumber to the dispersion relation, the ML estimates have to be obtained numerically. The modified estimators show somewhat improved performance with respect to sampling variability [7].

Unless in the singular case where D is a δ–function, the ML solution is not unique. The Fourier coefficients estimated in the above way constitute the start of a positive definite sequence and are thus, by the Bochner-Herglotz theorem, in accordance with a positive D. The extension of positive definite sequences is closely related to the trigonometric moment problem and Maximum Entropy (ME) spectral estimation [3]. It is therefore natural to seek uniqueness by maximising the entropy $-\int_0^{2\pi} \log[D(\theta)]d\theta$ subject to the constraints $D(\theta) > 0$, $\int_0^{2\pi} e^{-in\theta} D(\theta)d\theta = c_n = a_n + ib_n$, $n = 0, 1, 2$. The solution is

$$D(\theta) = \frac{1}{2\pi} \frac{\sigma_e^2}{|1 - \phi_1 e^{-i\theta} - \phi_2 e^{-i2\theta}|^2}, \tag{8}$$

where the parameters ϕ_1, ϕ_2 and σ_e^2 are obtained from the Yule-Walker equations [16]. The ME solution is the solution closest to the uniform distribution in the entropy metric subject to data consistency, *i.e.* c_1 and c_2, and although in this way optimal when no *a priori* information of the directional distribution is available, the method may also be extended to include prior knowledge [9].

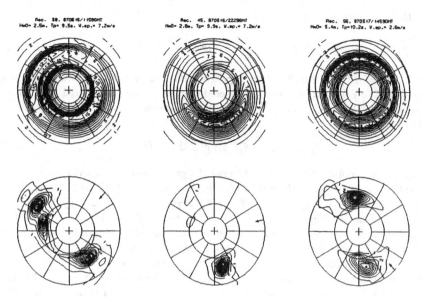

Figure 2: Examples of wavenumber spectra collected during the Labrador Sea Extreme Wave Experiment in 1987. Circles signify wavelengths of 100, 200 and 400m, respectively. Upper row: directional distribution computed by the Longuet-Higgins estimate ; lower row: MEM-estimate.

Heave/pitch/roll data buoys have routinely been monitoring the wave climate at the Norwegian oil fields since 1980. In practice, the time series from the buoys must be rotated into a fixed coordinate system and calibrated for hydromechanical and electronical transfer functions before the actual estimation of the spectra and the Fourier coefficients is carried out. Some examples of estimated wavenumber spectra are shown in Fig. 2. These data were acquired during the Labrador Sea Extreme Wave Experiment (LEWEX) in 1987 [25]. The upper row shows spectra computed using the Longuet-Higgins directional estimate [14]

$$\hat{D}(\theta) = \frac{1}{2\pi}[1 + \frac{2}{3}(\hat{a}_1\cos(\theta) + \hat{b}_1\sin(\theta)) + \frac{1}{3}(\hat{a}_2\cos(2\theta) + \hat{b}_2\sin(2\theta))], \qquad (9)$$

whereas the lower row displays the corresponding MEM estimates. Given c_1 and c_2, the Longuet-Higgins estimate is analogous to a Blackman-Tukey spectral estimate, whereas the MEM estimate corresponds to the Burg ME estimate.

Returning to Eqn. (5) for general arrays, little progress in using the ML-formulation directly has been reported. By factoring out the direction independent part of the transfer functions, i.e. writing $T_i(\omega, \theta) = R_i(\omega)h_i(\theta, \omega)$, the cross spectral matrix may be written $\Sigma = SR\Phi R^H$ where $R = diag(R_1, \cdots, R_N)$ and $\Phi = \int_0^{2\pi} h(\theta)D(\theta)h(\theta)^H d\theta$ (We omit the dependence on ω in the following).

The first attempts to extract spectra from spatial arrays considered Hermitian forms $\hat{D}(\theta) = \gamma^H(\theta)\hat{\Phi}\gamma(\theta)$ where the vector of weights γ was independent of Φ. Since the expectation of $\hat{D}(\theta)$ is approximately

$$\mathcal{E}(\hat{D}(\theta)) = \int_0^{2\pi} K(\theta, \theta') D(\theta') d\theta', \tag{10}$$

where $K(\theta, \theta') = \gamma^H(\theta) h(\theta') h(\theta')^H \gamma(\theta) = |\gamma^H(\theta) h(\theta')|^2$, the various procedures aim at approximating K to a δ-function. Several of these formulations may be reduced to minimising an expression $\gamma^H(\theta) \Xi \gamma(\theta)$ while $|\gamma^H(\theta) h(\theta)| = 1$. The solution, apart from a constant of modulus 1, is given by

$$\gamma_s(\theta) = \Xi^g h(\theta) / (h(\theta)^H \Xi^g h(\theta))^{1/2}, \tag{11}$$

where Ξ^g is a generalised inverse of Ξ. The use of a generalized inverse has turned out to be important in practice when working with heavily correlated series [11]. One example is maximising $K(\theta, \theta' = \theta))$ while $\int_0^{2\pi} K(\theta, \theta') d\theta' = \gamma(\theta)^H \Xi \gamma(\theta) = 1$, where $\Xi = \int_0^{2\pi} h(\theta) h(\theta)^H d\theta$. This gives the Longuet-Higgins directional estimate when applied to heave/pitch/roll data.

However, these linear estimators are today mostly replaced by *data adaptive* estimators. In 1969 Capon [4] suggested a method for array processing which later has become known as the Maximum likelihood method (MLM) for directional spectra. The term Maximum likelihood stems from the fact that the method may be interpreted as the maximum likelihood estimate for the direction of a plane wave recorded in a background of noise. The method was introduced in the context of ocean wave spectra by Davies and Regier [6] and is derived by minimising $\int_0^{2\pi} K(\theta, \theta') D(\theta') d\theta'$ while $K(\theta, \theta) = 1$, the solution of which is again given by Eqn. (11) with $\Xi = \Phi$. The procedure is repeated for each direction θ and the estimate, D_{MLM}, is scaled afterwards,

$$D_{MLM}(\theta) = \kappa \gamma^H(\theta) \Phi \gamma(\theta) = \kappa / h^H(\theta) \Phi^g h(\theta), \ \kappa^{-1} = \int_0^{2\pi} (h^H(\theta) \Phi^g h(\theta))^{-1} d\theta \tag{12}$$

In practice, the standard estimate is used for Φ since the expression remains meaningful as long as $\hat{\Phi}$ is positive semi-definite [12].

Contrary to the ME-distribution constructed from a given set of Fourier coefficients, the distribution produced by Capon's ML method is not in accordance with a cross spectrum computed from Eqn. (5). However, an interesting idea of iteratively improving the estimate has been suggested by Pawka [21]. Let $\Phi_D = \int_0^{2\pi} h(\theta) D(\theta) h(\theta)^H d\theta$ and consider the operator $M : D \to D_{MLM}$ defined by

$$D \to \Phi_D \to D_{MLM} = \kappa / h^H \Phi_D^g h. \tag{13}$$

The operator is not in general 1-1 since any function orthogonal to $\{h_i h_j^*\}_{i,j=1,N}$ may be added to D without affecting $M(D)$ as long as D is kept positive with an integral equal to 1.

One way of formulating this as an inverse problem would now be to introduce some kind of functional to be maximised together with the constraint $M(D) = \hat{D}_{MLM}$ where $\hat{D}_{MLM}$ is the ML estimate obtained from the data. So far, good experience with the *ad hoc* iterative scheme

$$D_{n+1} = \max[0, D_n + \omega_R(\hat{D}_{MLM} - M(D_n))], \ D_0 = \hat{D}_{MLM} \tag{14}$$

$$\int_0^{2\pi} D_{n+1}(\theta) d\theta = 1$$

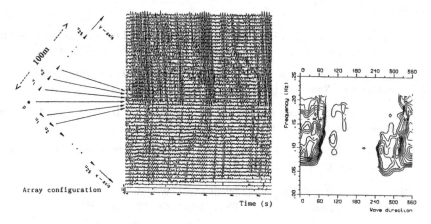

Figure 3: Time series of velocity recordings from an acoustic Doppler current meter recorded at the Odin field in the Norwegian Sea and corresponding directional spectrum [12].

has been obtained without taking any functional into account [12]. An in depth theoretical study of this iterative maximum likelihood method (IMLM) has not been carried out. Numerical experiments indicate that the relaxation parameter ω_R may be chosen around 1 . With real data, the convergence, measured in terms of $\| D_n - \hat{D}_{MLM} \|_\infty$, decreases geometrically in the beginning, but reaches a fluctuating level after a few iterations [12].

No practical algorithm for the maximum entropy estimate for general arrays has so far been developed. Algorithms which first estimate Fourier coefficients and subsequently use Eqn. (8) have not shown better performance than the simpler IMLM-algorithm discussed above. In fact, Fourier coefficients computed by a numerical integration of the IMLM-distribution typically show less sampling variability than directly estimated Fourier coefficients.

Fig. 3 shows an example of time series recorded by an acoustic Doppler current meter. This instrument records the wave induced water velocity component along two orthogonal, horizontal sonic beams in addition to the dynamic pressure at the instrument [12]. The velocity recordings were in this case every 4.6m such that the instrument in effect represents a spatial array with an extension more than 100m. An example of a spectrum is shown to the right.

4   Conclusions

With the development of advanced data acquisition systems, measurements of directional wave spectra, often in near real time, are becoming ever more important for ship routing and offshore operations, offshore engineering, and for calibration and control of numerical wave models. Although focusing on *in situ* measurements above, remote sensing techniques, of which the image producing synthetic aperture radar is the most promising [25], may turn out to give global monitoring systems by the end of the century. The increased availability and use of the spectra is also likely to induce further theoretical developments, bringing in methods from related areas like radar, radio astronomy and seismic processing [18].

References

[1] Barstow, S.F. and H.E. Krogstad: General analysis of directional ocean wave data from heave/pitch/roll buoys , Mod. Ident. and Control **5**, 47-70 (1984).

[2] Brillinger, D.V.: *Time Series. Data Analysis and Theory*, Holt, Rinehart and Wilson Inc., New York (1975).

[3] Burg, J.P.: *Maximum Entropy Spectral Analysis*, Ph.D. Thesis, Dept. of Geophysics, Stanford Univ. (1975).

[4] Capon, J.: High resolution frequency-wavenumber spectrum analysis, Proc. IEEE **57**(8), 1408-1418 (1969).

[5] Creamer, D.B. , F. Henyey, R. Schult and J. Wright: Improved linear representation of ocean surface waves, J. Fluid Mech. **205**, 135-161 (1989).

[6] Davies, R.E. and L.A. Regier: Methods for estimating directional wave spectra from multi-element arrays, J. Mar. Res. **35**, 453-477 (1977).

[7] Glad, I.K.: Statistical properties of directional spectrum estimators, Thesis, Dept. of Mathematics, NTH (in Norwegian) (1990)

[8] Hasselmann, K. : On the non-linear energy transfer in a gravity-wave spectrum, Part 1: General theory , J. Fluid Mech. **12**, 481-500 (1962). Part 2: Conservation theorems; wave-particle analogy; irreversibility, *Ibid.* **15**, 273-281 (1963). Part 3: Evaluation on the energy flux and swell-sea interaction for a Neumann spectrum, *Ibid.* **15**, 385-398 (1963).

[9] Jones, L.K.: Approximation-theoretic derivation of logarithmic entropy principles for inverse problems and unique extension of the maximum entropy method to incorporate prior knowledge, SIAM J. Appl. Math. **49**, 650-661 (1989)

[10] Kinsman, B.: *Wind Waves* , Prentice-Hall Inc., Englewood Cliffs, New Jersey (1965).

[11] Krogstad, H.E.: Maximum likelihood estimation of ocean wave spectra from general arrays of wave gauges, Mod. Ident. Control **9**, 81-97 (1988).

[12] Krogstad, H.E. , R.L. Gordon and M.C. Miller: High-resolution directional wave spectra from horizontally mounted acoustic doppler current meters, J. Atmos. Ocean. Techn. **54**, 340-352 (1988).

[13] Long, R.B.: The statistical evaluation of directional estimates derived from pitch/roll buoy data, J. Phys. Ocean. **9**, 373-381 (1980).

[14] Longuet-Higgins, M.S. , D.E. Cartwright and N.D. Smith: Observations of the directional spectrum of sea waves using the motion of a floating buoy, in *Ocean Wave spectra*, Printice-Hall, 111-136 (1963).

[15] Longuet-Higgins, M.S.: The statistical analysis of a random moving surface, Phil. Trans. Roy. Soc. **A 249**, 321-387 (1957).

[16] Lygre, A. and H.E. Krogstad: Maximum entropy estimation of the directional distribution in ocean wave spectra, J. Phys. Ocean. **16**, 2052-2060 (1986).

[17] Mardia, K.V. , J.T. Kent and J.M. Bibby: *Multivariate Analysis*, Acad. Press, London (1979).

[18] McDonough, R.N.: Application of the maximum likelihood method and the maximum entropy method to array processing, in *Nonlinear Methods in Spectral Analysis*, Topics in Applied Physics **34**, ed. S. Haykin. Springer, Berlin, Heidelberg, New York, 181-244 (1979).

[19] Mitsuyasu, H., A. Masuda and Y.-Y. Kuo: The dispersion relation of random gravity waves, Part 1, J. Fluid Mech. **92**, 717-730 (1979).

[20] Newman, J.N.: *Marine Hydrodynamics*, MIT Press, Cambridge, Massachusetts (1980).

[21] Pawka, S.S.: Island shadows in wave directional spectra, J. Geophys. Res. **14**, 1800-1810 (1983)

[22] Pierson, W.J.: *The theory and applications of ocean wave measurements systems at and below the sea surface on the land, from aircraft and from spacecraft*, National Aeronautics and Space Administration, CR-2646 , 389p. (1976).

[23] Phillips, O.: *The Dynamics of the Upper Ocean*, 2nd Ed. 1977, Cambridge University Press, Cambridge (1977).

[24] Phillips, O.: Wave interactions - the evolution of an idea, J. Fluid Mech. **106**, 215-227 (1981).

[25] Vachon, P.W. , R.B. Olsen, C.E. Livingstone, N.E. Freeman: Airborne SAR imagery of ocean surface waves obtained during LEWEX: Some initial results , IEEE Trans. on Geosc. and Rem. Sensing **26**, 548-561 (1988).

Address:
Harald E. Krogstad
SINTEF Industrial Mathematics
N-7034 Trondheim
NORWAY

OPTIMAL SHAPE DESIGN

Pekka Neittaanmäki

University of Jyväskylä, Department of Mathematics
P.O. BOX 35, SF 40351 Jyväskylä, Finland

1. Introduction

The design of the geometry (a geometrical layout) for the structure is the primary problem of designers of structural systems. In spite of graphical work stations and modern software for analyzing the stucture, finding the best geometry for the stucture by "trial and error" is still a very tedious and timeconsuming task. The goal in optimal shape design (structural optimization, or redesign) is to computerize the design process and therefore shorten the time it takes to design new products or improve the existing design. Structural optimization is already used in certain applications in the automobile, marine, aerospace industries and in designing truss and shell structures (with minimum weights). In general, however the structural optimization is just beginning to penetrate the industrial community. The integrated FEM (Finite Element Method) and CAD (Computer Aided Design) technologies within optimization loop will (hopefully quite soon) fully computerize the design loop.

Parameters chosen describe the design (geometry) of the system are called design variables. The design parameters can be either finite dimensional (vector) or distributed parameters. Optimal shape design problems can be divided roughly into three classes: optimal sizing, domain optimization and topology optimization.

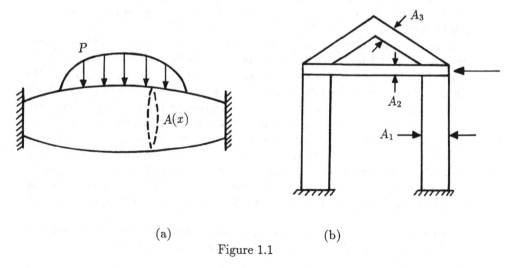

(a)   (b)

Figure 1.1

45

M. Heiliö (ed.), Proceedings of the Fifth European Conference on Mathematics in Industry, 45–54.
© 1991 B. G. Teubner Stuttgart and Kluwer Academic Publishers. Printed in the Netherlands.

Optimal sizing usually deals with structural optimization. We assume that the layout of the structure is given and we try to find optimal sizes of the structural members. The sizes of the members are chosen as the design parameters that can be of a vector or distributed type. In Figure 1.1 (a)–(b) we see two typical sizing problems in structural optimization: optimal sizing of a beam (distributed parameter) and of a frame (vector parameter), [**1,2,3,4,6**].

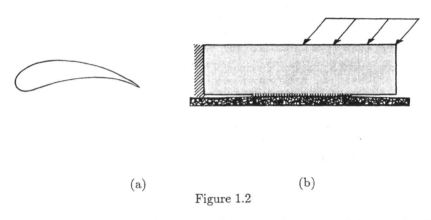

(a)   (b)

Figure 1.2

In domain optimization (or variable boundary optimization) the shape of the two– or three–dimensional domain is sought. Usually the problem is reduced to finding a vector function which defines the unknown boundary. In Figure 1.2 (a)–(b) we see two quite different domain optimization problems: finding an optimal shape of an airfoil and optimization of the contact surface of an elastic body on a rigid foundation, cf. [**3,4,6,8**].

Topology optimization deals with the search of optimal lay-out of the system. For example, find the optimal lay-out of the truss such that the weight of the truss is minimized and the truss can carry a given load without collapsing. On the other hand, in topology optimization, the mechanical body can be considered as a domain in space with high density of material, that is the body is described by the global density function that assigns the material to areas that are parts of the body. So one tries to distribute a minimum amount of material inside the structure in such a way that the structure is able to perform some mechanical task. Moreover, certain shaded areas are required to contain material. Topology optimization problems have an on-off nature and are therefore extremely difficult to solve in the distributed case. For some topology optimization type of problems see [**5**].

The previous division of optimal shape design problems into three classes is anything but strict. Moreover, it should be noted that the nature of the optimal shape design problem also depends on how the state of the system is modelled. For example, if the beam in Figure 1.1 (a) is modelled using the classical beam theory the problem is a sizing problem, but if the plane stress model is used the problem is a domain optimization problem.

2. SETTING OF THE PROBLEM

Let $\Omega \in \mathcal{O}$ (= set of admissible domains) be a domain for which we want to find an optimal design (an optimal geometrical layout). We suppose that $\mathcal{O}$ is a subset of some larger family $\tilde{\mathcal{O}}$; $\mathcal{O} \subseteq \tilde{\mathcal{O}}$.

With any $\Omega \in \tilde{\mathcal{O}}$ we associate a Hilbert space $V(\Omega)$ of functions, defined on Ω. In order to handle the situation mathematically, we introduce topologies in $\tilde{\mathcal{O}}$ and in $\{V(\Omega) \mid \Omega \in \tilde{\mathcal{O}}\}$. If Ω_n , $\Omega \in \tilde{\mathcal{O}}$, we have to define what it means that

$$(2.1) \qquad \Omega_n \xrightarrow{\tilde{\mathcal{O}}} \Omega .$$

Analogously, if $y_n \in V(\Omega_n)$, $y \in V(\Omega)$, Ω_n, $\Omega \in \tilde{\mathcal{O}}$, then we specify the convergence

$$(2.2) \qquad y_n \to y .$$

Let

$$(P) \qquad \Omega \in \mathcal{O} \to y(\Omega) \in V(\Omega)$$

be a mapping which with any domain $\Omega \in \mathcal{O}$ associates the solution of a state problem (given by equations, inequalities etc. in Ω) and let

$$(2.3) \qquad G = \{(\Omega, y(\Omega)) \mid \Omega \in \mathcal{O}\}$$

be its graph.

Finally, let $I(\Omega, y)$ with $\Omega \in \mathcal{O}$,and $y \in V(\Omega)$ be a cost function (criterion function), whose restriction on G will be denoted by $J(\Omega)$, i.e.

$$(2.4) \qquad J(\Omega) = I(\Omega, y(\Omega)).$$

The *abstract optimal shape design problem* is stated as follows:

$$(\mathbf{P}) \qquad \begin{cases} \text{Find } \Omega^* \in \mathcal{O} \text{ such that} \\ J(\Omega^*) \leq J(\Omega) \text{ for all } \Omega \in \mathcal{O}. \end{cases}$$

We will say that $(\Omega^*, y(\Omega^*))$ is an optimal pair for $(\mathbf{P})$.

THEOREM 2.1. *Assume that G is compact in the following sense: If $\{\Omega_n\}$, $\Omega_n \in \mathcal{O}$ is an arbitary sequence, there exist a subsequence $\{(\Omega_{n_k}, y(\Omega_{n_k}))\} \subset \{(\Omega_n, y(\Omega_n))\}$ and an element $(\Omega, y(\Omega)) \in G$ such that*[1] $\Omega_{n_k} \xrightarrow{\mathcal{O}} \Omega$, $y(\Omega_{n_k}) \to y(\Omega)$. *Moreover, let I be lower semicontinuous: If Ω_n, $\Omega \in \mathcal{O}$ with $\Omega_n \xrightarrow{\mathcal{O}} \Omega$ and if $y_n \in V(\Omega_n)$, $y \in V(\Omega)$ with $y_n \to y$, then*

$$(2.5) \qquad\qquad \liminf_{n \to \infty} I(\Omega_n, y_n) \geq I(\Omega, y) .$$

Then there exists at least one solution $\Omega^* \in \mathcal{O}$ of (**P**).

The proof of Theorem 2.1 is given in [**3**].

A large range of important optimal shape design problems which arise in structural mechanics, acoustics, electric fields, fluid flow and other areas of engineering and applied sciences can be formulated as in [**1,2,3,6,7,8**]. Typically,

$$J(\Omega) = \int_\Omega dx \text{ (minimization of the weight) },$$

$$J(\Omega) = \int_\Omega (y(x))^2 \, dx \text{ (minimization of displacements) },$$

$$J(\Omega) = \int_\Omega (\nabla y(x))^2 \, dx \text{ (minimization of stresses) },$$

$$J(\Omega) = \int_{\partial\Omega(u)} \frac{\partial}{\partial n} y(x) \, dx \text{ (minimization of contact stresses or boundary flux) }.$$

Sometimes we meet situations where the mapping $y : \Omega \to y(\Omega)$ is not differentiable (the state is described by a variational inequality, e.g.). To overcome this difficulty, one can apply some *regularization technique* to (P) in order to obtain a smooth problem [**3**].

If the variable part $\Gamma(u)$ of the boundary of Ω can be defined by a parameter u belonging to the set of admissible controls U_{ad}, the problem then is to find an optimal shape for $\Gamma(u)$. In this case one has the mappings

$$u \to \Omega(u) \to y(u) \to J(u; y(u))$$

and one gets the problem

$$(\mathbf{P(u)}) \qquad\qquad \min_{u \in U_{ad}} J(u; y(u)) .$$

Problems (**P**) and (**P(u)**) can be solved by the techniques in calculus of variations and mathematical programming. In applying nonlinear programming methods there arises two difficulties:

    – the mapping $u \to J(u; y(u))$ is not necessary differentiable [**3,7**]
    – the mapping $u \to J(u; y(u))$ is not necessary convex. [**4,7**]

---

[1] Topology in $\mathcal{O}$ is induced by the topology in $\tilde{\mathcal{O}}$.

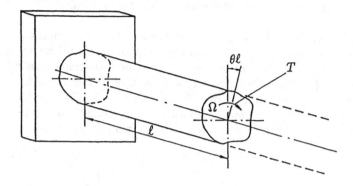

Figure 3.1

3. INDUSTRIAL APPLICATIONS

We shall present a collection of optimal shape design problems which would be interesting as well from an industrial point of view. For the technical background of these problems as well as numerical solution we refer to [3,4,7,8].

EXAMPLE 3.1. Maximization of the torsional rigidity of an elastic shaft.

Consider the torsion of the elastic shaft shown in Figure 3.1. Let Ω denote the cross-section of the shaft. A torque, T, is applied to the shaft at its free end, resulting in a unit angle of twist, θ. From the St Venant theory of torsion, the elastic deformation of the system is governed by the elliptic boundary value problem

$$(3.1) \qquad\qquad -\Delta u = 2 \quad \text{in } \Omega$$

$$(3.2) \qquad\qquad u = 0 \quad \text{on } \partial\Omega$$

where u is the Prandtl stress function. The torque–angular deflection relation is given by $T = GR\theta$, where G is the shear modulus of the material of the shaft and R is the torsional rigidity of the shaft given by

$$R(\Omega) = 2 \int_{\Omega} u(x)\, dx \ .$$

The problem of optimal shape design of Ω is given by

$$(3.3) \qquad\qquad \min_{\Omega \in \mathcal{O}} I(\Omega; u(\Omega)) \ ,$$

where

$$I(\Omega; u(\Omega)) = -R(\Omega) \ ,$$

$u(\Omega)$ solves (3.1), (3.2) and

$$\mathcal{O} = \{\Omega \subset \mathbf{R}^2 \mid \operatorname{means}(\Omega) \le A, \ A > 0 \text{ is given and } \Gamma \text{ is Lipschitz}\} \ .$$

EXAMPLE 3.2. Minimization of the weight of a thermal diffuser.

Consider the problem of finding an optimal shape for a minimum-weight thermal diffuser with *a priori* specifications on the input and output thermal power flux.

We assume that the thermal diffuser has a volume D symmetrical with respect to the z-axis (cf. Figure 3.2 (a)) whose boundary surface is made up of three regular pieces: the mounting surface Σ_1 (a disc perpendicular to the z-axis with its centre at $(0,0,0)$), the lateral adiabatic surface Σ_2 and the interface Σ_3 between the diffuser and the heatpipe saddle (a disc perpendicular to the z-axis with its centre at $(0,0,L)$).

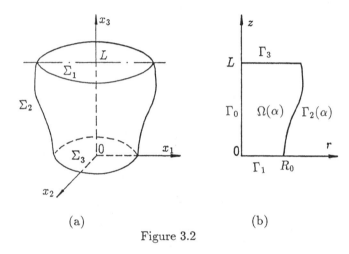

(a)   (b)

Figure 3.2

The temperature distribution over this volume $D(\alpha)$ is the solution of the stationary heat equation

(3.4) $$\Delta u = 0 \quad \text{in } D(\alpha)$$

with the following boundary conditions on the surface $\Sigma = \Sigma_1 \cup \Sigma_2 \cup \Sigma_3$ (on the boundary of D)

(3.5) $$\frac{\partial u}{\partial n} = q_{\text{in}} \quad \text{on } \Sigma_1$$

(3.6) $$\frac{\partial u}{\partial n} = 0 \quad \text{on } \Sigma \cup \Gamma_2$$

(3.7) $$u = u_3 \quad \text{on } \Sigma_3, \ u_3 = \text{constant},$$

where $\dfrac{\partial u}{\partial n}$ is the normal derivative on the boundary surface Σ. The parameter q_{in} appearing in (3.5) is the uniform inward thermal power flux at the source (positive constant). The radius R_0 of the mounting surface Σ_1 is fixed so that the

EXAMPLE 3.3.
Optimization of the shape of the poles of an electromagnet.

This problem is of interest for example in the manufacturing of very large electro-magnets and in magnetic tape storage on computers. Figure 3.3 (a) illustrates the two-dimensional approximation of the physical domain.

By symmetry we can restrict the design analysis to one-fourth of the domain only (see Figure 3.3 (b)). So the domain of interest, Ω, is given by $\Omega =]0, a[\times]0, b[$. It consists of three different subdomains, Ω_F, Ω_C, $\Omega_A = \Omega_{A,i} \cup \Omega_{A,e}$ of *ferric*, *copper* and *air* materials, respectively. Moreover, we have a subregion $D \subset \Omega_{A,i}$ where constant magnetic field is desired (Figure 3.3 (a)).

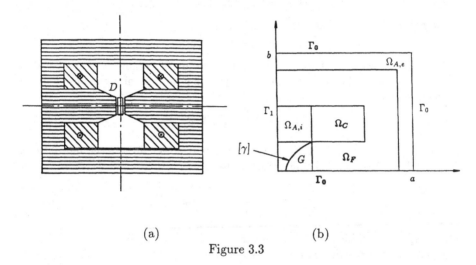

(a)   (b)

Figure 3.3

Our aim is to design a part of the boundary $\partial(\Omega_{A,i} \cap \Omega_F)$ described by a curve γ such that the graph of γ, denoted by $[\gamma]$, lies inside of a given set G (Figure 3.3 (b)). We can formulate the *shape design problem* as a minimization problem

$$\min_{\Omega_F \in \mathcal{O}} J(\Omega_F), \quad \mathcal{O} = \{\Omega_F \mid [\gamma] \in G\},$$

where J is the criterion function given by

$$J(\Omega_F) = \int_D |\nabla A - B_d|^2 \, dx,$$

where A is the electromagnetic potential in Ω and B_d is the desired field in the subregion D.

The electromagnetic potential satisfies the non-linear system

$$\nabla \cdot (v_r(|\nabla A|^2, x)\nabla A) = \mu_0 j \text{ in } \Omega, \quad A = 0 \text{ on } \Gamma_0, \frac{\partial A}{\partial n} = 0 \text{ on } \Gamma_1,$$

boundary surface Σ_1 is fixed in the design problem. Using the axial symmetry of our problem, one can consider the situation in $\mathbf{R}^2$ so that the class of shapes for the diffuser are characterized by a constant $L > 0$ and a positive function $\alpha(z)$, $0 \leq z \leq L$, with $\alpha(0) = R_0 > 0$. The domain, whose shape we are looking for, is (see Figure 3.2 (b))

$$\Omega(\alpha) = \{(z,r) \in \mathbf{R}^2 \mid 0 < z < L, \ 0 < r < \alpha(z), \ \alpha \in U_{\mathrm{ad}}\} \ ,$$

where

$$U_{\mathrm{ad}} = \{\alpha \in C^{0,1}([0,L]) \mid \alpha(0) = R_0\} \ .$$

Domain $\Omega(\alpha)$ is bounded by Γ_0, Γ_1, $\Gamma_2(\alpha)$ and Γ_3 (see Figure 3.2), where

$$\Gamma_2(\alpha) = \{(z,r) \in \mathbf{R}^2 \mid z \in [0,L], \ r = \alpha(z)\} \ .$$

In this case

$$\mathcal{O} = \{\Omega(\alpha) \mid \alpha \in U_{\mathrm{ad}}\} \ .$$

With the assumption that the diffuser is made up of homogeneous material of uniform density (no hollows) the design objective is to minimize the functional of the volume of $\Omega(\alpha)$:

$$J(\alpha) = \pi \int_0^L (\alpha(z))^2 \, dz \ ,$$

subject to the constraint on the outward thermal power flux at the interface Γ_3 between the diffuser and the heatpipe saddle:

$$\sup_{(z,r)\in\Gamma_3} -\frac{\partial u}{\partial n}(z,r) \leq q_{\mathrm{out}} \ ,$$

where q_{out} is a specified positive constant.

The optimal shape design problem reads

$(\mathbb{P})$
$$\begin{cases} \displaystyle \underset{\alpha \in U_{\mathrm{ad}}}{\text{Minimize}} \left\{ J(\alpha) = \pi \int_0^L \alpha(z)^2 \, dz \right\} \\[2mm] \displaystyle \text{subject to} \quad \sup_{(z,r)\in\Gamma_3} -\frac{\partial u(\alpha)(z,r)}{\partial n} \leq q_{\mathrm{out}} \ , \end{cases}$$

where $u = u(\alpha)$ is the solution to the state problem

$(3.4')$ $$\frac{1}{r}\frac{\partial u}{\partial r} + \frac{\partial^2 u}{\partial r^2} + \frac{\partial^2 u}{\partial z^2} = 0 \qquad \text{in } \Omega(\alpha)$$

$(3.5')$ $$\frac{\partial}{\partial n}u = q_{\mathrm{in}} \qquad \text{on } \Gamma_1$$

$(3.6')$ $$\frac{\partial}{\partial n}u = 0 \qquad \text{on } \Gamma_0 \cup \Gamma_2(\alpha)$$

$(3.7')$ $$u = u_3 \qquad \text{on } \Gamma_3 \ .$$

where $\mu_0 = 4\pi \cdot 10^{-7}\text{MKSA}^{-1}$, j is the x_3-component of the current density vector and

$$v_r(|\nabla A|^2, x) = \begin{cases} v_A = 1 \text{ MKSA for all } x \in \Omega_A \ , \\ v_C = 1 \text{ MKSA for all } x \in \Omega_C \ , \\ v_F = e + (f - e)\dfrac{|\nabla A|^{2\alpha}}{|\nabla A|^{2\alpha} + g} \text{ MKSA for all } x \in \Omega_F \ . \end{cases}$$

The constants α, e, f and g above can be fixed by experiments [7]. We choose $e = 5 \cdot 10^{-4}$, $f = 0.175775$, $g = 8758.756$ and $\alpha = 5.419241$.

EXAMPLE 3.4 Optimization of the header in paper machine.

A header is that part of paper machines, which has to deliver flow coming from pump to number of small pipes as uniformly as possible (Figure 3.4). It has been quite difficult to design the header just trying different kind of geometries and velocities.

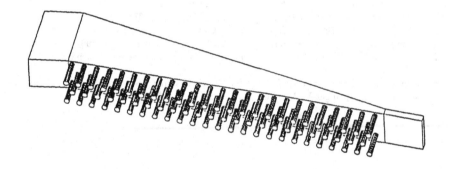

Figure 3.4

Flowrates in the pipes are effected by controlling recirculation and by changing the shape of the header. To the cost function is chosen the meandistribution of velocities at the end of the pipes. The flow can be modelled by the Navier-Stokes equations.

The design variables are the angle of inflow, the shape of the back wall, the shape of the pipes and recirculation. The shape of the header is assumed to be piecewise linear. When recirculation is changed, we also change velocity of inflow so that the mean value of velocities is constant.

REMARK 3.1. Above mentioned optimal shape design problems can in practice be solved by of FEM (finite element method) and NLP (nonlinear programming). In the discretization by FEM the boundary nodes of Ω lying on $\Gamma(\alpha)$ (which has to be redesigned) are so called design nodes. In the design process one tries to find optimal geometrical position for these design nodes through minimizing by NLP the criterion function. For applying NLP the FE-sensitivity analysis (computation of the gradient of the criterion function with respect to the variation of design nodes) must be performed. For details of practical realization of optimal shape procedure see [2,7,8].

REFERENCES

1. Arora, J., "Introduction to Optimum Design," McGraw-Hill, New York, 1989.
2. Brant, A.M., "Criteria and methods of structural optimization," Martinus Nijhoff Publishers, Dordrecht, 1986.
3. Haslinger, J. and P. Neittaanmäki, "Finite element approximation of optimal shape design. Theory and applications," J. Wiley & Sons, 1988.
4. Haug, E.J. and J.Cea (eds.), "Optimization of distributed parameter structures," Parts I–II, Nato Advances Study Institute Series, Series E, Alphen aan den Rijn: Sijthoff & Noordhoff, 1981.
5. Kohn, R.V. and Strang, G., *Optimal design and relaxation of variational problems I–III*, Comm. Pure Appl. Math. **39** (1986), 112–137, 139–182, 353–377.
6. Mota Soares, C.A. (Ed.), "Computer Aided Optimal Design: Structural and Mechanical Systems," NATO ASI Series, Series F, Vol 27, Springer Verlag, Berlin, 1987.
7. Neittaanmäki, P., *Computer aided optimal shape design*, to appear in Journal of Industrial Mathematics.
8. Pironneau, O., "Optimal shape design for elliptic systems," Springer Series in Computational Physics, Springer-Verlag, New York, 1984.

RANDOM FUNCTIONS IN PHYSICAL AND MECHANICAL SYSTEMS

J. VOM SCHEIDT
Technische Hochschule Zwickau
Fachbereich Mathematik/Informatik
Dr.-Friedrichs-Ring 2a
O-9541 Zwickau
BRD

ABSTRACT. Experience has shown that it is important to find a class of random functions having the two properties:
-The class of functions should possess relevant attributes which the user does not irritate and it should be able to provide the necessary parameters of these random functions.
-The mathematical theory with respect to this class of functions should have applicable results in the mind of the user. For example, it should be possible to obtain moments or distribution functions of the solution of differential equations containing such random functions.
The class of weakly correlated functions is one possibility to realize this concept. The weakly correlated random functions can also be characterized as functions without 'distant effect' or as functions of 'noise-natured character'. In this paper weakly correlated functions are applied to vibration problems and random temperature propagation.

The definition of a weakly correlated function can be illustrated by the following example:

$\{(\overset{1}{x},\overset{2}{x},\overset{5}{x}),(\overset{3}{x},\overset{4}{x})\}$ is the separation into maximum ε-neighbouring subsets of $(\overset{1}{x},\overset{2}{x},\overset{3}{x},\overset{4}{x},\overset{5}{x})$. The set $(\overset{1}{x},\overset{5}{x})$ is ε-neighbouring but it is not maximum ε-neighbouring relative to $(\overset{1}{x},\overset{2}{x},\overset{3}{x},\overset{4}{x},\overset{5}{x})$.

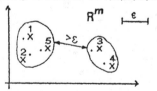

Figure 1. Example of ε-neighbouring sets

The definition of a weakly correlated function can be given in connection with the decomposition of the moments. In the shown example

M. Heiliö (ed.), Proceedings of the Fifth European Conference on Mathematics in Industry, 55–64.
© 1991 *B. G. Teubner Stuttgart and Kluwer Academic Publishers. Printed in the Netherlands.*

the fifth moment $<\prod_{p=1}^{5} f_\varepsilon(\overset{p}{x})>$ of a weakly correlated function has to decompose as

$$<\prod_{p=1}^{5} f_\varepsilon(\overset{p}{x})>=<f_\varepsilon(\overset{1}{x})f_\varepsilon(\overset{2}{x})f_\varepsilon(\overset{5}{x})><f_\varepsilon(\overset{3}{x})f_\varepsilon(\overset{4}{x})>$$

where $<.>$ denotes the expectation $E\{.\}$.

Initial value problems, boundary value problems, or eigenvalue problems of ordinary or partial differential equations which lead to a structure of the solution of the form

$$g(r_\varepsilon(x,\omega)) \qquad \text{with} \qquad r_\varepsilon(x,\omega)=\int_D F(x,y)f_\varepsilon(y,\omega)dy; \qquad D \subset R^m,$$

where $g(t)$ denotes a non-random function, are investigated in the further considerations.

It is possible to expand statistical characteristics as moments or distribution densities with respect to the correlation length ε. Some results will be given in case of $g(t)=t$.

Results of first order as to ε are the following statements:

1. It is

$$\lim_{\varepsilon \to 0} < \frac{1}{\sqrt{\varepsilon}^m} r_\varepsilon(\overset{1}{x},\omega) \frac{1}{\sqrt{\varepsilon}^m} r_\varepsilon(\overset{2}{x},\omega) > = \int_D F(\overset{1}{x},y)F(\overset{2}{x},y)a(y)dy \stackrel{.}{=} {}^2A_1$$

where $a(x)$ denotes the intensity of the weakly correlated function given by

$$a(x) = \lim_{\varepsilon \to 0} \frac{1}{\varepsilon^m} \int_{\{y:\, |y-x|<\varepsilon\}} <f_\varepsilon(x)f_\varepsilon(x+y)>dy.$$

The basic idea of this limit relation was also applied by Ornstein, Uhlenbeck, and van Lear to Brownian motion and Boyce to a Sturm-Liouville problem.

2. The convergence in distribution

$$\lim_{\varepsilon \to 0} \frac{1}{\sqrt{\varepsilon}^m} r_\varepsilon(x,\omega) = \xi(x,\omega)$$

can be shown where $\xi(x,\omega)$ denotes a random Gaussian function.

Results of higher order relative to ε can be formulated by the next two statements:

3. The k th moment of r_ε can be expanded by the following formulas for k even

$$<r_\varepsilon^k> = (k-1)!!({}^2A_1)^{\frac{k}{2}} \varepsilon^{\frac{mk}{2}} + \begin{cases} C_{2,1}^{even} \varepsilon^{\frac{k}{2}+1} + o(\varepsilon^{\frac{k}{2}+1}) & \text{for } m=1 \\[2mm] C_{2,m}^{even} \varepsilon^{\frac{mk}{2}+1} + o(\varepsilon^{\frac{mk}{2}+1}) & \text{for } m>1 \end{cases}$$

and for k odd

$$<r_\varepsilon^k> = 0 + C_{1,m}^{odd} \varepsilon^{\frac{m(k+1)}{2}} + C_{2,m}^{odd} \varepsilon^{\frac{m(k+1)}{2}+1} +o(\varepsilon^{\frac{m(k+1)}{2}+1})$$

where the first summands on the rigth-hands sides are the results of first order given by the limit theorems.

4. Expansions of the distribution density of

$$r_\varepsilon(\omega) = \int_D F(x) f_\varepsilon(x,\omega) dx$$

follow from the steps:

4.1 A transformation of the form $\tilde{r}_\varepsilon(\omega) = \dfrac{r_\varepsilon(\omega)}{\sqrt{{}^2A_1 \varepsilon^m}}$ is performed.

4.2 The distribution density $\tilde{p}(\tilde{u})$ of $\tilde{r}_\varepsilon(\omega)$ is expanded relative to $\sqrt{\varepsilon}$ up to terms of order $\sqrt{\varepsilon}^2$.

4.3 After retransformation the density $p(u)$ of $r_\varepsilon(\omega)$ can be written as

$$p(u) = \frac{1}{\sqrt{2\pi\varepsilon^m\,{}^2A_1}} \exp\left(- \frac{1}{2} \frac{u^2}{\varepsilon^m\,{}^2A_1}\right) \left\{ 1 + \right.$$

$$\left\{ \begin{array}{ll} \dfrac{1}{6} \dfrac{{}^3A_2}{\sqrt{{}^2A_1}^3} H_3\left(\dfrac{u}{\sqrt{\varepsilon\,{}^2A_1}} \right)\sqrt{\varepsilon} + [\ldots]\varepsilon \Big\} + o(\varepsilon) & \text{for } m=1 \\[4ex] \dfrac{1}{2} \dfrac{{}^2A_2}{{}^2A_1} H_3\left(\dfrac{u}{\sqrt{\varepsilon^m\,{}^2A_1}} \right)\varepsilon + \dfrac{1}{6} \dfrac{{}^3A_2}{\sqrt{{}^2A_1}^3} H_3\left(\dfrac{u}{\sqrt{\varepsilon^m\,{}^2A_1}} \right)\varepsilon^{\frac{m}{2}} \Big\} + o(\varepsilon) & \text{for } m>1 \end{array} \right.$$

where $H_p(u)$ are the Chebyshev-Hermite polynomials defined by

$$H_p(u) = (-1)^P \exp(\tfrac{1}{2}u^2) \frac{d^2}{du^2}\exp(-\tfrac{1}{2}u^2)$$

and ${}^2A_2$, ${}^3A_2$ denote statistical characteristics as ${}^2A_1$ given by statistical dates of $f_\varepsilon(x,\omega)$ and $F(x)$. The first summand corresponds to the second mentioned above.

Some generalisations can be given. For example, the definition of a weakly correlated connected random vector function $(f_{1\varepsilon}(x,\omega),\ldots,f_{s\varepsilon}(x,\omega))$ can be easily formulated. A partially weakly correlated random function $f_\varepsilon(x,y,\omega)$ relative to y is defined in a such way that the basic idea only refers to the variable y. The statements given above can be extended to random vectors of the form

$$(r_{1\varepsilon}(\omega),\ldots,r_{n\varepsilon}(\omega)) \text{ with } r_{p\varepsilon}(\omega) = \sum_{q=1}^{s} r_{pq\varepsilon}(\omega); \quad r_{pq\varepsilon} = \int_D F_{pq}(x) f_{q\varepsilon}(x,\omega) dx$$

and to functions of these random vectors $g(r_{1\varepsilon}(\omega),\ldots,r_{n\varepsilon}(\omega))$ where $g(t_1,\ldots,t_n)$ denotes a non-random function.

Firstly, we will apply these first order expansions to random eigenvalue problems of the form

$$M(\omega)u = \lambda N(\omega)u \qquad \qquad \text{on} \qquad D = [a,b];$$

$$\text{boundary conditions:} U_q[u] = 0 \qquad \text{for} \qquad q = 1,2,\ldots,2m$$

where $M(\omega)$ and $N(\omega)$ are given by

$$M(\omega)u = \sum_{k=0}^{p} (-1)^k \left[f_k(x,\omega)u^{(k)} \right]^{(k)} ; \quad N(\omega)u = \sum_{k=0}^{p'} (-1)^k \left[g_k(x,\omega)u^{(k)} \right]^{(k)}$$

with $p' < p$ and $U_q[u]$ by

$$U_q[u] = \sum_{k=0}^{2p-1} \left[\alpha_{qk} u^{(k)}(a) + \beta_{qk} u^{(k)}(b) \right].$$

Statements concerning the statistical characteristics of the eigenvalues $\lambda_k(\omega)$ and the eigenfunctions $u_k(x,\omega)$ can be obtained by the connection of theory of w.c. functions with the perturbation theory of Rellich and the theory of Ritz containing the approximation of the above eigenvalue problem by a eigenvalue problem as to matrices.

In order to demonstrate some typical applications the example

$$(f(x,\omega)u'')'' = \lambda(-g_1 u'' + g(x,\omega)u) \qquad \text{on} \qquad D = [0,1]$$

$$u(0) = u(1) = u''(0) = u''(1) = 0$$

is considered. This eigenvalue problem containes the problem of the eigenvibration of a simple supported bar

$$(EIu'')'' = \lambda \rho Au,$$

and the buckling problem of a bar,

$$(EIu'')'' = - \lambda u'',$$

in connection with the stability of bars. In this differential equation it is

     E: modulus of elasticity
     I: moment of inertial of the cross-sectional area
     ρ: mass per unit
     A: cross-sectional area.

With respect to numerical considerations, for example, we assume:

1. The cross-sectional area is supposed to be a circle with the radius $r(x,\omega)$. It follows

$$A(x,\omega) = \pi r^2(x,\omega) \qquad \text{and} \qquad I(x,\omega) = \frac{1}{2} \pi r^4(x,\omega).$$

2. We put $\langle r \rangle \equiv r_0$, $\langle \rho \rangle \equiv \rho_0$, and assume that

$$\bar{r}_\varepsilon(x,\omega) = r(x,\omega) - r_0; \qquad \bar{\rho}_\varepsilon(x,\omega) = \rho(x,\omega) - \rho_0$$

are independent and weakly correlated random normal distributed processes. The parameters γ, β are introduced by

$$\sigma_\rho = \gamma \cdot \rho_0; \qquad \qquad \sigma_r = \beta \cdot r_0$$

as a 'measure of the randomness' of $r(x,\omega)$ and $\rho(x,\omega)$. In particular, the correlation function of $\bar{r}_\varepsilon(x,\omega)$ can have the form

$$<\bar{r}_\varepsilon(x)\bar{r}_\varepsilon(y)> = \sigma_r^2 \begin{cases} 1 - \dfrac{1}{\varepsilon}\,y-x & \text{for } y-x < \varepsilon \\ 0 & \text{otherwise} \end{cases}$$

and $\bar{\rho}_\varepsilon(x,\omega)$ have a similar form.

We will now give an example for a result with respect to the eigenfrequencies $\omega_k = \sqrt{\lambda_k}$.

Supposing that $r(x,\omega)$ and $\rho(x,\omega)$ fulfill the inequalities

$$P(\ r(x,\omega)-r_0 \le 3\ \beta r_0) = 0.9972; \quad P(\ \rho(x,\omega)-\rho_0 \le 3\ \gamma\rho_0) = 0.9972$$

for $\beta=0.1$ and $\gamma=0.1$ then Fig.2 can be obtained. In the case of a random radius and a non-random mass per unit it is found that

$$P(239\ \tfrac{1}{s} \le \omega_1 \le 280\ \tfrac{1}{s}) = 0.9544,$$

and in the case of a random mass per unit and a non-random radius

$$P(244\ \tfrac{1}{s} \le \omega_1 \le 264\ \tfrac{1}{s}) = 0.9544.$$

Hence, the influence of $r(x,\omega)$ to the first frequency is stronger than the influence of $\rho(x,\omega)$.

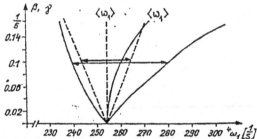

Figure 2. Comparison of the influence of $r(x,\omega)$ and $\rho(x,\omega)$ on the first frequency

We consider the case of bars loaded by a force F.

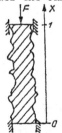

Figure 3. A loaded bar with random surface

Then $\lambda_1(\omega)$ denotes the maximal force by which the bar can be loaded without failure. The function

$$p_\beta(\bar{F}) = P(\ \bar{\lambda}_1(\omega) \leq \bar{F}) = P(\ \lambda_1(\omega) \leq F),$$

$$\lambda_1(\omega) = \lambda_1(\omega)\frac{4}{E\pi r_0^4}\ ;\qquad\qquad <\lambda_1> = \pi^2$$

gives the probability of failure if the bar is maximally loaded by a force F. These probabilities are plotted in Fig.4.

Since $\beta=0$ corresponds to the non-random case it is easey to see from

$$\lim_{\beta \to 0} p_\beta(\bar{F}) = \begin{cases} 0 & \text{for} \quad \bar{F} < \pi^2 \\ 1 & \text{for} \quad \bar{F} \geq \pi^2 \end{cases}$$

that our considerations agree with the deterministic case. Furthermore, it follows from

$$p_0(10) = 1 \qquad \text{and} \qquad p_\beta(10) < 1 \qquad\qquad \text{for } \beta > 0.$$

that bars with random areas fail less frequently if they are loaded by a force corresponding to $\bar{F}$.

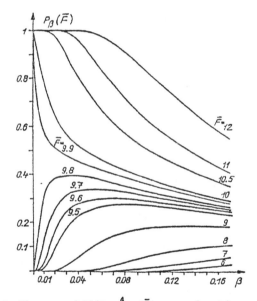

Figure 4. The propability $^4p_\beta(\bar{F})$ as a function of β

Now we deal with random vibrations of vehicles. A simple example is shown in Fig.5.

The spring force can be described by the relation $F_{sp} = F_{sp}(s)$ where s denotes the excitation and the damping force by $F_{da} = F_{da}(v)$ where v is the velocity, $v = s'$.

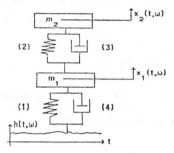

Figure 5. Two-mass vibration system

In our example the relations

(1) $F_{sp,1} = c_1(x_1 - h)$ (3) $F_{da,3} = a_1(x_2' - x_1') + a_2(x_2' - x_1')^2 + a_3(x_2' - x_1')^3$

(2) $F_{sp,2} = c_2(x_2 - x_1)$ (4) $F_{da,4} = 0$

lead to the differential equations

$$m_1 x_1'' + c_1(x_1 - h) - a_1(x_2' - x_1') - a_2(x_2' - x_1')^2 - a_3(x_2' - x_1')^3 - c_2(x_2 - x_1) = 0$$

$$m_2 x_2'' + a_1(x_2' - x_1') + a_2(x_2' - x_1')^2 + a_3(x_2' - x_1')^3 + c_2(x_2 - x_1) = 0$$

For general cases we have to investigate systems of differential equations of the form

$$Mz' + Nz + \sum_{k=2}^{m} B_k(z) = g(t,\omega)$$

where z is given by

$$z(t,\omega) = \left[x_1'(t,\omega), \ldots, x_n'(t,\omega), x_1(t,\omega), \ldots, x_n(t,\omega) \right]$$

and M, N are $2n \times 2n$-matrices. The inhomogeneous term has the form

$$g(t,\omega) = \sum_{q=0}^{2} P_q h^{(q)}(t,\omega)$$

with the $2n \times 2n$-matrices P_q. The random vector $h(t,\omega) = (h_p(t,\omega))_{1 \leq p \leq 2n}$ possesses the components

$$h_p(t,\omega) = \begin{cases} r(t + c_p, \omega) & p = 1, \ldots, P_a \\ 0 & p = P_a + 1, \ldots, 2n \end{cases}$$

where c_p denotes the temporal distance between the p th axle and he first axle. The cases $h_p \equiv 0$ or $h_p \not\equiv 0$ depend on the special construction of the considered vehicle. The random function $r(t,\omega)$ describes the surface of the road and it is put

$$r(t,\omega) = \int\limits_{-\infty}^{t} q(t-s)f_\varepsilon(s,\omega)ds$$

where $f_\varepsilon(t,\omega)$ is a weakly correlated random process and $q(t)$ a non-random function. Finally, the non-linear terms $B_k(z)$ are given by

$$B_{k,q}(z) = \sum_{q_1,\ldots,q_k=1}^{2n} b_{q,q_1\ldots q_k} z_{q_1}\ldots z_{q_k}\quad.$$

In applications the measured surface of the road is approximated by $r(t,\omega)$. For this purpose the spectral density of the measured surface is approximated by the first order spectral density of $r(t,\omega)$,

$$S_r^1(\alpha) = \frac{a\varepsilon}{\pi} \int\limits_0^\infty \cos(\alpha t) \int\limits_0^\infty q(u)q(t+u)dudt,$$

by means of parameters contained in $q(t)$ and a, ε. Then, the measured spectral density of the surface of the road corresponds to the spectral density of this process.

By application of results of the weakly correlated theory correlation relations

$$C_{pq}^{kh}(t_1,t_2) = <(z_p^{(k)}(t_1) - <z_p^{(k)}(t_1)>)\cdot(z_q^{(h)}(t_2) - <z_q^{(h)}(t_2)>)>$$

can be expanded with respect to ε up to terms of second order. Expansions up to the first order correspond to expansion results up to first order of the linear differential equation belonging to the given system ($B_k \equiv 0$). It is possible to show that

$$C_{pq}^{kh}(t_1,t_2) = C_{pq}^{kh}(|t_2-t_1|)$$

up to terms of second order. Hence, real spectral densities

$$S_{pq}^{kh}(\alpha) = \frac{1}{\pi} \int\limits_0^\infty \cos(\alpha t)C_{pq}^{kh}(t)\, dt$$

can be considered. Properties of these spectral densities often give information to the user about the investigated vibration system. Furthermore, assertions concerning to the distribution of $x_p^{(k)}(t,\omega)$ can be calculated. A numerical analysis of technological vibration systems can be provided by means of computer programmes made. These results can be extended to systems with a great number of masses.

The last application is concerned with problems of heat conduction influenced by random sources of heat. An example of such a problem contains the investigation of temperature propagation in brakes of vehicles. We deal with the boundary-initial value problem:

$$\frac{\delta u}{\delta t} = \alpha(\frac{\delta^2 u}{\delta x^2} + \frac{\delta^2 u}{\delta y^2})\qquad\text{on}\qquad G = \{(x,y):\ x \leq R;\ 0\leq y\leq L\}$$

initial condition: $u(0,x,y) = u_0(x,y)$

boundary conditions: $\frac{\delta u}{\delta y}(t,x,y) \Big|_{(\delta G)_1} = P(t,x,\omega)$

$$\left[\lambda \frac{\delta u}{\delta n}(t,x,y) + \alpha_p \{u(t,x,y) - u_R(t,x,y)\} \right]_{(\delta G)_p} = 0 \quad \text{for } p = 2,3,4$$

where α denotes the temperature conductivity, λ the termal conductivity number, α_p, $p=2,3,4$, the heat transition numbers, u_R a non-random boundary temperature, and $P(t,x,\omega)$ a random function on $(\delta G)_1$ with respect to a given surface power.

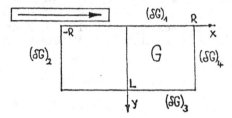

Figure 6. Domain G for the investigated problem

Assuming $g(x,y) \subset C_0^\infty(G)$ it follows

$$U(t,\omega) \overset{\cdot}{=} (\bar{u}(t,\ldots,\omega),g) = \int_0^t \int_{-R}^R F(t-s,x)\bar{P}(s,x,\omega)ds\,dx$$

where $\bar{P}$ is defined by $\bar{P} = P - \langle P \rangle$ and $\bar{u} = u - w$. The non-random function w is given by the solution of the averaged problem concerning the above problem. The averaged problem is obtained by substituting random quantities by their mean values. The function $F(t,x)$ can be determinded as

$$F(t,x) = -\alpha \sum_{k,p=0}^\infty v_{2p}(0)\exp(-\alpha\lambda_{kp}t)\, v_{1k}(x)\, (f_{kp},g)$$

where λ_{kp} and $f_{kp}(x,y) = v_{1k}(x) \cdot v_{2p}(y)$ are the eigenvalues and eigenfunctions of the eigenvalue problem

$$-\Delta f = \lambda f; \qquad -f_n \Big|_{(\delta G)_1} = 0; \qquad [\lambda f_n + \alpha_q f]_{(\delta G)_q} = 0, \quad q=2,3,4.$$

In the further investigations $\bar{P}(t,x,\omega)$ is assumed to be weakly correlated or partially weakly correlated, respectively. Using expansions as shown in the first part of the paper correlation relations, spectral densities, conditional probabilities, and distribution functions of the solution $\bar{u}(t,x,y,\omega)$ can be obtained.

For example, it is possible to calculate the dependence between $\bar{u}(t_1,P_1)$

and $\bar{u}(t_2,P_2)$ (see Fig.7).

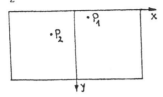

Figure 7. Half-plane with coordinate axis

This dependence behaviour is strongly connected with heat stresses. Fig.8 shows the results of a comparison between electro-analogical methods and the application of the weakly correlated theory. The results measured by means of physical methods are denoted by a cross (x) and this figure contains our calculated values for different correlation functions R_q of $\bar{P}(t,x,\omega)$ with $\sigma^2=(\frac{1}{\lambda_m}P_m)^2 p(1-p)$.

The best fit to the measured values would be obtained by an intensity

$a_{best} = 2.004 \cdot 10^{11} \frac{K^2}{ms}$ and these values are plotted by a hatched line.

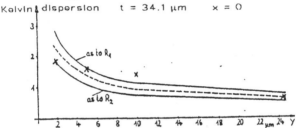

Figure 8. Dispersion as function of y for different intensities and measured values

The physical values were assumed to be $\alpha = 1.278 \cdot 10^{-5} \frac{m^2}{s}$; $P_m = 1.26 \cdot 10^7 \frac{W}{m^2}$

the temporal correlation length $t_{cor} = 65 \cdot 10^{-6}$ s, and the spatial correlation length $x_{cor} = 5$ μm.

References:
   [1] Boyce, W.E. (1966) 'Stochastic nonhomogenous Sturm-Liouville problems',J. Frankl. Institute 282, 206-215.
   [2] vom Scheidt, J., Purkert, W. (1983) 'Random Eigenvalue Problems', Akademie Verlag Berlin; North Holland New York /Amsterdam/Oxford.
   [3] vom Scheidt, J. (1990) 'Random Equations of Mathematical Physics' Akademie Verlag Berlin; Wiley.

Mathematical Models in Chemical Engineering

Hj. Wacker, T. Kronberger, A. Ortner, L. Peer*

1. Introduction

Let me start with two observations concerning the impact of mathematics on chemical engineering. Firstly, costs for man power increased from 25% of product development costs (man power - software - hardware) in the seventies to 75% now. Therefore it has become increasingly attractive to use commercial software packages to reduce the time and costs resp. for developing new products. Secondly, both the NSF [5] in USA and the E.F.Ch.E. [16] in Europe organized a forecast for the development in the nineties. With respect to mathematics the NSF showed optimism especially concerning the use of optimization. The E.F.Ch.E. was more sceptic on mathematical modelling but favorized dynamic models and optimization.

Concerning computer supported solutions there is a connection between numerical approaches and hardware. The producers of supercomputers, like CRAY, have become interested in the chemical engineering market. Flowsheets of plants lead to large scale nonlinear equations (steady state) [15] and differential algebraic equations (dynamic case) [12]. In many practical problems there arise even PDEs. Up to now the largest part of all available software is based on the sequential modular approach [8] which can be run even on workstations. There is an increasing tendency to use the equational approach [9] especially for dynamic systems. This approach is highly efficient but needs some storage. Therefore at least mini supercomputers, like CONVEX, might be useful. In case of large reaction systems, influence of fluid dynamic phenomena etc. even larger computers might be needed. For a state of the art see e.g. [3], [6], [14]

2. Plate Distillation Columns - Dynamic Case

In [1] we gave a detailled description of a steady state plate distillation column and the numerical treatment of the resulting nonlinear algebraic systems. How near to reality do we get using a model of this type? In 1985 Krishnamurthy and Taylor, published a model [11] and offered a comparison with experimental data. This model is slightly more refined than that we used in [12] as it takes into consideration also the case of phase transition thus describing implicitly a kind of stage efficiency. The model does not assume equilibrium. Phase transition change mass and energy balances and mass transfer coefficients are needed. The MERQ - equations (material balances - energy balances - rate equations - eqilibrium equations) were solved by Newton's method and compared to reality. The absolute difference between predicted and measured mole fractions are mostly smaller than 4% . The comparison, however, concentrates on wellknown hydrocarbons where sufficiently checked data are available. In other cases we were told that, roughly, 10% - 15% error was to be accepted.

---

* all Univ. Linz, Math. Deptm.

M. Heiliö (ed.), Proceedings of the Fifth European Conference on Mathematics in Industry, 65–74.
© 1991 *B. G. Teubner Stuttgart and Kluwer Academic Publishers. Printed in the Netherlands.*

Here we confine ourselves to present a short description of a dynamic plate distillation column. For details see [12].

2.1. The Mathematical Model for a Dynamic Column

Contrary to our model for the steady state [1], we do not have equilibrium on the trays any more. To describe the mass transfer between liquid and vapor phases we apply the two film theory where a third auxiliary phase (the film) is assumed to exist between the two phases.

In contrast to the steady state model we have to consider vapor and liquid holdups in addition to the flow rates. We obtain balance equations for both phases.

It should be noted, that for the first (evaporator) and last (condenser) tray the following equations have to be slightly modified.

Mass balances (see Fig. 1):

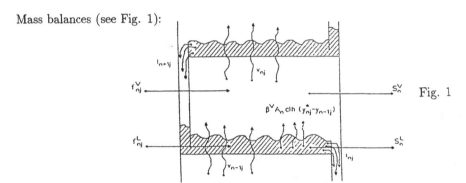

Fig. 1

Vapor phase - bulk:

$$\frac{dh_{nj}^V}{dt} = f_{nj}^V - S_n^V y_{nj} + V_{n-1} y_{n-1j} - V_n y_{nj} + \beta^V A_n clh \left(y_{nj}^* - y_{n-1j} \right) \quad (1)$$

$$n = 2, 3, \cdots, N-1$$

Liquid phase - bulk:

$$\frac{dh_{nj}^L}{dt} = f_{nj}^L - S_n^L x_{nj} + L_{n+1} x_{n+1j} - L_n x_{nj} + \beta^L A_n clh \left(x_{nj}^* - x_{nj} \right) \quad (2)$$

$$n = 2, 3, \cdots, N-1$$

Film

$$\beta^V \left(y_{nj}^* - y_{n-1j} \right) = -\beta^L \left(x_{nj}^* - x_{nj} \right) \quad (3)$$

Energy balances:

$$\frac{dE_n^V}{dt} = F_n^V e_n^{f,V} - S_n^V e_n^V + V_{n-1} e_{n-1}^V - V_n e_n^V + Q_n^V + \Delta E_n^{ev} \tag{4}$$

$$\Delta E_n^{ev} := \sum_{j=1}^{M} \beta^V A_n clh \left(y_{nj}^* - y_{n-1j}\right) \cdot \left[e_n^V - e_n^L\right]$$

$$\frac{dE_n^L}{dt} = F_n^L e_n^{f,L} - S_n^L e_n^L + L_{n+1} e_{n+1}^L - L_n e_n^L + Q_n^L - \Delta E_n^{ev} \tag{5}$$

$$n = 2, 3, \cdots, N - 1$$

Equilibrium Relations: the film theory model assumes thermodynamical equilibrium in the film

$$y_{nj}^* = k\left(y_{n1}^*, \cdots, y_{nM}^*, x_{n1}^*, \cdots, x_{nM}^*, T_n^*, p_n\right) \cdot x_{nj}^* \tag{6}$$

$$n = 1, \cdots, N$$

For the variables temperature, pressure, and flow rate we have algebraic equations: The temperature is defined by the holdup and the enthalpy. The user may prescribe the pressure in the evaporator. The pressure drop of each tray allows to compute the pressure in connection with the pressure of the tray below. Finally the flow rates are defined by the tray-geometrics.

As in the case of the steady state we use a tray to tray ordering both for the equations and the variables. For the Jacobian we get a block tridiagonal structure.

2.2. Numerical Solution

We get a system of semi-explicit differential algebraic equations of dimension $N(4M + 8)$ where, roughly, half of the equations are algebraic ones.

$$\frac{dx_1}{dt} = f_1(x_1, x_2, t) \qquad\qquad x_1(0) = x_0 \tag{7}$$

$$0 = f_2(x_1, x_2, t)$$

Usually our specification leads to an index 2 problem. We use the trapezoidal rule for the differential part and solve the resulting nonlinear systems for $x_1(t_{n+1})$, $x_2(t_{n+1})$ by Newton's method [12]. It can be proved that index 2 problems can be solved by a regularization technique [7]. Compare also [4].

2.3. Example: Dynamic Alcohol Distillation

In Fig. 2 we have a steady state distillation column separating methanol, ethanol and water. After three minutes a linear change in the composition of the feedstream and the reflux ratios starts. The change takes 15 minutes. The final feed and reflux data are given in Fig. 2.

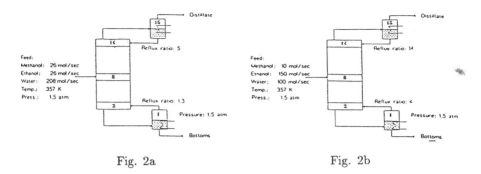

Fig. 2a   Fig. 2b

There is no further change in any parameters afterwards. It is known that after
one or two hours a new steady state should be obtained. We confine ourselves to
present the composition of the distillate as a function of time (Fig. 3). As one can
see in that figure the distillate gets into a steady state condition after about one
hour.

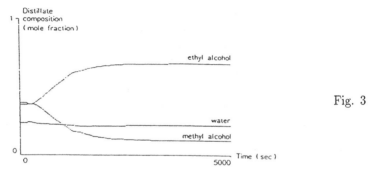

Fig. 3

For more details and another example see [12].

3. Optimization of Flowsheets

3.1. Combined Simulation - Optimization of Flowsheets

In [10] it was described how flowsheets are simulated using the sequential
modular approach. Optimization both for the steady state and the dynamic case
is performed by the same strategy. Instead of using two independent programms
one for simulation and one for optimization we can combine both and thus reduce
the numerical effort considerably.

The simulation process is described by an operator equation:

$$F(x,p) = 0 \tag{8}$$

where p is a known parameter vector and $x = x(p)$ the vector of the state variables. In the case of a plate distillation column F is either a nonlinear algebraic system (steady state) or a system of differential algebraic equations (dynamic case). For the objective we get:

$$\phi(x,p) = \underset{p}{Min!} \tag{9}$$

Inequality constraints are reduced to equational constraints by an active index set strategy. Equality constraints are included into (8). (8) is solved by Newton's method and (9) by a Quasi Newton technique. The latter is based on the gradient of ϕ :

$$\frac{d\phi}{dp} = \phi_x x_p + \phi_p$$

For the calculation of x_p we exploit identity (8) and get the linear system

$$F_x x_p + F_p = 0 \tag{10}$$

for x_p. We can arrange that F_x is exploited both for Newton's method and for the solution of (10) at the same iteration point. A $L - U$ decomposition, which may be based on a sparse solver, is already available from the solution of (8).

In the case of a dynamic column we have to solve the differential algebraic system:

$$\dot{x}(t) = f(x(t), y(t), u(t), t) \qquad x(0) = x_0$$
$$0 = g(x(t), y(t), u(t), t) \tag{11}$$

We assume the index to be 2 at most. We even may reduce the index to 1 by using a suitable regularization [7]. W.l.o.g. we proceed as follows. We differentiate (11) w.r.t. the parameter vector $u(t)$, interchange the order of differentiation and get:

$$\dot{x}_u = f_x x_u + f_y y_u + f_u \qquad \frac{\partial x_j(0)}{\partial u_k} = \begin{cases} 1 & x_j = u_k \\ 0 & \text{else} \end{cases}$$
$$0 = g_x x_u + g_y y_u + g_u \tag{12}$$

In 2.2. we mentioned that the numerical simulation of (11) is done by the trapezoidal rule and Newton's method for $x(t_{n+1}), y(t_{n+1})$. For given $u(t_{n+1})$ we get:

$$\cdot \begin{pmatrix} I - \frac{\Delta t_n}{2} f_x & -\frac{\Delta t_n}{2} f_y \\ g_x & g_y \end{pmatrix} \Bigg|_{(x^{(k)}(t_{n+1}), y^{(k)}(t_{n+1}), u(t_{n+1}), t_{n+1})}$$

I. e. we solve the linear differential algebraic system (12) for (x_n, y_n), using a system matrix the $L - U$ decomposition of which is already at hand. For the control variable we use an approximation:

$$u(t) = \sum_{k=1}^{r} u_k \alpha_k(t)$$

The resulting finite dimensional optimization problem is solved by Quasi Newton.
The gradients $\nabla_u x, \nabla_u y$ we get from (12).

3.2. Example: Optimization of an Oxygen Plant - Steady State

In [10] we described the numerical simulation of an oxygen plant, see Fig. 4.

Fig. 4 Oxygen Plant

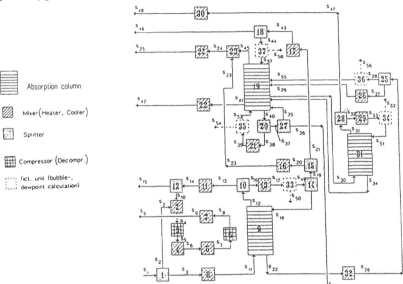

The flowsheet consists of 37 units:

> plate absorption columns (AC): 3; mixers (M); coolers (C); heaters (H); splitters (Sp): 27;
> Streams: feed: 1; product stream: 12; connecting streams: 45

We want to maximize the purity (= percentage) of oxygen in product stream S_{42}.

Constraint:

For simulation we use the sequential modular approach. We have the following connecting streams cut: $S_{12}, S_{30}, S_{31}, S_{34}, S_{40}$. This gives 25 conditions.

There are 19 restrictions, 15 equality constraints, 4 unequality constraints, e.g.:

$T_7 = 195$ (temperature of $S_7 : 195K$)
$Sp_{14}^{49} + Sp_{14}^{19} = 1$ (splitter 14 splits up into S_{19} and S_{49})
$G9_{29} = 0.9999$ (vapor rate of stream $S_{29} = 0.9999$)
$G8_{37} = 10.012$ (total flux of $S_{37} = 10.012 mol/s$)
$G4_{49} - G4_{50} = 0$ (enthalpy of S_{49} = enthalpy of S_{50})   etc.

In addition we have 44 box constraints for the free parameters.
In total there are 47 variables: 25 resulting from cut streams and 22 free parameters.

We give some numerical results:

	Sp_{10}^{13}	Sp_{15}^{20}	S_{25}^{27}	$S_{19,1}^V$ [mol/s]	$S_{19,12}^V$ [mol/s]	Q_{23} [kJ/s]	Objective [%O_2]
initial guess	0.20	0.80	0.10	385.8	985	8000.0	84.4
solution	0.12	0.88	0.01	402.0	975	4058.4	97.7

$S_{19,12}^V$ stream stripped from absorber 19 at tray 12
Q_{23}: heat duty to $H23$

We needed 38 iterations; computing time (COMPAREX 7/78): 9'17". For more details see [15].

3.3. Example: Waste Gas Purification Plant

The following idea to deal with waste gas avoids the production of gypsum. The waste gas ingredients SO_2, NO_x are converted into the corresponding acids - sulfuric and nitrogen acids, by oxydation.

Fig. 5 : Scheme of a waste gas purification plant (ELIN)

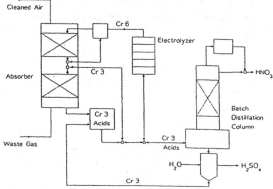

The polluted air is cleaned and leaves the system as top product of the absorber. The acids leave the absorber as liquid bottom product and are batch distilled. First H_2O, then HNO_3 is distilled. H_2SO_4 is received as bottom product.

We first tried to apply the rigorous dynamic model we described in Chapter 2. It was not possible, however, to get the necessary data, e.g. mass transition parameters β etc. Therefore we had to reduce the complexity of our model taking into account only mass balances. For ease of presentation we only give the model and the results for a strongly simplified model.

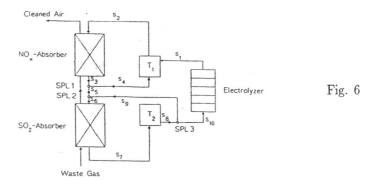

Fig. 6

We obtain the following system

Electrolyzer: $\quad s_1(t) \quad = \quad s_{10}(t) + \gamma_{Cr3N} s_{10,Cr3N}(t) + \gamma_{Cr3S} s_{10,Cr3S}(t)$

Tank T_1: $\qquad \dfrac{dm_1(t)}{dt} \quad = \quad s_1(t) + s_4(t) - s_2(t)$

Absorber (NO_x): $\quad s_3(t) \quad = \quad s_2(t) + \gamma_{NO} s_{NO}$

SPL 1: $\qquad\quad s_3(t) \quad = \quad s_4(t) + s_5(t) \quad$ (splitter)

SPL 2: $\qquad\quad s_6(t) \quad = \quad s_5(t) + s_9(t) \quad$ (mixer)

Absorber (SO_2): $\quad s_7(t) \quad = \quad s_6(t) + \gamma_{SO_2} s_{SO_2}$

Tank T_2: $\qquad \dfrac{dm_2(t)}{dt} \quad = \quad s_7(t) - s_8(t)$

SPL 3: $\qquad\quad s_8(t) \quad = \quad s_{10}(t) + s_9(t)$

Consequently we get a nonlinear system of twelve ODEs.
The following figure shows the concentration profile of the chromium compounds in the tanks T_1, T_2.

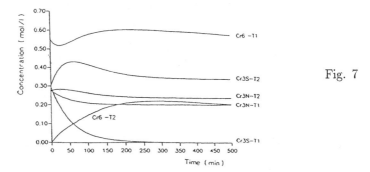

Fig. 7

4. Concluding Remarks

The future development for mathematical research is mainly directed to dynamic problems. Of special interest are models which are described by PDEs - e.g. heat exchangers, tubulare reactors, packed columns - as the field of partial differential algebraic equations is almost untouched and extremely promising. A second direction concerns optimization of dynamic problems leading to problems of optimal control for PDEs.

Both direction are in accordance with the intention of the chemical engineers. To make cooperation fruitful some data collecting will be necessary.

To support cooperation between mathematican and chemical engineers GAMM is going to install a working party. In the European frame ECMI is prepared to act as a platform for joint research of chemical engineers and mathematicans.

Nomenclature

A	tray area, m^2	N	number of trays
clh	clear liquid hight, m	Q	heat duty, kJ/s
e	specific enthalpy, kJ/mol	s	mass stream, mol/s
E	enthalpy, kJ	S	sidestream, mol/s
f	feed, mol/s	T	temperature, K
F	total feed, mol/s	v	vapor flow, mol/s
h	holdup, mol	V	total vapor flow, mol/s
H	total holdup, mol	x	liquid mole fraction
k	equilibrium constant	y	vapor mole fraction
l	liquid flow, mol/s	β	mass transfer coefficient, $mol/(m^3 s)$
L	total liquid flow, mol/s	γ_i	stoechiometric vector of reaction i,
m_k	vector of components in tank k, mol		$i \in \{Cr3N, Cr3S, NO, SO_2\}$
M	number of components		

Acknowledgements: part of this work has been supported by the Austrian NSF project P6674, VOEST and ELIN. In addition we thank DECHEMA, ÖMV, RAG and the CD Laboratory 'Modelling of Reactive Systems' for providing us with information.

REFERENCES

[1] D.Auzinger, L.Peer, Hj.Wacker, W.Zulehner: Numerical Calculation of Separation Processes, in: Case Studies in Industrial Mathematics, ECMI II, (ed. Engl/Wacker/Zulehner), Kluwer/Teubner, 1988, pp.131-154

[2] C.O.Bennett, J.E.Myers: Momentum, Heat, and Mass Transfer, McGraw-Hill, 1960, pp.1-810

[3] L.T.Biegler: Chemical Process Simulation: A Concise Survey, Chem. Eng. Proc., Oct. 1989, pp.50-61 (references on request)

[4] K.E.Brenan, S.L.Campbell, L.R.Petzold: Numerical Solution of Initial Value Problems in Differential-Algebraic Equations, Elsevier Science Publishing Co., Inc. (1989)

[5] D. Luss Cullen (Chairman): Frontiers in Chemical Engineering - Research needs and opportunities, Report of the National Research Council, National Academy Press (1988)

[6] R.Eckermann (ed.): Modern Computer Techniques and their Impact on Chemical Engineering, 26th Tutzing Symposium 1988, (1989), DECHEMA Monographs Vol.115

[7] H.Engl, M.Hanke, A. Neubauer: Tichonov Regularization on Nonlinear Differential Algebraic Equations, Math.Inst. Univ. Linz, Report No. 385, pp.1-17 (1989) (submitted)

[8] J.C.Fagley, B.Carnaham: The sequential-clustered method for dynamic chemical plant simulation, Comp. Chem. Eng., Vol.14, No.2, pp.161-178 (1990)

[9] H.P.Hutchinson, D.J.Jackson, W.Morton: The Development of an Equation Oriented Flowsheet Simulation and Optimization Package I, II, Comp. Chem. Eng. Vol.10, No.1, (1986) pp.19-29, 31-47

[10] F.Kokert, L.Peer, Hj.Wacker: Mathematical Simulation and Optimization of Chemical Plants, in: Proceedings of the Third European Conference on Math. in Ind. (eds. J. Manley et al.), Kluwer/Teubner, pp.75-90 (1990)

[11] R.Krishnamurthy, R.Taylor: A non equilibrium stage model of multicomponent separation processes: Part I, II, AIChE J. Vol.31, No.3, pp.449-465

[12] Th.Kronberger: Numerical Simulation of Steady State and Nonsteady State Distillation Columns with Nontheoretical Trays, Diploma Thesis, Math.Inst. Univ.Linz (Austria), (1990)

[13] M.D.Lu, R.L.Motard: Computer-Aided Total Flowsheet Synthesis, Comp. Chem. Eng., Vol.9, No.5, (1985), pp.431-445

[14] Oil & Gas Journal: Computers in petroleum in the 90s, A Special Supplement Mar 12, 1990, 32/3 - 32/34

[15] L.Peer: Simulation and Optimization of Chemical Plants (in German), Ph.D. Thesis, Math.Inst.Univ.Linz (Austria), (1990)

[16] J.Villermaux, K.Elgeti, R.Westerterp: The strategy of Chemical Engineering for the next 10 years,-A general Survey of the Reports from the Working Parties of the European Federation of Chem. Engineers, (Report) pp.1-5, (1989)

Address: Johannes Kepler Universität Linz
                Altenbergerstr. 69
                A-4020 Linz
                A U S T R I A

Mini-symposium:

OPTIMAL SUPPLY AND DISTRIBUTION OF ENERGY

Organiser: Hansjörg Wacker

USE OF PATTERN RECOGNITION IN ELECTRICITY LOAD FORECASTING

J. Banim and P.F. Hodnett

INTRODUCTION

This paper is closely related to the paper in these proceedings by Hyde and Hodnett[3] where it is shown that the total load, Y(t), at time, t, of an electricity system, can be written in the form

$$Y(t) = N(t) + W(t) + S(t) + R(t)$$

where N(t) represents the normal load component, W(t) represents the weather-sensitive load component, S(t) represents the load component due to special events and R(t) represents the random load component. Here we describe the use of pattern recognition methods for predicting W(t) when N(t) has been determined by the Hyde and Hodnett[3] procedure, S(t) is removed and R(t) is approximately normally distributed with zero mean and constant variance σ^2 and is therefore negligible. Hyde and Hodnett[3] used a linear regression method to predict W(t).

Two pattern recognition methods were tested. In one of these a nearest neighbour search technique as described in Jabbour et. al.[4] assumes that the recurrence of some weather pattern always reproduces the same weather load, W(t). A historical database of weather patterns and associated weather loads, W(t) is searched for a "nearest neighbour" to a given weather pattern and the resulting associated load, W(t) is the prediction for the weather load. Since the initial results with this method gave less accurate predictions for W(t) than those given by the linear regression procedure of Hyde and Hodnett[3], the method is not further described here. The second method called a learning machine (or matrix) method, used by Dillon et.al[1] and Fu Shu-Ti[2], gave predictions for W(t) more accurate than given by linear regression and is described in what follows.

M. Heiliö (ed.), Proceedings of the Fifth European Conference on Mathematics in Industry, 77–81.
© 1991 B. G. Teubner Stuttgart and Kluwer Academic Publishers. Printed in the Netherlands.

LEARNING MACHINE (MATRIX) METHOD

Past weather and load data is used to train a learning
machine (a matrix of numbers). Once trained, the learning
machine is used for prediction. The data on weather and
load, used to train the machine is called the training set.
Let a weather pattern be represented by a vector
$X = (x_1, .., x_n)$ where $x_1, ..., x_n$ are values of n weather
variables. As indicated in Hyde and Hodnett[3], the
weather influence is adequately represented by the six
variables identified in that paper so that here, n = 6.
The response to X is the weather sensitive load, W(t). The
objective is to establish a link between the weather
pattern, X and the weather sensitive load, W(t). To this
end the range of the weather load, W(t), is divided into
sub-intervals. The link between the weather load, W(t) and
the weather pattern, X is represented by a learning matrix,
V where each row corresponds to a weather load subinterval.
Similarly each weather variable, x_n, is divided into M
subintervals and each column of V corresponds to a
subinterval of one of the weather variables. The
subinterval widths for both weather load and weather
pattern are determined by the required accuracy and in this
paper are chosen as follows. The subinterval width for
W(t) is taken to be the typical range of W(t) divided by 50
i.e. the number of subintervals, M is 50 and since the
typical range for W(t) is (-100MW, 100MW) the subinterval
width is 4MW. Similarly the range of each of the weather
variables, x_n is divided into 50 subintervals so that, for
example, for the temperature (one of the weather
variables), which has a typical range of (-3°C, 15°C), the
subinterval width is 0.36°C. The learning matrix, V, is
therefore of the following form.

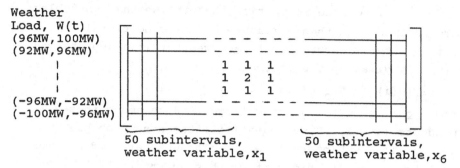

```
Weather
Load, W(t)
(96MW,100MW)
(92MW,96MW)
    |
    |
    |
(-96MW,-92MW)
(-100MW,-96MW)
```

50 subintervals, weather variable, x_1 50 subintervals, weather variable, x_6

It is necessary to construct 24 learning matrices, one for each hour of the day.

TRAINING THE LEARNING MACHINE

For a particular input weather pattern, X and associated weather load, W(t) six columns ($j_1, j_2, \ldots j_6$) and one row (i) of the matrix, V are activated. Thus six positions, (i,j_1), (i,j_2) .., (i,j_6) are identified in the matrix, V. To record the association between this weather pattern and weather load the matrix numbers at these six positions are <u>increased</u> by two while at the eight <u>adjacent</u> row and column positions to each of (i,j_1), ... (i,j_6) the numbers are <u>increased</u> by one. This number pattern is shown symbolically in the matrix above. Initially, the numbers in <u>all</u> matrix positions are set at one. This procedure continues for each element of the training data set and when complete the learning machine is "trained".

FORECASTING WITH THE LEARNING MACHINE

When the training stage is complete, corresponding to <u>each</u> weather load subinterval, W(i) for i = 1 to 50, a probability of occurrence, $P_j^s\{W(i)\}$, corresponding to subinterval, j of weather variable, s is defined by

$$P_j^s\{W(i)\} = v_{ij}^s / \sum_{i=1}^{i=50} v_{ij}^s$$

where v_{ij}^s is the number in matrix row i, column j of weather variable, s. Then, for an input weather vector of six weather variables (at an hour to be forecast), six weather subintervals $j_1, \ldots, j_6$ are activated with associated probabilities $P_{j_1}^1\{W(i)\}, \ldots, P_{j_6}^6\{W(i)\}$ and <u>each</u> i

in 1 to 50. On the assumption that the weather variables
are independent* then a measure of the conditional
probability of occurence of each weather load subinterval,
W(i), is given by the product

$$P^1_{j_1}\{W(i)\} \cdot P^2_{j_2}\{W(i)\} \cdot \ldots \cdot P^6_{j_6}\{W(i)\}$$

for each i in 1 to 50. The largest value among these
products (i = 1,...,50) identifies the most probable
weather load subinterval, W(i). The weather load forecast
is taken to be the midpoint of this subinterval.

The learning matrix is continually updated by adding
new weather and load data as it becomes available.

INITIAL RESULTS

Initial results with this procedure are very
encouraging. When the forecast weather load, W(t), from
here is added to the forecast normal load, N(t), given by
Hyde and Hodnett[3] to give a forecast total load, Y(t),
the difference between the forecast load and actual load
for the test period, May-June (1988), shows a mean absolute
error of 26MW with a standard deviation of 18MW.

This is more accurate than the results reported in
Hyde and Hodnett [3] for the same test period. Detailed
results for this period given in Table 1 show the forecast
load and actual load for each of the 24 hours of Tuesday,
31 May, 1988. These results may be compared with the
results reported by Hyde and Hodnett [3] for the same day
where it is seen that the results here are more accurate
with, for example, a maximum error over the day (here) of
34MW and in Hyde and Hodnett [3] a maximum error of 62MW.
For the results shown in Table 1 the average error over the
day is 13MW with a standard deviation of 8MW. These
results were generated with four weeks of training data
i.e. approximately 30 sets of data.

----

* Since the six weather variables used are not independent
  their uncorrelated principal components are used. This
  necessitates a linear transformation of the weather
  variables.

Table 1 - Comparison of actual loads and forecasted loads (in MW), with forecasting errors, for Tuesday 31 May 1988.

Hour	Load	Forecast	Error	Hour	Load	Forecast	Error
1	1065	1058	7	13	1782	1767	15
2	1015	986	29	14	1661	1650	11
3	974	953	21	15	1685	1677	8
4	940	932	8	16	1656	1637	19
5	935	926	9	17	1716	1706	10
6	923	911	12	18	1727	1693	34
7	1072	1059	13	19	1585	1583	2
8	1417	1417	0	20	1481	1465	16
9	1735	1735	0	21	1453	1448	5
10	1744	1736	8	22	1508	1500	8
11	1773	1755	18	23	1476	1457	19
12	1774	1759	15	24	1274	1259	15

ACKNOWLEDGEMENT

This project is jointly funded by EOLAS (Irish Science and Technology Agency) and the ESB under the Higher Education-Industry Cooperation Scheme operated by EOLAS

REFERENCES

[1]   Dillon, T.S., Morsztyn, K. and Phua, K., 1975, Proc. 5th Power Systems Computation Conference, Vol.1, pp.1-16, Cambridge, England.

[2]   Fu, Shu-Ti, 1981, Proc. 7th Power Systems Computation Conference, pp.581-587, Lausanne, Switzerland.

[3]   Hyde, O. and Hodnett, P.F., 1990, Proc. 5th European Symposium on Mathematics in Industry, Kluwer Academic Publishers, Dordrecht.

[4]   Jabbour, K., Riveros, J.F.V., Landsbergen, D., and Meyer, W., 1988, IEEE Trans. Power Syst., 3, 908-913.

Department of Mathematics
University of Limerick, Limerick, Rep. of Ireland

Overview of Optimal Operations Planning at the
Swedish State Power Board

by

H.Brännlund & D.Sjelvgren

1. The planning process

The Swedish State Power Board (SSPB) production resources consists mainly of hydro-
and nuclear plants and only a minor share of fossil fueled thermal power. Power is exchanged
with other utilities in Sweden and with other Nordic countries. This power exchange is expected
to grow in the future and will increase the power industries need for efficient planning tools.

Today activities associated with operations planning at SSPB comprise of a wide range of
complex problems and time horizons. In order to make the task of overall optimal operations
planning manageable, a sequence of smaller subproblems have been identified. The characteristic
of each subproblem is closely coupled to the length of the associated planning horizon. The
objective of the different subproblems in the process can be summarized as follows:

LONG TERM PLANNING , time horizons from 1-5 years. Planning of large hydro storages
as well as fuel management and maintenance scheduling for the nuclear plants. The randomness of
the reservoir inflows are particularly important in this planning phase.

SEASONAL PLANNING , time horizon of maximum 12 months, usually, associated
mathematical models have a time increment of 1 week. Is aiming at determining the amount of
water to be discharged from hydro reservoirs per week. The results from this planning activity
form the basis for the next stage, namely, the weekly planning.

WEEKLY PLANNING, makes use of a more detailed representation of the electric power
systems components. Time increments may vary from 1 hour up to a couple of hours.
Furthermore, small hydro reservoirs with a cycle of 1 day to 1 week are added to the system
model. Results from this planning activity are used in the daily planning.

DAILY PLANNING, has a very tight coupling to the power system control centre and the
real-time process. A feasible generation schedule is computed, hour by hour, taking into account
start-up and shut-down costs of thermal units, security constraints associated with the transmission system, coordination of hydro- and thermal generation.

2. Modelling

In case the power system involved consists of large proportion of hydro energy, the modelling becomes particularly complex. At SSPB, models have been developed to represent riversystems with cascaded hydroplants. The dynamics of the hydro storages are described by the
following difference equations:

M. Heiliö (ed.), Proceedings of the Fifth European Conference on Mathematics in Industry, 83–88.
© 1991 *B. G. Teubner Stuttgart and Kluwer Academic Publishers. Printed in the Netherlands.*

$$x_i(k+1) = x_i(k) - u_i(k) - s_i(k) + \sum_{j \supseteq I_{ui}} \{ u_j(k-\tau_{uij}) + s_j(k-\tau_{sij}) \} + w_i(k) \quad, \forall\, i \in I, k \in K \quad (1)$$

where x:s denotes stored water volume, u:s are discharges, s:s stands for spillages, w is natural local inflow , τ is the water delay time, I is the set of hydroplants and I_{ui} is a subset of plants lying immediately upstream of plant i and finally k denotes the time interval. It is important to notice that eq. (1) has a network structure which implies that a riversystem can be represented by a set of nodes and directed arcs as illustrated in figure 1.

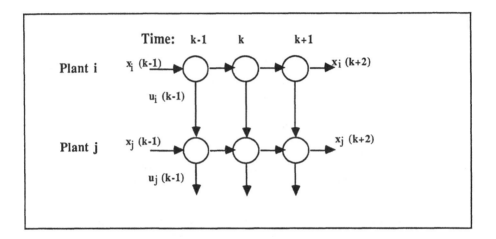

Figure 1. A network representation of two hydroplants in cascade.

In addition to eq (1) there are many other important constraints that are due to time dependent upper and lower limits on reservoir volumes and turbine discharges, limitations on rapid changes of turbine discharges and limitations on rapid headvariations etc., cf. ref [4].

$$\underline{x}_i(k) \le x_i(k) \le \bar{x}_i(k) \quad, \qquad \forall\, i \in I, k \in K \tag{2}$$

$$\underline{u}_i(k) \le u_i(k) \le \bar{u}_i(k) \quad, \qquad \forall\, i \in I, k \in K \tag{3}$$

$$\left| h_i(k) - h_i(k-1) \right| \le \delta_{hi}(k) \quad, \qquad \forall\, i \in I, k \in K \tag{4}$$

$$\left| u_i(k) - u_i(k-1) \right| \le \delta_{ui}(k) \quad, \qquad \forall\, i \in I, k \in K \tag{5}$$

where $h_i(k)$ is the effective head, $\delta_{hi}(k)$ is the allowed headvariation and $\delta_{ui}(k)$ is the allowed variation in turbine discharge. The constraints defined by eq. (4) are solution dependent since the reservoir head is a nonlinear function of the reservoir content.

$$p_i(k) = \eta_i(h,u) \cdot h_i(k) \cdot u_i(k) \qquad (6)$$

where $p_i(k)$ is the power generated at plant i during time period k and η_i is the generating efficiency.

In longterm- and seasonal planning, piecewise linear approximations of the hydroplant generating characteristic can be used [3]. A higher degree of modelling accuracy is required in short-term planning applications and in this case polynomial functions can be used [4,5].

The thermal plants are in the seasonal planning application represented by a probabilistic model [7]. Thus, each plant is described as a multi- state and multi- block unit with different cost of generation and availability data. Representation of nuclear plants includes in addition to the generation cost and plant availability model, also a modeling of the complicated schemes under which generation control at these plants can be done.

In the short term planning situation, more accurate information on the status of the power system is available and the models used are usually purely deterministic. If a major disturbance occurs, e.g. the tripping of a large nuclear plant, it is assumed that the planning model can be rerun.

A very important point in operations planning applications with respect to hydropower systems, concerns determination of proper reservoir levels at the end of the planning horizon. The different endpoint strategies used at SSPB in the context of seasonal planning include a specification of the desired level in the reservoirs at termination, specification of total energy stored in the reservoirs along one river or specification of total energy stored in the whole system. A fourth option that is available is to specify a desired level of the marginal value of hydropower at termination point of the study period. The marginal water values when calculated for individual reservoirs also play an important role in serving as a link for passing information from the seasonal planning model to the weekly- and daily planning models.

3.   Problem formulation and solution techniques

Having defined appropriate models for the components of the power system, the operation planning problem is formulated as a mathematical program. The objective of this optimization problem is to minimize the cost of thermal and nuclear generation subject to load requirements and restrictions associated with available production resources.

> minimize Total operating cost
> subject to
> * Loadbalance constraints
> * Restrictions on hydro power
> * Restrictions on thermal power
> * Transmissionsystem constraints

The above problem is a large scale optimization problem and in order to be able solve it effectively, some sort of decomposition scheme is necessary [2-4]. Through decomposition the mixed hydrothermal operation planning can be divided into smaller hydro- and thermal subproblems.

The hydrosubproblem can be formulated as a general non-linear programming problem with equality constraints in terms of network flow variables,

$$\min F(x)$$
subject to   (P1)
$$Ax = b$$
$$x \in X$$

where the objective function $F(x)$ is non-linear and represents the hydropower utility function. The vector x contains all system vectors, i.e. the state variables x and the control variables u while the systemmatrix A represents the hydrological couplings between plants along the same river and has a network structure.

At the SSPB we have applied large scale nonlinear programming techniques, specialized to take advantage of the network flow structure of the reservoir dynamics equations, refs. [3,4,8]. Problem (P1) is solved by successively computing feasible search directions d that will improve the current solution. The basic steps of the applied methods are;

S1.   Given a feasible solution x^i, generate a feasible direction of descent d that satisfies
        the conditions $Ad=0$ and $\nabla F(x) \, d \le 0$.
S2.   Determine the steplength α by a linesearch procedure in the search direction d.
S3.   Update the solution $x^{i+1} = x^i + \alpha \, d$.
S4.   Check if convergence has been reached, if not , return to step *S1*.

Due to the large scale of (P1) it is necessary that efficient solution techniques are used. For the algorithm *(S1)-(S4)* above, generation of the descent direction d is critical. One technique for determining a search direction is to solve the following problem (P2).

$$\min [\nabla F(x^i)]^t \, d$$
s.t.   (P2)
$$Ad=0$$

where $\nabla F(x^i)$ is the gradient vector of the objective function, evaluated at the point x^i. Due to the networkstructure of A, it is possible to solve (P2) as a minimum cost networkflow problem using specialized networkprogramming techniques. The problem (P2) can furthermore be decomposed into one subproblem for each river since no couplings between different riversystems exists.

The above described techniques have been implemented in an informationsystem developed at SSPB for use in seasonal planning where problems on the order of 10 000 equations and 20 - 30 000 variables are solved.

Other techniques that have been evaluated for solving (P1) is to apply a reduced gradient technique for determination of search directions in step *(S1)*. Reduced gradient methods are particularly efficient on the problems of type (P1) since the linear constraint set can be handled using same techniques as in linear programming methods. In the hydro scheduling application it is possible to further increase the efficiency by specializing the algorithm to exploit the network structure of the constraints.

In a straight forward implementation of a reduced gradient algorithm, search directions are given by the steepest descent, i.e. the negative gradient direction. More advanced techniques to generate descent directions is to use conjugate gradients or memoryless quasi-Newton updates [9]. In these calculations full use can be made of the inherent network properties of the problem.

The transmission security constraints are important to the planning done with respect to a weekly or daily horizon. In the Swedish power system, these restrictions mainly affect the hydropower production and hence a set of linear inequalities must be added to problem (P1),

$$[g_i]^t X \leq c_i \qquad i = 1, 2, \ldots, T \tag{7}$$

where g_i is the vector of coefficients corresponding to the i:th transmission constraint and c_i is a constant. The inequalities (7) do not have the desirable network properties and thus need special attention. Two different techniques have been developed for handling these general, linear constraints within a network programming framework.

First an approach based on basis partitioning will be discussed. In order to apply this method the inequality constraints (7) are converted into a set of equalities. The basis matrix B', for the resulting system of linear equations can always be partitioned in the following way, cf. ref [8]

$$B' = \begin{bmatrix} B & C \\ D & F \end{bmatrix} \begin{matrix} \}n \\ \}T \end{matrix} \tag{8}$$

where B is a square nxn submatrix that has a pure network structure and the T lower rows of B' are associated with the additional transmission security constraints. It is now possible to derive an inverse to B' on a compact form that enable us to take advantage of the special properties of the submatrix B and reduce the required computational labour.

Another technique that we have developed for solving problem (P1) augmented with inequalities (7) is based on a gradient projection scheme and an active set strategy, cf. ref. [10]. This is a promising approach since we in a straightforward manner are able to make use of the fact that in a normal operating situation, only a small number of the security inequality constraints will be binding. The basic idea of the method is to compute the reduced gradient with respect to the equality constraints associated with the reservoir dynamics. This gradient is then projected on the set of active security constraints in order to obtain a descent direction that will not violate any restrictions. The non-binding security constraints need only be monitored in order to avoid violation of any bounds.

4. References

[1] Pereira, M.V.F., "Optimal scheduling of Hydrothermal Systems - An overview", IFAC Symposium on Operation of Electric Energy Systems, Rio de Janeiro, Brazil, July 22-25, 1985, pp. 1-9.

[2] Ea, K and Monti, M., "Daily Operational Planning of the EDF Plant mix - Proposal for a new method", Proc. of PICA conf., San Fransisco, pp. 94-100, 1985.

[3] Sjelvgren, D. et. al., "Optimal Operations Planning in a Large Hydro-Thermal Power System", IEEE Transactions PAS-102, No.11, 1983.

[4]   **Brännlund, H., Sjelvgren, D.** et al., "Optimal Short Term Operation Planning of a large Hydrothermal Power System based on a nonlinear Network Flow Concept", IEEE Trans. Vol. PWRS-1, no.4, pp. 75-82, 1986.

[5]   **Brännlund, H.** et al., "Optimal Generation Scheduling of large Hydro-thermal Power Systems by Network Programming Methods", Proc. of PSCC, Lisbon, pp. 56-62, 1987.

[6]   **Andersson,T., Sjelvgren, D.** , and **Andersson, S.** , "Information System for Optimal Operation Planning and Associated Forecasting Techniques", Proceedings of Applied Optimization Techniques in Energy Problems, June, 1985, Linz, Austria, G.G. TEUBNER, Stuttgart, 1985.

[7]   **Andersson, S.** and **Sjelvgren, D.** , "A Probabilistic Production Costing Methodology for Seasonal Operations Planning of a Large Hydro and Thermal Power System", IEEE Winter Power Meeting, New York, 1986.

[8]   **Kennington, J.L.** and **Helgason, K.V**, "Algorithms for Network Programming", John Wiley and Sons, New York, 1981.

[9]   **Gill, P.E.** and **Murray, W.**, "Practical Optimization", New York, NY, Academic Press, 1981.

[10]   **Brännlund, H., Sjelvgren, D.** et al., "Short Term Generation Scheduling with Security Constraints", IEEE Trans. Vol. PWRS-3, no.1, 1988.

Authors adress: The Swedish State Power Board
                            S-162 87   Vällingby
                            Sweden

APPROXIMATE SOLUTION OF A SYSTEM OF RESERVOIR POWER PLANTS BY VARIATIONAL TECHNIQUES *

A. Gangl, J. Gutenberger and Hj. Wacker

1. Introduction

The operation of a system of hydroelectrical power plants is a complex problem as the plants may be coupled both electrically (i.e., minimum power output) and hydraulically (i.e., the water outflow from one plant may be a significant part of the inflow to one or more downstream plants).

In this paper we present a daily release-policy for a system consisting of two hydro units in series. The discharge from the upstream reservoir is assumed to flow directly into the downstream reservoir with no time lag. A reasonable criterion of performance is the "rated" energy production, that is, we wish to maximize the energy production weighted by a time dependent tariff over a 24h planning period $[0, T]$. The underlying model leads to the following control problem

$$(P) \qquad \text{Maximize} \qquad \int_0^T a(t)\{\sum_{i=1}^2 f_i(V_i(t))Q_i(t)\eta_i(Q_i(t), f_i(V_i(t)))\}\, dt \qquad (1)$$

subject to

$$\dot{V}_1(t) = Z_1(t) - Q_1(t) \qquad (2)$$
$$\dot{V}_2(t) = Z_2(t) + Q_1(t) - Q_2(t) \qquad (3)$$
$$V_i(0) = V_i(T) = V_i^{max} \qquad (4)$$
$$V_i^{min} \le V_i(t) \le V_i^{max} \qquad (5)$$
$$0 \le Q_i(t) \le Q_i^{max} \qquad (6)$$

with bounded state variables $V_i \in C[0, T]$ and control variables $Q_i \in L_\infty(0, T)$. Here $a(t)$ represents the tariff, $V_i(t)$ the volume of the i-th reservoir, $Q_i(t)$ the total discharge through the turbines of plant i and $Z_i(t)$ the natural inflow to reservoir i. The functions f_i and η_i describe the head (i.e., the height difference between the surface of the reservoir and the turbines) and the efficiency of the i-th plant, respectively. Note that the efficiency in general depends on the head and discharge.

A straightforward approach to solve this infinite dimensional problem is given by discretizing (P). This attack, however, leads to a large-scale optimization problem with a nonlinear, nonconvex objective and a large number of linear constraints. One needs sophisticated numerical methods (cf. [1], [4]) to cope with these difficulties.

---

*This work has been supported partly by the "Bundesministerium für Wissenschaft und Forschung" (BMfWF) and the "Fonds zur Förderung der wissenschaftlichen Forschung" (FWF) P6674, P7780.

M. Heiliö (ed.), Proceedings of the Fifth European Conference on Mathematics in Industry, 89–93.
© 1991 B. G. Teubner Stuttgart and Kluwer Academic Publishers. Printed in the Netherlands.

Alternatively, one may try to reduce the complexity of the problem by analytically determining the structure of the optimal solution. Afterwards the problem can be solved with less numerical effort. I.e., one may either transform (P) into an equivalent finite dimensional optimization problem or one may apply multiple shooting to a suitable boundary value problem (cf. [6]).

2. Predetermination of the optimal structure

Concerning the optimal control of a single plant results have been obtained by Gfrerer [2] by variational methods and by Phú [8] by the so-called method of region analysis. In this paper we generalize the results of Gfrerer to two plants in series. We will meet the following assumptions:

(A1) f_i is continuously differentiable with $f_i'(V_i) > 0$

(A2) $\eta_i(Q_i, f_i(V_i))$ is constant

(A3) Z_i is known in advance and $\;0 < Z_1(t) < Q_1^{max}\;$ and $\;0 < Z_2(t) < Q_2^{max} - Q_1^{max}$

(A4) $f_1'(V_1)Z_1(t)\eta_1 - f_2'(V_2)Q_2^{max}\eta_2 > 0\quad$ for all $t \in [0, T]$ and V_i satisfying (5)

(A5) $a(t) = \begin{cases} a_H & \text{for } t \in (0, \xi] \\[2mm] a_L & \text{for } t \in (\xi, T] \cup \{0\} \end{cases} \qquad (a_H > a_L > 0)$

Remarks:

- (A1) is not very restrictive as it holds for most reservoirs.

- (A2) is a simplification. For realistic, nonconstant efficiency functions the existence of a solution of (P) cannot be guaranteed.

- We confine ourselves to the situation (A3) that there is no spilling water. The case of spilling water was analysed by Lindner [5].

- (A4) is met to reduce the number of possible cases in our analysis.

- In Austria the tariff is a piecewise constant function with at most two different values a day. 0 corresponds to 6am and ξ to 10pm.

In the sequel we briefly outline how to determine the structure. Applying the Maximum Principle of the calculus of variation (see [3]) we obtain necessary conditions for an optimum $(V_i^*, Q_i^*)_{i=1,2}$. Our procedure is essentially based on (A2), under which the control variables appear linearly. Consequently, the optimal solution consists of extremal (bang-bang) and singular subarcs only and it remains to analyse in which sequence these subarcs have to be combined.

Under (A3) the optimal solution of the second storage is of the same structure as for a single plant with inflow $Z_2(t) + Q_1^*(t)$. Using the results in [2] we obtain

$$Q_2^*(t) = \begin{cases} Z_2(t) + Q_1^*(t) & t \in (0, t_{2A}] & \text{singular} & (V_2^* = V_2^{max}) \\ Q_2^{max} & t \in (t_{2A}, \xi] & \text{extremal} \\ 0 & t \in (\xi, t_{2B}] & \text{extremal} \\ Z_2(t) + Q_1^*(t) & t \in (t_{2B}, T] & \text{singular} & (V_2^* = V_2^{max}) \end{cases} \tag{7}$$

Our analysis yields the following result for the first storage

$$Q_1^*(t) = \begin{cases} Z_1(t) & t \in (0, t_{1A}] & \text{singular} & (V_1^* = V_1^{max}) \\ Q_1^{max} & t \in (t_{1A}, t_{1C}] & \text{extremal} \\ Z_1(t) & t \in (t_{1C}, t_{1D}] & \text{singular} & (V_1^* = V_1^{min}) \\ 0 & t \in (t_{1D}, t_{1B}] & \text{extremal} \\ Z_1(t) & t \in (t_{1B}, T] & \text{singular} & (V_1^* = V_1^{max}) \end{cases} \tag{8}$$

where $0 \leq t_{1A} \leq t_{2A} \leq t_{1C} \leq t_{1D} \leq \xi$ and $\xi \leq t_{iB} \leq T$, i=1,2. Summing up the structure is fixed by (at most) 6 variables. It might happen that some singular subarcs in (7) and (8) vanish. E.g., if the capacity of the first reservoir is sufficiently large, V_1^* does not reach the lower bound V_1^{min} and $t_{1C} = t_{1D}$.

3. Numerical solution

Replacing $Q_1^*(t)$ in (7) by (8) difficulties arise since we do not know a priori whether $t_{1B} \geq t_{2B}$ (see fig. 1) or $t_{1B} \leq t_{2B}$ (see fig. 2). Therefore, to find the global optimum, we have to perform a case collecting. For each case we are able to transform (P) into an equivalent 6-dimensional optimization problem with 2 equality and 10 inequality constraints. We confine ourselves to the case $t_{1B} \geq t_{2B}$.

Maximize $J(t_{1A}, t_{2A}, t_{1C}, t_{1D}, t_{2B}, t_{1B})$ subject to

$$0 \leq t_{1A} \leq t_{2A} \leq t_{1C} \leq t_{1D} \leq \xi \leq t_{2B} \leq t_{1B} \leq T$$

$$V_1(t_{1C}) \geq V_1^{min}, \qquad V_2(\xi) \geq V_2^{min} \tag{9}$$

$$\int_{t_{1D}}^{t_{1B}} Z_1(t)dt = V_1^{max} - V_1(t_{1C})$$

$$\int_{\xi}^{t_{2B}} Z_2(t)dt = V_2^{max} - V_2(\xi)$$

where

$$J : = \eta_1 \left\{ a_H \left(f_1(V_1^{max}) \int_0^{t_{1A}} Z_1(t)dt + Q_1^{max} \int_{t_{1A}}^{t_{1C}} f_1(V_1(t))dt + f_1(V_1(t_{1C})) \int_{t_{1C}}^{t_{1D}} Z_1(t)dt \right. \right.$$

$$+ \quad a_L f_1(V_1^{max}) \int\limits_{t_{1B}}^{T} Z_1(t)dt \Big\} + \eta_2 \Big\{ a_L f_2(V_2^{max}) \Big(\int\limits_{t_{2B}}^{T} Z_2(t)dt + \int\limits_{t_{1B}}^{T} Z_1(t)dt \Big)$$

$$+ \quad a_H \Big(f_2(V_2^{max}) \{ \int\limits_{0}^{t_{1A}} Z_1(t) + Z_2(t)dt + \int\limits_{t_{1A}}^{t_{2A}} Z_2(t) + Q_1^{max}dt \} + Q_2^{max} \int\limits_{t_{2A}}^{\xi} f_2(V_2(t))dt \Big) \Big\}$$

$$V_1(t) := V_1^{max} + \int\limits_{t_{1A}}^{t} Z_1(\tau) - Q_1^{max}d\tau \text{ for } t \in [t_{1A}, t_{1C}] \text{ and}$$

$$V_2(t) := V_2^{max} + \int\limits_{t_{2A}}^{t} Z_2(\tau) + Q_1^{*}(\tau) - Q_2^{max}d\tau \text{ for } t \in [t_{2A}, \xi]$$

(9) has been successfully solved with an algorithm due to Schittkowski (cf. [7]).
The second case can be treated analogously.

References

[1] H. Gfrerer: Globally Convergent Decomposition Methods for Nonconvex Optimization Problems, Computing 32, 199-227 (1984).

[2] H.Gfrerer: Optimization of Hydro Energy Storage Plants by Variational Methods, ZOR 28, B87-B101 (1984).

[3] I.V. Girsanov: Lectures on the Mathematical Theory of Extremum Problems, Lecture Notes in Economics and Math. Systems 67, Springer, Berlin (1972).

[4] J. Gutenberger, A. Gangl and Hj. Wacker: Decomposition Methods with Superlinear Convergence for Power Plant Systems, submitted to ECMI (proceedings of the Lahti conference, eds. M. Heilio et al.), Teubner, Stuttgart, pp 1-4 (1991).

[5] E. Lindner: Models for the Optimal Control of Storage Power Plants, VWGÖ, Wien (1987).

[6] H. Maurer and W. Gillessen: Application of Multiple Shooting to the Numerical Solution of Optimal Control Problems with Bounded State Variables, Computing 15, 105-126 (1975).

[7] K. Schittkowski: On the Convergence of a Sequential Quadratic Programming Method with an Augmented Lagrangian Line Search Function, Math. Operationsforschung u. Statistik, Optimization 14, 197-216 (1983).

[8] H.X. Phú: On the Optimal Control of a Hydroelectrical Power Plant, Systems & Control Letters 8, North Holland, 281-288 (1987).

Institut for Mathematics, University of Linz
Altenbergerstr. 69, A-4040 Linz / AUSTRIA

APPENDIX:

Optimal reservoir levels

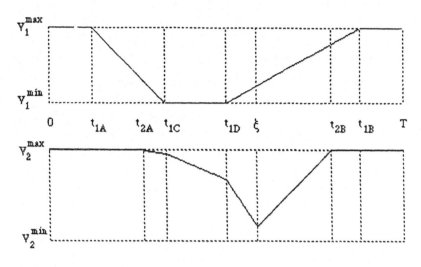

fig.1 (CASE $t_{1B} \geq t_{2B}$)

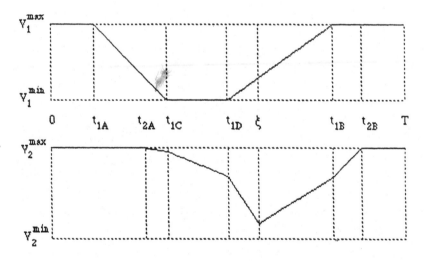

fig.2 (CASE $t_{2B} \geq t_{1B}$)

DECOMPOSITION METHOD WITH SUPERLINEAR CONVERGENCE FOR POWER PLANT SYSTEMS

J. Gutenberger, A. Gangl and Hj. Wacker

1. Introduction: To optimize the yearly income of a system of storage power plants leads to a problem of optimal control (cf. [2]). Discretization w. r. t. time of the mathematical model results in a large-scale nonlinear optimization problem which can be solved by a decompostion-convexification technique developed by Gfrerer [4]. In this paper we try to reduce the numerical effort by increasing the rate of convergence w. r. t. the convexification variable.

2. The method: The resulting numerical model has the form

$$(P) \qquad \text{minimize} \quad f(x)$$
$$\text{subject to} \quad Ax=b, Kx \geq q$$

where $f: R^n \to R$ is a given function, twice continuously differentiable, nonlinear and nonconvex. A and K are $m \times n$ ($m < n$) and $k \times n$ matrices resp. with full ranks. Artificial splitting of some variables provides separability: $f(x) = \sum_{i=1}^{s} f_i(x^i)$, $x=(x^1,...,x^s)$, $f_i: R^{n_i} \to R$, $\sum_{i=1}^{s} n_i = n$. For the inequality constraints $Kx \geq q$ we get: $K_i x^i \geq q_i$, $i=1,...,s$, (K_i $k_i \times n_i$ matrix).

(P) is solved by a primal dual method which decomposes the problem. In order to extend the decomposition technique to nonconvex problems, Gfrerer [4] considered the following convexification:

$$(CP) \qquad \text{minimize} \quad f(x) + \tfrac{1}{2} (x-y)^T C (x-y) , \qquad y \in R^n,$$
$$\text{subject to} \quad Ax=b, Kx \geq q$$

$C \in L(R^n)$, pos. def., is chosen such that (CP) is strictly convex w.r.t. x and such that C preserves the separable structure. In Bertsekas [1] C is chosen as a multiple of the identity. A major drawback of this approach is the local convergence. Gfrerer has shown that the "amount of convexification" should be as small as possible to fasten convergence. Therefore a priori information of the Hessian of f should be included. For details see [3],[5].

Let $x(y)$ denote the unique solution of problem (CP) which is solved by duality. We consider the dual problem:

$$(D) \qquad \text{maximize} \quad \phi(\lambda) \quad ,\lambda \in R^m$$

95

M. Heiliö (ed.), Proceedings of the Fifth European Conference on Mathematics in Industry, 95–98.
© 1991 B. G. Teubner Stuttgart and Kluwer Academic Publishers. Printed in the Netherlands.

where $\phi(\lambda) = \min \{ f(x) + {}^1\!/_2 (x\text{-}y)^T C (x\text{-}y) + \lambda^T (Ax\text{-}b) \mid Kx \geq q \}$

$$= \sum_{i=1}^{s} \min \{ f_i(x^i) + {}^1\!/_2 (x^i\text{-}y^i)^T C_i (x^i\text{-}y^i) + \lambda^T (A_i x^i\text{-}b) \mid K_i x^i \geq q_i \}.$$

Note that only the equality constraints are treated by duality, whereas the inequality constraints, which do not destroy separability of the problem, are treated directly. It is a well known result of duality theory, that under certain convexity assumptions a maximum $\lambda(y)$ of (D) yields the unique solution $x(y) = x(y, \lambda(y))$ of (CP). For that the following functional φ is well defined:

$$\varphi(y) = f(x(y)) + {}^1\!/_2 (x(y)\text{-}y)^T C (x(y)\text{-}y)$$

A fixed point of $x(.)$ minimizes φ and fulfills the first order necessary conditions for a local minimizer of f. The new value of the convexification variable is defined by the iteration (cf. [4])

(G) $y^{k+1} := x(y^k) ,$

which may be regarded as a descent method with fixed step size. Since the rate of convergence is only linear, we tried to accelerate the method by applying Newton's method to minimize φ.

(IM) $y^{k+1} := y^k - [\nabla^2 \varphi(y^k)]^{-1} \nabla \varphi(y^k)$

where $\nabla \varphi(y) = (y\text{-}x(y))^T C,$

$\nabla^2 \varphi(y) = C - C[M - MA^T(AMA^T)^{-1}AM]C, \quad M = Z(Z^T(\nabla^2_x f(x(y)) + C)Z)^{-1}Z^T,$

Z nxk_a denotes the zero-space of the matrix K_a, consisting of those rows of K corresponding to the active constraints of $Kx \geq q$ at $x(y)$. (k_a ... number of active constraints).

Performing one Newton step means to solve a high dimensional system of linear equations. This is done by the method of conjugate gradients, where the main numerical effort is to multiply the system-matrix $\nabla^2 \varphi$ with vectors. Based on the sparsity of the matrices M, C and A this can be done quite efficiently. (Note: the matrix $\nabla^2 \varphi$ is full).

Alternatively one may use the inverse of $\nabla^2 \varphi$ explicitly:

$$\nabla^2 \varphi(y)^{-1} = C^{-1} + [N - NA^T(ANA^T)^{-1}AN], \quad N = Z(Z^T(\nabla^2_x f(x(y)))Z)^{-1}Z^T .$$

However computational experience has shown, that this approach is not favorable due to stability problems ($(Z^T(\nabla^2_x f(x(y)))Z$ ... indefinite !).

One major drawback is the fact, that changes in the active index set cause jumps in the Hessian $\nabla^2\varphi$. In our practical problem (optimal control of storage power plants) this happens very often. The implementation of a line search algorithm together with regularization of $\nabla^2\varphi$ has helped us to overcome these difficulties to some extent:

$$y^{k+1} := y^k - \beta[\nabla^2\varphi(y^k)+\alpha I]^{-1} \nabla\varphi(y^k)$$

$\beta \le 1$ ... line search parameter, α ... regularization parameter.

3. Practical example

The method has been applied to a storage power plant system consisting of three units.

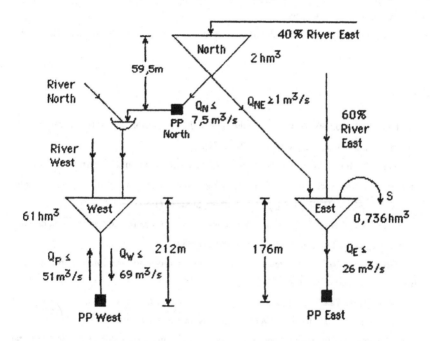

Water from reservoir North can be released either to reservoir East and/or via a pipe system to a small plant North. This water and the water of river North is then collected in a small pool and afterwards led to reservoir West. Unit West additionally allows for pumping, that is water can be transferred back to the reservoir. At East we have to consider spillage water due to the small capacity of the reservoir.

The aim is to maximize the monetary values of the energy produced in the system.

For a more detailed description and numerical results see [3], [5] and [6].

Performing a yearly optimization we get a problem of $9 \cdot N$ variables and $25 \cdot N$ constraints. N denotes the number of discretization points.

We compare our improved method (IM) with Gfrerer's technique (G).

N=60: (-> 540 variables, 1500 constraints)

	number of iterations	number of evaluations of the dual function	CPU - time (minutes)
(G)	404	1359	10.42
(IM)	73	520	5.42

We see that the number of iterations and evaluations of the dual function could be reduced considerably. The CPU-time is halved.

For higher dimension (N≥120) the additional effort for calculating a Newton step cancels the advantage of fewer iterations.

References:

[1]   D. Bertsekas: Convexification procedures and decomposition methods for nonconvex optimization problems. JOTA 29, 169-197 (1979).

[2]   A. Gangl, J. Gutenberger and Hj. Wacker: Approximate solution of a system of reservoir power plants by variational techniques, submitted to ECMI (proceedings of the Lahti conference, eds. M. Heilio et al.), pp 1-5, Teubner, Stuttgart (1991).

[3]   A. Gangl: Short term optimization of an extension of a storage power plant system, diploma thesis, University Linz, Institute of Mathematics (1989).

[4]   H. Gfrerer: Globally convergent decomposition methods for nonconvex optimization problems, Computing 32, 199-227 (1984).

[5]   J. Gutenberger: Yearly optimization and comparison of two extensions of a storage power plant system, diploma thesis, University Linz, Institute of Mathematics (1989).

[6]   Hj. Wacker et al.: Optimal extension of a system of hydro energy storage plants, to appear in proceedings of the fourth european conference on mathematics for industry, eds. Hj. Wacker et al., pp 1-6, Teubner, Stuttgart (1990)

Institute for Mathematics
University of Linz
Altenbergerstr. 69
A-4040 LINZ / AUSTRIA

AN ELECTRICITY LOAD FORECASTING SYSTEM

O. Hyde and P.F. Hodnett

INTRODUCTION

The Irish Electricity Supply Board (ESB) operates an
isolated electricity network with a peak load of
approximately 2350 MW, using a variety of types of
electricity generation i.e. coal, oil and peat fired
stations, hydro electricity and a pumped storage station.

In this paper, a system to forecast short term demand
for electricity is developed based upon the development of
a forecasting procedure which uses both past load data and
weather conditions.

LOAD COMPONENTS

The load demand, although highly variable, is not
entirely random, and the variations in the demand for power
in the short term can be attributed to (a) hour of day,
(b) day of week, (c) season of year, (d) special days, (e)
weather influences, (f) special events, (g) random unknown
factors. A number of authors including Debs [1], Jabbour
et al [2], Lijesen and Rosing [3], Rahman and Bhatnagar [4]
have examined these factors in detail. Total load is then
assumed to consist of four separate components. First,
there is the normal load component, incorporating factors
(a)-(d), which is basically the weather-independent
component of the load. A long-term factor which has to be
accounted for is load growth, from year to year reflecting
increases in customer demand and number of customers.

A second significant component of the load is the
weather-sensitive part where changes in total load occur
due to changes in weather conditions.

A third factor is that occasions arise where unusual
or special events affect the load. The load is affected in
a special way by events such as (a) school closures, (b)
industrial disputes, (c) system disturbances (lightning,
gales etc), (d) elections, (e) major television events.

Finally, there is a random component of the load which

99

M. Heiliö (ed.), Proceedings of the Fifth European Conference on Mathematics in Industry, 99–103.
© 1991 B. G. Teubner Stuttgart and Kluwer Academic Publishers. Printed in the Netherlands.

is not attributable to either normal load, weather
influences or special events.

MODEL IDENTIFICATION

In analysing the load, a linear relationship between
the total load and the above components is assumed. The
total load of the system at a point in time, t, is written
in the form

$$Y(t) = N(t)+W(t)+S(t)+r(t) \ldots \ldots (1)$$

where Y(t) represents total load, N(t), represents normal
load components, W(t), represents weather-sensitive load
component, S(t), represents load component due to special
events, r(t), represents random load component at time t.

As time t is measured periodically (every hour), we
employ an index k to specify the kth total hourly load
measured from the specified initial time. This gives

$$Y(k) = N(k)+W(k)+S(k)+r(k) \ldots \ldots \ldots (2)$$

The following are the steps involved in identifying an
overall model of the form given in equation (2). The first
step is the collection of historical data of hourly load
and weather information. The quality of the data is
checked, with guide-lines set for the treatment of missing
data. Then, load variations due to special events are
removed, as they do not represent normal load behaviour,
giving

$$Y(k) = N(k)+W(k)+r(k) \ldots \ldots \ldots (3)$$

Total load is now composed of normal load and weather load,
both of which can be modelled separately, and of a random
load component, r(k).

The effects of load growth are removed from the load data
before analysis of the data begins by normalizing the
yearly load by the method used in [2].

Weather Load Model

The year is divided into six bimonthly periods of similar
weather conditions to account for seasonal trends in
weather patterns, as follows, January-February; March-
April; May-June; July-August; September-October; November-
December.

For each period, an hour when normal load is approximately constant is chosen (i.e. any variation in load behaviour is due to weather influences alone), for example, 3 p.m. on mid-weekdays. For this selected data, equation (3) gives

$$Y(k) = N+W(k)+r(k) \dots\dots\dots\dots\dots (4)$$

where k refers to a measurement sample rather than a specific time, in this instance. It is found that the random component, r(k) is approximately normally distributed with zero mean and constant variance σ^2 and is therefore neglected later. Regression analysis is used to test the significance of possible weather variables on load variation and to identify those weather factors which are necessary and sufficient to represent the effects of weather on load variation.

Normal Load Model

To develop the normal load model, the data is now further subdivided into smaller periods, consisting of a particular hour on a particular day of the week (or special day) in one of the six bimonthly periods. Equation (3) is written (with r(k) neglected) as follows, with k again referring to a measurement sample (for a single period)

$$N(k) = Y(k)-W(k) \dots\dots\dots\dots\dots (5)$$

Since the values of Y(k) are known from the data and since W(k) can now be determined using the weather load model therefore estimates for the normal load components N(k) can be calculated from equation (5) over the sample of measurements for each period. Averaging N(k) given by equation (5), gives an estimate of the normal load component which is used to forecast future normal loads for that period. The process is repeated for every hour of every day in every bimonthly period, creating the normal load model.

Combining Models

Electrical load can now be forecasted for any time period provided accurate weather forecasts can be obtained for that time. The forecast electrical load is the sum of the

individual forecasts from the normal load and weather load
models. Load growth has been incorporated into the model
through destandardizing the forecast load. Adjustments can
also be made to the forecast load to take into account any
special events that may be foreseen.

INITIAL RESULTS

A historical data base consisting of 7 years of load and
weather data was created. Through regression analysis a
total of 15 potential weather variables in 6 locations
throughout the country were tested for significance and
through this process the following weather load model was
identified (at k^{th} hour)

$$W(k) = \alpha_1 T(k) + \alpha_2 T(k-24) + \alpha_3 Sn(k) + \alpha_4 V(k) + \alpha_5 H(k) \qquad \ldots (6)$$

where, T represents dry bulb temperature in $^{\circ}C$, Sn
represents sunshine duration in hours, V represents wind
velocity in knots, H represents relative humidity (%) and
α_i (i = 1 to 5) are the corresponding parameters for these
weather variables.

With respect to location weather in the Dublin area only
was found to be significant reflecting the fact that more
than half the electricity demand is in the Dublin area.
The five weather variable parameters α_i (i = 1 to 5) are
allowed to have different values for each of the six
bimonthly periods of the year, but each parameter, α_i, is
assumed to be constant throughout a particular bimonthly
period.

Initial test results with the model are encouraging with
the model results giving a good approximation to the real
ESB load data. For example for the test period, May-June
(1988), the difference between the model predictions and
real load data shows a mean absolute forecasting error of
2.1% and a standard deviation of 1.6%. For these model
tests the weather load model used is

$$W(k) = -0.015T(k) - 0.015T(k-24) - 0.065Sn(k) + 0.009V(k)$$
$$+0.003H(k) \qquad \ldots (7)$$

The normal load model provides separate forecasts for each
day of the week (7 days) and for the special days

(a) first Monday in June, public holiday and (b) Ascension Thursday and Corpus Christi Thursday both Church holydays treated as one day. There is therefore a total of nine (seven ordinary days plus two special days) forecast days in the May-June (1988) bimonthly period. Detailed results for this period are shown in Table 1 which shows the forecast load and actual load for each of the 24 hours of Tuesday, 31 May, 1988.

TABLE 1 - Comparison of actual loads and forecasted loads, (in MW) with forecasting errors, for Tuesday 31st May 1988.

Hour	Load	Forecast	Error		Hour	Load	Forecast	Error
1	1065	1061	4		13	1782	1748	34
2	1015	989	26		14	1661	1600	61
3	974	959	15		15	1685	1640	45
4	940	940	0		16	1656	1629	27
5	935	933	2		17	1716	1654	62
6	923	927	-4		18	1727	1721	6
7	1072	1042	30		19	1585	1592	-7
8	1417	1363	54		20	1481	1479	2
9	1735	1682	53		21	1453	1449	4
10	1744	1737	7		22	1508	1510	-2
11	1773	1729	44		23	1476	1474	2
12	1774	1767	7		24	1274	1241	33

ACKNOWLEDGEMENT

This project is jointly funded by Eolas (Irish Science and Technology Agency) and the ESB under the Higher Education - Industry Cooperation Scheme operated by Eolas.

REFERENCES

[1] Debs, A.S., 1988, "Modern Power Systems Control and Operation", Klumer Acad.Publ., Boston, U.S., Chapter 9.

[2] Jabbour, K., Riveros, J.F.V., Landsbergen, D., and Meyer, W., 1988, IEEE Trans. Power Syst., 3, 908-913.

[3] Lijesen, D.P. and Rosing, J., 1971, IEEE Trans. PAS, 90, 1757-1767.

[4] Rahman, S., and Bhatnagar, R., 1988, IEEE Trans. Power Syst., 3, 392-399.

Department of Mathematics
University of Limerick, Limerick, Rep. of Ireland

DAILY LOAD PREDICTION VIA
CLUSTER ANALYSIS AND LEARNING MACHINES

EWALD H. LINDNER

1. Introduction

The aim of this project is to develop a software package for predicting the load of an electric power-supply company for a time period of 24 hours. This short-term prediction is intended for assisting the load dispatchers in doing this preliminary stage of the daily unit commitment problem. The companies which we have in mind are situated in Central Europe, their peak load is about 1000 MW, their supply area amounts to 10000 km^2 resp. 1 million persons. (In Austria) load dispatchers are still not eager to use forecasting methods supported by resp. prognoses calculated by computers but prefer their own manual ways of predicting to them. Therefore, we report on a method which is similar to that. The representatives of the clusters, and the learning machine, will model the box board slices, and the overlaying, resp., of that manual method.

This project was stimulated by Siemens Austria AG, which also wanted a model where the load to be predicted is not split into parts (nominal load, residual load etc.), but treated as a whole and yields a 24 hourly prediction in one step, so that the load curve is smoother compared with a time series model. The results of our method will be compared with an adaptive and learning load forecasting method (Zarzer[9], [10]), which combines elements of time series analysis and regression analysis.

For a learning machine approach, compare Dillon e.a. [3] and one, where the load is split into parts predicted separately, see e.g. Fu Shu-Ti [5]. For comparing different forecasting models, which do not include pattern recognition, we refer to Deistler e.a. [1,2]. Edwin and El Sayed [4] at first present a time series model for the load prediction and secondly analyze the dependence of the cost functional (essentially for thermal power plants) on the forecasting error.

2. Data, variables, and the model

Our model was applied to two electric power-supply companies. In both cases just the electric load and the (air) temperature were recorded every 30 minutes, but no information

M. Heiliö (ed.), Proceedings of the Fifth European Conference on Mathematics in Industry, 105–108.
© 1991 B. G. Teubner Stuttgart and Kluwer Academic Publishers. Printed in the Netherlands.

on precipitation, humidity, wind velocity, wind's direction, clouding, etc. was available. These missing quantities obviously influence the electric load, on the other hand they are hardly predicted with sufficient accuracy for a time period of 24 hours due to informations of the local hydrografic service. For details cp. König and Tröger [6], and Edwin and El Sayed [4].

Because of the data recording process the electric load for day i is represented by a 48-dimensional vector $L(i)$. Hence, a single prediction implies to calculate also a 48-dimensional vector. Furthermore the term "day" will mean the so-called energetic day, which starts two hours earlier than the calender day, i.e. at 10 p.m.

In a first step we condensed information on temperature to mean, lowest and highest temperature $T(i), T_{\min}(i)$, and $T_{\max}(i)$, resp., of day i. Afterwards, we determined the number of past days to be taken into account in our model. We used the standard method for time series and considered the partial correlation matrices. The partial correlation matrix $R_{yy \cdot x}$ is the correlation matrix of the stochastic variable $y - \hat{y}$, where $\hat{y}$ is the best linear approximation of y by x, cp. Schlittgen and Streitberg [7]. In our cases y is the joint distribution of the load at a certain time point and the temperature of the past N_P days and x is the joint distribution of the temperature of the past N_T days, $N_T < N_P$. Hence, the coefficients of this matrix give us a first order information on the remaining information. We stopped this empirical determination, whenever the partial correlation was less than 2 percent. For the data of Pfalzwerke, FRG, we ended up with $N_T = 4$ and $x = (T(i - j), j = N_T, \cdots, 0$ by $-1, T_{\max}(i), T_{\min}(i))$. The elements of x will be called weather characteristics.

For a standard learning machine model (Dillon e.a. [3]) the elements of the event output vector are the load at a certain time point, discretized in a certain number of intervals. In our case we cluster the daily load curves in two ways, namely, the unscaled load curves into N_{US} and the multiplicatively by the mean load scaled load curves into N_S clusters. Both times we at first apply a divisive hierarchical and then a partitioning cluster algorithm, that is DISMEA and KMEANS by Späth [8]. So the rows of the learning matrix LM correspond with (the centers of) the clusters of the unscaled load curves. The columns of LM correspond with the weather characteristics, which are shifted by a constant factor each and discretized into N_W intervals of constant length. Finally we add another block of N_S columns which correspond with the clusters of the scaled load curves. They represent "typical days", i.e., the qualitative behaviour of the load curves. In fact, each block of columns for one weather characteristic may be regarded as a single learning machine. LM is initialized by zero and learns from the past system behaviour. For day $i = -N_D, \cdots, -1$ let $p(i)$ be the index of the cluster, load curve $L(i)$ belongs to, and $q(i)$ be the index of the interval, $T(i - N_T)$ is assigned to. For sake of simplicity let us just consider this first weather characteristic. Then $\mathrm{LM}(p(i), q(i))$ is increased by 1. Hence, LM finally contains the absolute relativities. The conditional probability, that load curve cluster j will occur at day 0, if $T(-N_T)$ were assigned to interval k, is

$$\mathrm{Prob}\,(p(0) = j | q(0) = k) \;=\; \mathrm{LM}\,(j, k) / \sum_{l=1}^{N_{US}} \mathrm{LM}\,(l, k)$$

Let $\bar{L}_j$ be the center, i.e. the mean value, of cluster j. Then

$$L_1 := \sum_{j=1}^{N_{US}} \bar{L}_j \cdot \text{LM}(j,k) / \sum_{l=1}^{N_{US}} \text{LM}(l,k)$$

is a forecast for the load, if $T(-N_T)$ were assigned to interval k. We finally modify all above summations in that way, we consider just indices j, and l, iff there is a non zero entry in row j, and l, and the column according to the type of day 0. The same way we treat the remaining weather characteristics, yielding $L_2, \cdots, L_{N_T+3}$. Finally, we end up with the prediciton

$$L_P := \sum_{m=1}^{N_T+3} L_m \rho_m$$

where the ρ_m are either set or determined by linear least squares. In this case we minimize the error between forecast and the load taken place within the past 8 to 17 days. In any case, we need a forecast of the temperature and the type of the day which the load is to be predicted for.

3. Results for Pfalzwerke, FRG

The data base contained 183 days and was supplied by Siemens Austria AG. We obtained $N_T = 4, N_{US} = 27, N_S = 9$, shifted the temperature data by its minimum and then discretized these data in $N_W = 7$ intervals of $4°$ C length each. The first 4 days were not fed in LM, as the relevant weather data were incomplete. So we used $N_D = 160 - N_{LS}$ days as fill in, and N_{LS}, ranging from 8 to 17 days, as period for adapting $\rho_m, m = 1, \cdots, 7$. 19 days were used for evaluating the following error objectives, which are compared with two methods Z1, and Z2 of Zarzer [9], [10]. Method Z1 combines elements of time series analysis and regression analysis. Method Z2 consists of Z1 plus an error learning method, which is again based on Z1.

	error in percent (standard deviation in percent)		
	this method	Z1	Z2
mean error	4.46(4.78)	5.23	1.99
error of work	2.01(2.11)	4.18	1.31
maximum error	16.31(6.10)	15.04	5.57
minimum error	0.09(0.10)	0.27	0.11
error at load maximum	2.08(1.56)	3.06	1.51
error at load minimum	5.56(5.28)	8.27	2.65

Hence, this method is superior to Zarzer's basic method Z1, but inferior to his two stage method. By also applying a further error learning method as for Z2 we would equalize the results of Z2 at least. As the above three implemented methods were not honored at all by our local electric power-supply company this improved version was not implemented.

References

[1] Deistler M., Fraißler W., Petritsch G., Scherrer W.: Ein Vergleich von Methoden zur Kurzfristprognose elektrischer Last, Archiv für Elektrotechnik 71 (1988), S. 389-397

[2] Deistler M., Fraißler W., Petritsch G., Scherrer W.: Methoden zur Kurzfristprognose elektrischer Last, Research Report No. 82, Institut für Ökonometrie, OR and Systemtheorie, TU Wien, 1987, S. 1-59

[3] Dillon T.S., Morsztyn K., Phua K.: Short term load forecasting using adaptive pattern recognition and self organizing techniques, paper 2.4/3 in: Conference-Proceedings, Vol.1, of the 5th Power System Computation Conference, Cambridge, Sept. 1975, Queen Mary College (ed.), University of London

[4] Edwin K.W., El Sayed M.A.H.: Lastprognosen für die Tageseinsatzplanung von Kraftwerkssystemen, etz Archiv 2 (1980) 6, S. 171-176

[5] Fu Shu-Ti: A pragmatic approach for short-term load forecasting using learning-regression method, paper VI/13, in: Proceedings of the 7th PSCC, Lausanne, July 1981, p. 581-587

[6] König S., Tröger D.: Kurzfristige Vorhersage wetterabhängiger Summenlasten, Elektrizitätswirtschaft 74 (1975) 1, S. 21-24

[7] Schlittgen R., Streitberg B.H.J.: Zeitreihenanalyse, R. Oldenburg Verlag, München, 1984

[8] Späth H.: Cluster-Analyse-Algorithmen, 2. verb. Aufl., R. Oldenburg Verlag, München, 1977

[9] Zarzer E.A.: Ein adaptives Regressions-Prognosemodell zur Tageslastprognose, ÖZE 36 (1983) 1, S. 1-4

[10] Zarzer E.A.: An adaptive and learning forecasting method, in: Hj. Wacker (ed.): Applied Optimization Techniques in Energy Problems, B.G.Teubner Stuttgart, 1985, p.471-485

Address: Institut für Mathematik
                      Johannes Kepler Universität Linz
                      A-4040 Linz/Auhof, Altenbergerstr. 69
                      AUSTRIA

THE OPTIMAL SCHEDULING OF HYDRO-ELECTRIC POWER GENERATION

P. Oliveira, S. Mckee, C. Coles

Department of Mathematics
University of Strathclyde
Glasgow, Scotland, U.K.

Abstract. The weekly optimization of a power system is presented. The mathematical model, a mixed integer (linear) programming problem, is discussed and the solution of the unit commitment and economic dispatch problem is determined on an hourly basis. The system contains 840 integer variables and 1344 continuous variables, with a sparsity of 0.11%. This report focuses on a branch-and-bound technique for the solution of the problem.

1. Introduction

The two main decisions [Cohen and Sherkat (1987), Tong and Shahidehpour (1989)] in power scheduling are: unit commitment, which indicates which of the generating units are committed (on or off) in every time interval of the scheduling horizon; economic dispatch, which is the allocation of the system load among the generating units to meet the demand. The objective of a least cost solution implies the simultaneous consideration of these two decisions.

2. The System

The system we consider is a lumped model of the Scottish Hydro-Electric plc system. It consists of five generating units: one nuclear, two thermal (oil and coal fired), one conventional hydro, and a pump-storage unit. The inflows to the hydro units are assumed constant over a 24 hour period. This information is given in energy units, assuming that there are no great variations in reservoir heads. The inflow to the pump-storage unit is assumed to be 1% of the hydro unit inflow;

M. Heiliö (ed.), Proceedings of the Fifth European Conference on Mathematics in Industry, 109–114.

also lower and upper limits are prescribed for both generation and pumping. The demand is assumed constant over each 1 hour period. The particular system considered in this paper has a large hydro component and the costs for producing and for pump-storage were supplied by Scottish Hydro-Electric plc.

3. The Mathematical Model

The nuclear component is of the "must-run" type and so, for optimal scheduling, this component may be neglected. The time interval over which the system operates is divided into T equal time periods.

3.1 THE THERMAL SYSTEM

There are N thermal units. Each unit has an associated integer variable which denotes whether the unit is committed (on) or not (off)

$$\alpha_i^t = \begin{cases} 1, & \text{if unit } i \text{ is on during period } t, \\ 0, & \text{otherwise;} \end{cases}$$

Note that this implies that x_i^t, the power output, $\underline{x}_i \alpha_i^t \leq x_i^t \leq \bar{x}_i \alpha_i^t$. The costs are:

(a) Running costs per unit

$$F_i \alpha_i^t + V_i x_i^t, \tag{1}$$

where F_i is the fixed cost, and V_i is a linear approximation to the fuel cost over the range $[\underline{x}_i, \bar{x}_i]$ and is zero for $x_i^t = 0$.

(b) Start-up cost per unit

$$U_i \beta_i^t, \tag{2}$$

where U_i is the start-up cost for unit i, and β_i^t is an integer variable defined as

$$\beta_i^t = \begin{cases} 1, & \text{if unit } i \text{ is started in period } t, \\ 0, & \text{otherwise.} \end{cases}$$

$$\beta_i^t \geq \alpha_i^t - \alpha_i^{t-1}.$$

(c) Shut-down cost per unit

$$D_i \gamma_i^t, \tag{3}$$

where D_i is the shut-down cost for unit i, and γ_i^t is an integer variable defined as

$$\gamma_i^t = \begin{cases} 1, & \text{if unit } i \text{ is shut-down in period } t, \\ 0, & \text{otherwise.} \end{cases}$$

$$\gamma_i^t \geq \alpha_i^{t-1} - \alpha_i^t.$$

(d) If the minimum down time is Ξ_i for unit i, then necessarily the following restriction holds

$$\gamma_i^t + \sum_{j=t+1}^{t+\Xi_i-1} \beta_i^j \leq 1, \tag{4}$$

where Ξ_i is an integer greater than 1.

3.2 THE HYDRO SYSTEM

The hydro system consists of K hydro units, and L pump-storage units.

3.2.1 Hydro plants. The continuity equation gives the following relation

$$v_k^t = v_k^{t-1} + f_k^t - y_k^t - s_k^t, \tag{5}$$

where

v_k^t is the kth reservoir storage at period t,

y_k^t is the discharge of kth hydro plant at period t,

f_k^t is the influx to k^{th} hydro plant reservoir at period t, and $f_k^t \geq 0$.

s_k^t is the spillage from k^{th} reservoir at period t, and $s_k^t \geq 0$.

The operation of the hydro units must be within the reservoir limits, and operating limits of the turbines. The $0-1$ variable δ_k^t denotes whether the hydro unit is committed in period t

$$\delta_k^t = \begin{cases} 1, & \text{if unit } k \text{ is on during period } t, \\ 0, & \text{otherwise.} \end{cases}$$

The operating costs of the hydro units are taken to be linear

$$H_k y_k^t + S_k s_k^t, \tag{6}$$

where H_k and S_k are, repectively, the value of the discharge and spillage.

3.2.2 *Pump-storage plants.* From continuity we get

$$r_l^t = r_l^{t-1} + g_l^t - q_l^t + p_l^t, \tag{7}$$

where

r_l^t is the lth reservoir storage at period t,

q_l^t is the discharge from the lth unit at period t,

p_l^t is the pumped water to the lth reservoir at period t,

g_l^t is the influx to the lth reservoir at time t, and $g_l^t \geq 0$.

The operation of the pumped-storage units must be within the operating limits of the reservoir, turbines and pumps. Due to the design of the plant there cannot be generation and pumping at the same time. This can be expressed as $q_l^t p_l^t = 0$. In order to overcome this nonlinearity, two integer variables are introduced,

$$\mu_l^t = \begin{cases} 1, & \text{if generating in period } t, \\ 0, & \text{otherwise.} \end{cases}$$

$$\nu_l^t = \begin{cases} 1, & \text{if pumping in period } t, \\ 0, & \text{otherwise.} \end{cases}$$

Therefore $q_l^t p_l^t = 0$ can be expressed as, $0 \leq \mu_l^t + \nu_l^t \leq 1$. The cost of operating the lth pump-storage unit is

$$G_l q_l^t - P_l p_l^t, \tag{8}$$

where G_l is the value of the equivalent quantity of water used for generating each unit of power while P_l is the value of the water pumped to the reservoir.

3.3 THE DEMAND AND RESERVE CONSTRAINTS

In this study we shall consider satisfying the demand (d^t), while allowing just enough excess capacity to cover the constant reserve (R) imposed by the pool system operating in the United Kingdom. Therefore

$$x_i^t + y_k^t + q_l^t - \Theta_l p_l^t \geq d^t, \tag{9}$$

$$\bar{x}_i \alpha_i^t + \bar{y}_k \delta_k^t + \bar{q}_l \mu_l^t - \Theta_l p_l^t \geq d^t + R, \tag{10}$$

where Θ_l is the efficiency of the pumping process.

3.4 THE MIXED INTEGER MODEL

In summary the scheduling can be modelled as the mixed integer linear programming problem

$$\min_{\alpha_i^t, x_i^t, \delta_k^t, y_k^t, \mu_l^t, q_l^t, \nu_l^t, p_l^t} \sum_{t=1}^{T} \{ \sum_{i=1}^{N} (U_i \beta_i^t + F_i \alpha_i^t + V_i x_i^t + D_i \gamma_i^t) + \sum_{k=1}^{K} (H_k y_k^t + S_k s_k^t)$$

$$+ \sum_{l=1}^{L} (G_l q_l^t - P_l p_l^t) \}, \qquad (11)$$

for $i = 1, \ldots, N$, $k = 1, \ldots, K$, $l = 1, \ldots, L$, $t = 1, \ldots, T$.

subject to the constraints given above.

4. The Solution

The branch-and-bound approach, [Fletcher (1987)], considers a sequence of simpler linear programming problems derived from the original problem. The top node is a relaxation of the original problem which is relatively easy to solve. The sucessors of the top node are a set of problems (also relaxations of the original problem) each having a disjoint space; the union of the solution spaces of these sucessors is the solution space of the top node.

The Sciconic/VM package was used to solve the problem by the branch and bound method. The package allows the user to define the strategy of searching the tree.

5. Conclusions

Several runs have been made for different operating conditions, (see Table 1), which have validated the model, and have supplied insight into the costs of the different policies.

The results we have obtained show that it is not possible to carry out a full search of the branch and bound trees in reasonable time. However, the improvement of the sucessive solutions, seems to indicate that a solution within 2% of the optimum can be obtained in 1 to 2 hours of CPU time on a VAX 11/785. Nevertheless, the structure of the underlying matrix suggests that the implementation of Benders' decomposition technique [Benders (1962)] and Lagrangian relaxation [Muckstadt and Koenig (1977)] should lead to substantial reduction in computation times [Habibollahzadeh and Bubenko (1986), Merlin and Sandrin (1983)] and this is currently being studied.

Table 1
Numerical Results - 2 to 7 days

N.D.	Int.	N.S.	Br.	Prim.	Dif.	CPU
2	$3.39E5$	1	137	$3.39E5$	0.0	0 : 09 : 48
3	$4.94E5$	1	205	$4.94E5$	0.0	0 : 07 : 32
4	$6.49E5$	1	224	$6.49E5$	0.0	0 : 17 : 18
5	$8.00E5$	3	728	$7.99E5$	0.1	4 : 28 : 00
6	$9.28E5$	1	726	$9.26E5$	0.3	2 : 50 : 00
7	$1.06E6$	1	783	$1.05E6$	0.8	4 : 40 : 00

N.D.-number of days; Int.-integer solution $\mathcal{L}$; N.S.-number of solutions found
Br.-number of branches; Prim.-primal solution $\mathcal{L}$; Dif.-% difference
CPU-computation time

Acknowledgements

The authors would like to acknowledge Scottish Hydro-Electric plc (in particular, Donald Isles) for supplying the data used in this report. The first two authors would like to acknowledge useful discussions with Professor Wacker and his group at the Johannes Kepler University at Linz. The first author was funded by the British Council and the University of Minho, Portugal, the assistance of which is gratefully acknowledged.

References

Bauer, W., Gfrerer, H., Lindner, E., Schwarz, A. and Wacker, Hj. (1987). Optimization of the Gosau hydro energy power plant system. *Mathematical Engineering in Industry*, Vol. 1, 3, 169-190.

Benders, J. F. (1962). Partitioning procedures for solving mixed-variables programming problems. *Numerische Mathematik*, 4, 238-252.

Cohen, A. I. and Sherkat, V. R. (1987). Optimization-based methods for operations scheduling. *IEEE Proc.*, Vol. 75, 12, 1574-1591.

Fletcher, R. (1987). Practical Methods of Optimization. 2nd. Ed. John Wiley.

Habibollahzadeh, H. and Bubenko, J. A. (1986). Application of decomposition techniques to short-term operation planning of hydrothermal power system. *IEEE Trans. Power Syst.*, Vol. PWRS-1, 1, 41-47.

Merlin, A. and Sandrin, P. (1983). A new method for unit commitment at Electricite de France. *IEEE Trans. Power App. Syst.*, Vol. PAS-102, 5, 1218-1225.

Muckstadt, J. A. and Koenig, S. A. (1977). An application of lagrangian relaxation to scheduling in power-generation systems. *Operations Research*, Vol. 25, 3, 387-403.

SCICONIC/VM User Guide (1981). Scicon Computer Services Ltd.

Tong, S. K. and Shahidehpour, S. M. (1989). An overview of power generation scheduling in the optimal operation of a large scale power system. *Proc. Power Industry Computer Applications*.

GREEDY ALGORITHM FOR MAINTENANCE SCHEDULING IN ELECTRIC POWER SYSTEM

ZS. SOÓS and L. Varga

ABSTRACT.   An effective method has been developed to solve the
maintenance scheduling problem of thermal power plants in an electric
power system. The basis of the algorithm is the observation that the
model for scheduling unit maintenances with some simplification can be
formulated as a parallel machine scheduling problem. On this basis a
modified and fitted version of the largest processing time (LPT)
algorithm has been applied to solve the general maintenance scheduling
problem. Computational experience has shown that it has significant
advantages.

1. Introduction

Scheduling of generating equipment for preventive maintenance plays an
important factor in power system reliability. The maintenance schedule
is a discrete optimization problem. The exact discrete optimization
methods need enormous computer storage and excessive computation time
so they are not suitable for systems with large number of units. We
have to be content with heuristic algorithms which are able to
generate reasonably good feasible schedules within a short running
time.
      In this paper a reserve-levelling technique is presented. We
point out that there is connection between the theory of scheduling
and the reserve levelling approach. Based on this connection an
algorithm has been developed to solve the maintenance scheduling
problem.

2.   The model of maintenance scheduling and its equivalence to a certain parallel machine problem

First of all the time horizon can be divided into equal increments or
periods (e.g. months, weeks). Let $C_1, C_2, \ldots, C_n$ denote the capacity of
the generating units. The *total installed capacity* of the power system
is $C = \sum\limits_{i=1}^{n} C_i$ assumed to be constant over all scheduling periods. Let P_j

M. Heiliö (ed.), Proceedings of the Fifth European Conference on Mathematics in Industry, 115–118.
© 1991 B. G. Teubner Stuttgart and Kluwer Academic Publishers. Printed in the Netherlands.

be the predicted peak load (demand) in the j-th time period. The gross reserve or *maintenance space* in a particular period is $M_j = C - P_j$, $j = 1, 2, \ldots, m$. Let the maintenance of n_j units be scheduled in period j, the available *reserve space* in this period is M_j reduced by the capacities of the units being removed from the system for maintenance

$$R_j = M_j - \sum_{k=1}^{n_j} c_{j(k)}$$ where j(k) is the index of the unit in the k-th

position in sequence scheduled in the j-th period.

In general the reliability indices are proportional to the system reserve, so that we have to maximize the minimum system reserve over the time periods of the scheduling horizon:

$$R_{opt} = m\ a\ x\ (m\ i\ n\ R_j)$$
$$\scriptstyle 1 \le j \le m$$

Under the following simplifying assumptions, this simplified maintenance scheduling model is equivalent to the parallel machine scheduling problem where the objective is to minimize the maximum completion time.
The assumptions are as follows:

    i.) $P_j = P$ = constant, then the maintenance space is $M = C-P$;

    ii.) the maintenance length is one period for any generating unit;

    iii.) any generating unit can be scheduled for maintenance in any time period.

Since the maintenance space M is constant, a smaller planned outage will provide a larger reserve. Therefore the task is equivalent to minimize the maximum of the planned outages over the periods, that

is to minimize $\max\limits_{1 \le j \le m} L_j$, where $L_j = \sum\limits_{k=1}^{n_j} c_{j(k)}$ that is the sum of

the unit capacities scheduled for maintenance in j-th period. Then the task is to find L_{opt} where

$$L_{opt} = m\ i\ n\ (\ m\ a\ x\ L_j\ ),$$
$$\scriptstyle 1 \le j \le m$$

that is to minimize the maximum loss of capacity over the time periods.

Let us set up the following *equivalences*:
-   *the i-th time period corresponds to the i-th machine.*
-   *the capacities of the generating units correspond to the processing times of the jobs on the machines.*

Observe that due to these equivalences, the completion time on the j-th machine is L_j, j=1,2,..,m, and the objective of the parallel machine scheduling problem is

$$m\ i\ n\ (\ m\ a\ x\ L_j\ ),$$
$$\scriptstyle 1 \le j \le m$$

that is the same as the objective for the maintenance scheduling problem. The proof of the equivalence is completed and clear.

3. The algorithm

In this section the algorithm is presented derived from a greedy type method - the so-called LPT (Largest Processing Time) algorithm - used to solve parallel machine problem.

The developed LPT strategy to solve the general maintenance scheduling problem as follows:

Step 0: Compute the values of the maintenance space for every time period, $j = 1, 2, \ldots, m$.

Step 1: Order the set of units into a sequence nonincreasing with respect to the capacity, that is $i_j > i_k$ then $C_{i_j} \leq C_{i_k}$.

Place the units with the same capacities in nonincreasing order of maintenance length.

Step 2: Take the units in the order of step 1. Suppose maintenance of the next unit requires h time periods. Consider all possible time intervals of length h. Let these begin at period k, where $k = 1, 2, \ldots, m-h+1$. Find the value of k for which

$$\min_{k \leq i < k+h} R_i$$

is a maximum. The maintenance should begin in this period.

Repeat this step until all generating units are scheduled for maintenance.

The selecting rule used in the second step allows the handling of the more than one period long maintenances. For a one period long one the rule is equivalent to the original LPT strategy. Our procedure is similar to the levelize reserve algorithm but uses special selecting rule.

4. The computer program and results

An iteractive, user-friendly program has been developed on an IBM PC for planning the preventive maintenance of the generating units.

In an electric power system there are some specific additional restrictions, namely that every generating unit generally has two separate maintenance periods. There is, moreover, a bound for the time interval between repair and maintenance. These stipulations do not modify the algorithm introduced in the previous section. They involve some restrictions in the maintenance assignment rule.

Other programs are available to solve the problem, such as the MASCO and ICARUS, but they are unable to handle multi-period maintenances and the constraints mentioned above.

The MASCHE (MAintenance SCHEduling) package can be applied for scheduling the maintenance periods of those thermal power plants where the units are separately connected (boiler-turbine). These power plants supply the main portion of total capacity in the Hungarian Electric Power System. Other power plants (so-called non- scheduled power plants) are considered with their available capacities reduced by the predicted maintenance outages. These are a small part of the external sources. The most convenient number of the yearly equidistant time intervals was 52 in the case of electric power system in Hungary.

We performed the revision of the maintenance scheduling plan for the years 1989 and 1990 by the MASCHE program and obtained a more uniform and in the critical time periods *larger* system reserve compared with traditional methods used in the past.

References

Yamayee, Z. A. (1982) 'Maintenance Scheduling: Description, Literature Survey and Interface with Overall Operations Scheduling', IEEE Transactions on Power Apparatus and Systems   Vol PAS-101, No 8. Aug. 2770-2779.

El-Sheikhi F. A. and Billinton R. (1985) 'Maintenance Considerations in Generating Capacity Reliability Assessment- Methodology and Application' Maintenance Management International 5, 63-76.

Contaxis, G. C., Kavatza, S. D. and Vournas, C. D. (1989) 'An Interactive Package for Risk Evaluation and Maintenance Scheduling'   IEEE Transactions on Power Systems, Vol.4. No.2, May 389-395.

Syslo, M., Deo, N. and Kowalik, J. S., (1983) Discrete Optimization Algorithms with Pascal Programs,   Prentice Hall, Englewood Cliffs New Jersey.

Expansion Planning for Electrical Generating System, International Atomic Energy Agency, Vienna, 1984.

Zsolt SOÓS
Computer and Automation Institute of Hungarian Academy.
H-1111 Budapest, XI. Kende u. 13-17.
Hungary

László VARGA
Hungarian Electricity Board,
H-1011 Budapest, Vám u. 5-7,
Hungary

CONTRIBUTED PAPERS

ANALYTICAL AND COMPUTER DESIGN OF SPIROID GEARS

V. ABADJIEV   D. PETROVA

ABSTRACT. Spiroid skew-axis type gears are used to connect
non-intersecting and non-parallel shafts. Their analytical
design consists of solving two basic problems: synthesis
of the primary circles and synthesis of the tooth surfaces.
The optimal parameters of the worm and gear blancs as well
as the geometric parameters of their teeth are obtained as
a result of the successful synthesis by multi-criteria
computer optimization.

1. Introduction

Spiroid skew-axis type gearing is based on established
concepts. The small gear is a conic worm with non-symmetric helical teeth and the large one is a conical face-
type gear with non-symmetric spiral teeth. Spiroid worms
usually have from one to six threads. Reduction ratios
as great as 360:1 for a single spiroid-gear mesh are
practical. The lower limit of reduction ratios is 10:1.
Normally the spiroid worm is a driver but in the lower
ratios either member can be a driver. The surfaces of the
spiroid gear teeth are generated by Olivier's second
principle. Thus, spiroid gear teeth are cut by a hob whose
form, proportion and parameters are like the ones of the
spiroid worm. In result, a conjugate linear contact
between the tooth surfaces is realized easy. The spiroid
gear-drives have larger gear-pair efficiency values (up
to 97 per sent) than the cylindric worm gear-drives and
similar ones to those of double-enveloping worm-gear drives
Spiroid gears have many more teeth in simultaneous contact
and therefore, greater torque transmission capacity [1,2,3].
The described features ensure their application in heavy
industry, hoisting and hauling devices, etc.

2. Analytical Design of Spiroid Gears

The analytical design of spiroid gears includes two basic
problems: synthesis of the primary circles and synthesis

M. Heiliö (ed.), Proceedings of the Fifth European Conference on Mathematics in Industry, 121–126.
© 1991 B. G. Teubner Stuttgart and Kluwer Academic Publishers. Printed in the Netherlands.

of the tooth surfaces. When designing spiroid gears the
question of the pitch point place choice is essential.
This point is a point both of the primary circles and of
the conjugate tooth surfaces. Therefore, the choice of its
place influences on blanks and teeth proportions and on
the gears quality.

2.1. SYNTHESIS OF THE PRIMARY CIRCLES.

The concept of primary circles and its applications to
the theory of gearing have been defined in [4]. Let two
skewed axes 1-1(spiroid worm axis) and 2-2(spiroid gear
axis) be given. Their mutual position is completely
determined by the angle δ

and the offset a_w (Fig.1).
Let a point P be given too.
The primary circle H_i^a lies
in a plane which is perpendicular to the axis i-i
and H_1^a and H_2^a are externally tangent at P. It
means that they have one
common point P and are
situated in different
half-spaces with respect
to the plane T_m which
contains the tangents to
H_1^a and H_2^a at P. Eight
parameters determine the
radii and the mutual
position of H_1^a and H_2^a :
r_i, a_i, θ_i - semi-polar
coordinates of P in the
system S_i (i=1,2) and two
angles δ_1 and δ_2 which
define the orientations of
the planes of H_1^a and H_2^a
with respect to the
straight-line m-m which is
perpendicular to T_m. Ex-
pressing the radius-vector

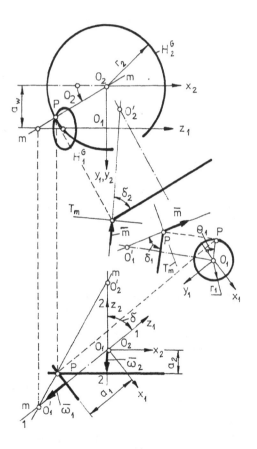

Figure 1

O_1P and the unit vector of m-m

both by r_1, a_1, θ_1, δ_1 and by r_2, a_2, θ_2, δ_2 one obtains the following set of equations

$$r_1\cos\theta_1 = a_2\sin\delta + r_2\cos\theta_2\cos\delta$$
$$r_1\sin\theta_1 = a_w - r_2\sin\theta_2$$
$$a_1 = r_2\cos\theta_2\sin\delta - a_2\cos\delta \qquad (1)$$
$$\cos\delta_1\sin\theta_1 = \cos\delta_2\sin\theta_2$$
$$\sin\delta_1 = \sin\delta_2\cos\delta + \cos\delta_2\cos\theta_2\sin\delta$$

To solve (1) one should consider 5 among 10 unknowns δ, a_w, δ_1, r_1, a_1, δ_2, r_2, a_2, θ_1, θ_2 as free parameters. If a spiroid gear-pair is considered the first 5 unknowns are chosen as free parameters. It should be mentioned that $\delta_1\in[0,\frac{\pi}{2})$, $\delta_2\in(0,\frac{\pi}{2}]$, $\theta_1\in[0,\frac{\pi}{2})$, $\theta_2\in(0,\frac{\pi}{2})$, $r_1>0$, $r_2>0$. The analytical solution of system (1) and the conditions which the free parameters should fulfill so that (1) to be solved have been discussed in detail in [4].

For example, if $\delta=\frac{\pi}{2}$, $\delta_1>0$ and $a_1>(a_w-r_1)tg\delta_1$ then the parameters of orthogonal spiroid gears can be obtained by the formulae that follow:

$$cotg\theta_2 = (r_1tg\delta_1 + a_1)/a_w,$$

$$\sin\theta_1 = tg\delta_1 tg\theta_2, \quad a_2 = r_1\cos\theta_1,$$

$$\cos\delta_2 = \cos\delta_1\sin\theta_1/\sin\theta_2,$$

$$r_2 = (a_w - r_1\sin\theta_1)/\sin\theta_2.$$

When we design spiroid gears primary circle H_i^σ is treated as a circle of a pitch cone with taper angle δ_i (i=1,2). If we know the speed ratio $i_{12}=\omega_1/\omega_2$ we can define the position of the pitch helix passing through the pitch point P (ω_i-angular velocity).

2.2 SYNTHESIS OF THE TOOTH SURFACES

Since the technology of the spiroid gears manufacture is based on Olivier's second principle we are interested exclusively in the synthesis of the thread surfaces Σ_1^j (j=1,2). Σ_1^j is a linear conic helicoid with a constant axial parameter. The generation of Σ_1^1 used as low-side thread surface of a spiroid worm with right-hand threads is shown in Fig. 2. Σ_1^j is a convolute helicoid in the most common case and is generated by the straight-line L^j which performs a composite right-hand helical motion consisting of [5]:

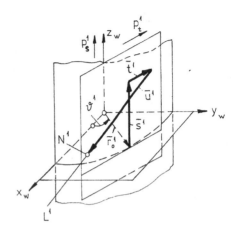

a) helical motion along the worm axis $O_w z_w$ with a parameter p_s^j and b) helical motion with a parameter p_t^j in a plane which tangent to the cylinder having a radius r_o^j and this motion is performed perpendicularly to the axis $O_w z_w$. The obtuse angle between L^j and $O_w z_w$

Figure 2

is ξ^j. The equation of Σ_1^j and Σ_2^j in the coordinate system firmly connected with the worm appear to be

$$x_w = r_o^j \cos\vartheta^j + (u^j \sin\xi^j - p_t^j \vartheta^j)\sin\vartheta^j$$
$$y_w = r_o^j \sin\vartheta^j - (u^j \sin\xi^j - p_t^j \vartheta^j)\cos\vartheta^j \qquad (2)$$
$$z_w = p_s^j \vartheta^j + u^j \cos\xi^j$$

where u^j and ϑ^j are parameters of Σ_1^j.
j=1 and the upper signs refer to low-side thread surfaces while j=2 and lower signs apply for high-side ones. A linear helical surface is developable (involute) if the parameter of distribution $h^j = p_s^j + (r_o^j + p_t^j)\cot g\xi^j = 0$. One obtains Archimedean helicoids if $r_o^j = 0$. Synthesis of Helicon skew-axis type gears is accomplished upon assuming $\delta = \pi/2$, $\delta_1 = 0$ and $p_t^j = 0$.

The right-hand helical surfaces $\Sigma_1^j (j=1,2)$ correspond to the angular velocity vectors directions shown in Fig. 1.

3. Computer Design

The computer program aims at calculating: a)the optimal geometric and constructive parameters of the spiroid worm and gear; b)the constructive parameters of the hob; c)the gears quality parameters [6].
Let us describe some estimation criteria.

3.1. STANDARD AXIAL MODULE

The decrease in the cutting tools nomenclature involves necessity of designing a hob with a standard axial module (since the hob parameters are functions of this module). The criterion reads that the calculated module should coincide with given exactness with one of the standard modules which form a set in the program system [7].

3.2 ELIMINATION OF THE ORDINARY NODES

When the lubricant oil film is broken a semi-dry and dry friction between the tooth surfaces is present. As result the loading capacity, the efficiency and the durability of the gears decrease. One of the causes is the existence of ordinary nodes. They are the common points of infinitesimal-ly close contact lines. Such a point is characterized by an anular relative velocity vector when the normal vector to the tooth surface in this point exists. Analytical relations guaranteeing non-existence of ordinary nodes in the region of mesh are obtained.

3.3. ANGLE OF NORMAL FORCE TRANSMISSION α_{12}

This angle depends on the gear helix angle and on the low side tooth normal pressure angle. Smaller values of α_{12} involve better transmission of the forces in the region of mesh and greater efficiency values.

3.4. ELIMINATION OF THE POINTS OF INTERFERENCE

The point of interference is characterized by an anular relative velocity vector when the normal vector to the tooth surface in this point does not exist. These points cause an undercutting of the gear surface and the interfe-rence of the both tooth surfaces in working meshing. The gear tooth surface dimensions must be restricted so that the points of interference to be eliminated in the region of mesh.

Requirements treating the sliding velocity value, the
efficiency etc., are taken into account also.
The input data of the program system are initial and final
values and steps of variation of the free parameters, set
of standard modules, constructive and kinematic constants,
key parameters corresponding to each of quality criteria
with values of 1 or 0 depending on whether a given criterion will be taken in consideration or not.
The output information is written in a table containing
geometric, constructive and technological parameters that
are necessary to make the design drawings for producing the
hob and gears.

REFERENCES

1. Saari, O. (1956) 'The Mathematical Background of Spiroid
   Gears', Industrial Mathematics, No 7.
2. Nelson, W. (1961) 'Spiroid Gearing. Part 1-Basic Design
   Practices', Machine Design, Vol. 33, No 4
3. Dudley, D. (1962) Gear Handbook. The Design, Manufacture
   and Application of Gears, Mc Craw-Hill Book Company,
   INC, New York-Toronto-London.
4. Abadjiev, V. and Petrova, D. (1989) 'Synthesis of
   Geometric Primary Circles of Externally Meshed Skew-
   Axes Gears', Proc. of Automation Design'89 Conference
   of ASME, Vol. H0509C.
5. Abadjiev, V. (1984) 'On the Synthesis and Analysis of
   Spiroid Gears', Ph.D. Thesis, Sofia, Bulgaria(in bulg.)
6. Abadjiev, V. and Petrova, D. (1985) 'Program System for
   Designing Spiroid Gears', Proc. of Aktuelni Problemi
   masinskih elemenata i konstrukcija, Vol. 1, Ohrid,
   Yugoslavia, pp. 65-76.
7. Abadjiev, V. and Petrova, D. (1985) 'Automated Synthesis
   of Standardized Spiroid Gears', Proc. of 5th National
   Congress on Theoretical & Applied Mechanics, Vol. 2,
   Sofia, Bulgaria, pp. 111-116.

Institute of Mechanics and Biomechanics
Bulgarian Academy of Sciences
Acad. G. Bonchev str., bl. 4
1113 Sofia
Bulgaria

TESTING THE EASE OF LEARNING OF INDUSTRIAL SYSTEMS

B SERPIL ACAR
Engineering Design Institute
Loughborough University of Technology
Loughborough, Leics.
LE11 3TU
U.K.

Introduction

Increasing use of Computer Aided Engineering systems in manufacturing industry is a sign of appreciation of their importance in terms of quality, accuracy, and cost of design and manufacture. Choosing an efficient system from the overcrowded market, however, is a very difficult task. Once the short list of the systems with facilities to serve the purposes is produced, companies usually reconsider the details of the systems and then decide which one is 'best' for them. Some of these facts are quantitative, measurable, such as purchase, maintenance and running costs. On the other hand, some of them are qualitative such as: systems being easy to learn, easy to use, and the customer has no measures to compare the systems. Consequently, most of these systems are declared to be 'easy to learn' by vendors but in reality little attempt has ever been made to quantify such claims.

In this paper an account of a new mathematical technique is given which makes use of learning theory in order to objectively compare human-computer interface of any industrial system. As an example, the technique is applied to compare the ease of learning of user interfaces of two Computer Aided Design systems.

Calculation of Learning Times

Learning is defined as change of behaviour as a result of practice. One approach to be able to measure learning is to find a relationship which enables performance to be defined as a function of practice. This approach leads to the fitting of learning curves, related to the likelihood of a correct response and performance observed during the experiments.

Replacement learning is one of the well-known learning models used to generate learning curves. In this model, some erroneous habits are replaced by correct responses, while the total number of entries remains constant. In this model, the learning process is defined as a transfer of elements from the 'unknown' state to the 'learned' state

The replacement model learning curve, the relationship between performance y and practice (trials) x is given in [1] as

M. Heiliö (ed.), Proceedings of the Fifth European Conference on Mathematics in Industry, 127–130.
© 1991 B. G. Teubner Stuttgart and Kluwer Academic Publishers. Printed in the Netherlands.

$$y = a - (a - b)(1 - \theta)^{x-1} \tag{1}$$

where a denotes the asymptotic level of performance

b denotes the initial level of performance

θ denotes the learning parameter

In order to calculate the time spent to reach a certain level of knowledge and to make comparisons of reaching the same level of different systems; let us normalise the data set by dividing each data point by the expert's performance. Let us define unit time as "time spent by an expert to complete a pre-defined unit work". Performance values, number of unit work completed per unit time should now be between 0 and 1 for a learner.

Let us also specify a level of performance α as a proportion of the target, say 60%. From equation (1) we can find the trial value j for any desired level α. i.e. the corresponding trial that the subject has reached the level α is

$$j = \frac{\log \dfrac{\alpha - a}{b - a}}{\log (1-\theta)} + 1 \tag{2}$$

The trial number j can also be used to find the cumulative unit time spent to reach level x. In this case, we are interested in the number of unit time spent for one complete unit of work, i.e. 1/y. Therefore, the sum of number of unit time spent between the first and j^{th} trials, integral of 1/y between 1 and j, gives the cumulative time spent during the trials after the first unit trial to reach level α.

The time spent between first and j^{th} trial is

$$I_{(1,j)} = \int_{1}^{j} \frac{1}{a - (a-b)(1-\theta)^{x-1}} \, dx$$

$$= \frac{x-1}{a} - \frac{1}{a \log(1-\theta)} \log (a - (a-b)(1-\theta)^{x-1}) \Big|_{1}^{j}$$

$$= \frac{1}{a} \left\{ (j-1) - \frac{1}{\log(1-\theta)} \log \left(\frac{a - (a-b)(1-\theta)^{j-1}}{b} \right) \right\} \tag{3}$$

The value of equation (3) is meaningful (positive) if the value of b, the initial value for proportion of target/unit time, is less than α (j >1). Otherwise, the number of trials to reach the level α would be taken as 1, i.e. j = 1, which makes the value of (3) zero. In each case the total time spent is found by adding the time spent in the first trial to the equation (3). That is:

$$I = I_1 + I_{(1,j)}$$

(4)

$$\text{where } I_1 = \frac{1}{b}$$

and substituting (2) in (3) gives:

$$I_{(1,j)} = \begin{cases} 0 & ; b \geqslant \alpha \\[2em] \dfrac{1}{a \log(1-\theta)} \log \dfrac{b(\alpha-a)}{\alpha(b-a)} & ; b < \alpha \end{cases}$$

Example Application

The technique developed in Section 2 is applied to compare learning times of a prototype front end to a commercial Constructive Solid Geometry modeller [2] with the commercial modeller itself. The prototype system provides a familiar interface to the design/manufacturing engineer so that his/her ideas may be expressed in terms of sufficiently high level design/manufacturing features. Hence, concepts of geometric primitives and set theoretic operations are replaced by the new and very important tool in the product modelling: 'features' [3].

A series of experiments details of which can be found in [5] have been conducted with a group of subjects inexperienced in CAD but with some experience in conventional engineering drawings, the results are processed, learning curves are produced. Figures 1 and 2 show learning curves of two features, step and countersink, for both systems.

In Figure 1, the chosen level of knowledge, $\alpha = 0.60$ is reached on trial $j = 1.6$ and $j_2 = 10.7$ when the prototype and the commercial systems are used respectively. Shaded areas, learning times in terms of unit time is calculated as 2.94 and 25.59 respectively.

Countersink example in Figure 2 is a typical example of reaching well above the level $\alpha = 0.60$ during the first trial. In this case time spent during the first trial, I, of equation (4) is taken as I, since $1_{(1,j)} = 0$. The shaded areas are calculated as 1.42 and 15.47 for feature-based prototype and commercial systems respectively. When the statistical analysis is completed [4], it was clear that the feature-based interface has been far superior to the commercial CAD systems interface as reflected in the examples of this section.

Conclusion

The mathematical technique to measure and compare learning times is presented. The technique fills a gap by making use of systematic measuring, fitting learning curves and calculating learning times, to quantify verbal claims such as "easy to use" which are usually based on *ad hoc* techniques. The use of expert times provides a very useful tool in the objective comparison of different systems. Result of an application of this technique to the data from extensive learning experiments is given as an example to show the superiority of a feature-based CAD system. The technique can be applied to compare the ease of learning of the user interface of any industrial system as long as the tasks are fully defined.

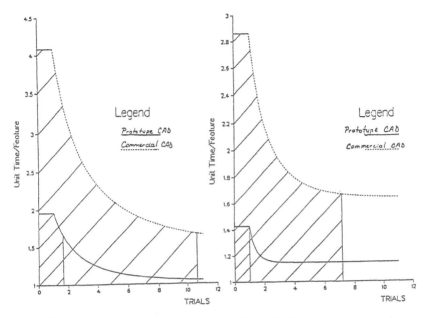

Figure 1 Step   Figure 2 Countersink

References

[1]   Restle, F., and Greeno, J.G., (1970), "Introduction to Mathematical
        Psychology", Addison-Weslwy, USA.
[2]   Acar, B.S.,Case,K.,Bennaton J., and Hart,N.,(1986),"Evaluation of a
        New Approach to Design and Process Planning". Proceedings of the 4th
        Conference on UK Research in Automated Manufacture, I.Mech.E., London,
        U.K.
[3]   Stiny, G., (ed) (1988), Report on the Workshop on Features in Design and
        Manufacturing, University of California, Los Angeles, CA.
[4]   Case, K., and Acar, B.S., (1988), "Computer Aided Production Engineering
        Linking CAD, Process Planning and Tool Design Experimentation",
        Final Report to SERC, Loughborough University of Technology.
[5]   Case K. and Acar B.S.,(1989), "Learning Studies in the Use of Computer
        Aided Design System for Discrete Part Manufacture",Behaviour and Information
        Technology, V8, pp53,368.

Address for Correspondence: **Dr B Serpil Acar** , Engineering Design Institute,
Loughborough University of Technology, Loughborough, Leics. LE11 3TU, England.

MATHEMATICAL MODELLING OF TWO-PHASE FLOW IN BOILERS

C.J. ALDRIDGE

ABSTRACT: A simple mathematical model for heated steam-water flow in a boiler is reported, and some preliminary results are presented.

1. Introduction

The flow of a heated steam-water mixture through a boiler commonly occurs in industrial situations, yet the formulation of a mathematical model for such a process is beset with difficulties. A complete description is not possible for a number of reasons: the tube contains two distinct phases, which may be arranged within the channel in various patterns (called flow régimes), and typically the flow is turbulent. Hence models based on *averaged* equations have been proposed, for example by Drew (1983). However, such models require a large number of constitutive relations to be prescribed; these may be difficult to determine, and generally lead to complicated equations. On the other hand, simpler modelling approaches can result in an ill-posed equation set, as noted by Seward (1988, p44).

In order to obtain a fast but efficient solution procedure, we show how a complicated model may be reduced, using asymptotic methods, to a much simpler system. A brief description is given here; for further details, see Aldridge and Fowler (1991).

2. System Model

Two-phase heated flows evolve through different régimes, which are commonly distinguished as bubbly, slug, churn and annular flow (see Hewitt, 1982). Figure 1 illustrates these régimes. In idealising the flow, we distinguish two 'boundaries', at $z = r$ and $z = s$. Subcooled water flows from $0 < z < r$, and at $z = r$ nucleate boiling begins. For $r < z < s$ the flow is two-phase, passing through the various régimes. 'Dryout' occurs at $z = s$, and superheated steam flows for $z > s$.

Within the two-phase region, our model is based on the one-dimensional averaged equations derived by Drew and Wood (1985). We choose a *two-fluid* model, consisting of seperate balance laws for the gas and liquid phases, plus jump conditions at the interface. Previous studies by Davies and Potter (1967) and Zuber and Findlay (1965) have employed the simpler homogeneous and drift-flux models, both of which are likely to give an inferior description than a two-fluid model, particularly in annular flow. Conserving mass, momentum and enthalpy for each phase, we have six hyperbolic equations (if we include both sound speeds); depending on the application, the time scales may be very different,

M. Heiliö (ed.), Proceedings of the Fifth European Conference on Mathematics in Industry, 131–134.
© 1991 B. G. Teubner Stuttgart and Kluwer Academic Publishers. Printed in the Netherlands.

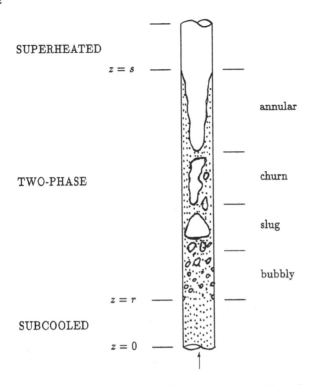

Figure 1: Flow régimes in a boiler tube

resulting in a stiff system, and the prescibed boundary data itself may be ill-posed.

Within the single-phase regions, the flow is described by standard one-dimensional (averaged) conservation laws.

3. Simplification

The model then obtained is rather complicated, and we seek a simpler description, which represents the fundamental features of the flow, but is easier to analyse. The model must therefore be non-dimensionalised, using data from typical operating conditions, and the sizes of terms examined. By neglecting those terms adjudged to be small, we obtain a simplified system. This procedure has, for example, been carried out for a model of a boiler tube, and from this simplified model we *deduce*:

(i) the phasic enthalpies are very close to equilibrium values,

(ii) the region of bubbly/slug/churn flow is short, and

(iii) within the annular region, we effectively have a drift-flux type model for conservation of momentum for the liquid.

4. Summary

The simplified model is relatively easy to solve numerically, and a FORTRAN code is cur-

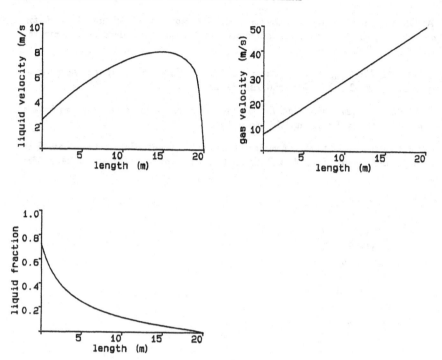

Figure 2: Steady state solutions in the annular régime

rently being completed to do this. Here, we just report some preliminary results: Figure 2 shows steady state solutions in the annular region, assuming constant fluid properties and friction factors.

In conclusion, we have produced a model describing the flow in three different regions of a boiler, and, using asymptotic techniques, simplified the system into a form which may readily be solved. The ideas outlined above are likely to be very fruitful in providing efficient methods of solving industrial problems.

References

[1] C.J. ALDRIDGE AND A.C. FOWLER. A two-fluid, transient model for two-phase flow in a boiling tube (in preparation). 1991.

[2] A.L. DAVIES AND R. POTTER. Hydraulic stability: An analysis of the causes of unstable flow in parallel channels. In *Proc. Symp. Two-Phase Flow Dynamics, Eindhoven, EUR 4288e*, pages 1225–1266, 1967.

[3] D.A. DREW. Mathematical modelling of two-phase flow. *Annual Review of Fluid Mechanics*, 15:261–291, 1983.

[4] D.A. DREW AND R.T. WOOD. *Overview and Taxonomy of Models and Methods for Workshop on Two-Phase Flow Fundamentals*. Nat. Bureau of Standards, Gaithersburg, Maryland, Sept. 22–27, 1985.

[5] G.F HEWITT. Flow regimes. In Hetsroni, G., editor, *Handbook of multiphase systems*, chapter 2.1, pages 2.3–2.43. McGraw-Hill, 1982.

[6] P.E. SEWARD. *A two-fluid model for the analysis of gross flow instabilities in boiling systems*. PhD thesis, University of Oxford, 1988.

[7] N. ZUBER AND J. FINDLAY. Average volumetric concentration in two-phase flow systems. *Transactions of the ASME, Series C, Journal of Heat Transfer*, **87**:453–468, 1965.

C.J. ALDRIDGE
University of Oxford
Mathematical Institute
24-29 St. Giles'
Oxford OX1 3LB
United Kingdom

STATIONARY PIPE FLOW OF POWER-LAW FLUIDS

Gunnar Aronsson

1. Introduction

Fluids exhibiting non-newtonian behaviour occur frequently in food industry and in many chemical process industries, like for example paint production, pharmaceutical industry, etc. An important subclass of all non-newtonian fluids consists of the so-called power-law fluids. These are fluids such that their constitutive behaviour is described with sufficient accuracy by a certain power "law" involving the velocity gradient and stress tensors.

Now the flow of a power-law fluid is already known explicitly for a number of situations with simple geometry. In other cases, approximations must be used. Here, the steady laminar flow in a pipe with arbitrary cross-section is considered. A mathematical model for the flow problem will be given, and a finite element method for numerical solution will be suggested.

2. Remarks on power-law fluids

Concerning power-law fluids and the validity of their constitutive law, we refer to:

I) [B-A-H], pp. 172–175. (Observe in particular the discussion on p. 175)

II) [SCH], pp. 138–139. (Discussion on p. 139)

III) [T-G], pp. 167–169

IV) [BÖH], p. 59

V) [SKE], pp. 7–12

VI) [A-M], p. 55

The power law is also known as the *law of Ostwald and de Waele*. Clearly, it is not a law at all, but an attempt at empirical curve fitting with great simplicity, which turned out to be quite successful.

It is understood that the fluid considered here is incompressible. Further, thermal effects are neglected. It is convenient to start from the power law as stated in [A-M], p. 55:

$$\mathbf{\tau}' = 2KS^{\frac{n-1}{2}}\mathbf{D}.$$

Here, $\mathbf{\tau}'$ = deviatoric (=extra) stress tensor, $K > 0$ and $n > 0$ are material constants, $\mathbf{D}$ = rate of strain tensor (also called rate of deformation tensor), $S = 2tr(\mathbf{D}^2) = 2\,\mathbf{D} : \mathbf{D}$. Note that $\mathbf{D} = \frac{1}{2}(\nabla\mathbf{V} + (\nabla\mathbf{V})^{\top})$, where $\nabla\mathbf{V}$ = velocity gradient tensor.

M. Heiliö (ed.), *Proceedings of the Fifth European Conference on Mathematics in Industry*, 135–138.

3. Analysis of the flow problem

We will consider a steady laminar flow through a straight pipe with an arbitrary cross-section Ω. The flow is driven by a pressure difference and end effects are neglected, i.e. the pipe is "long". It is understood that the flow is everywhere in the direction of the pipe. Such a flow problem is considered in some detail in [BÖH], pp. 156–159. An important difference is that we consider here only power-law fluids, and another difference is that gravity will not be neglected here. It is seen from below that (in the notation of [BÖH]) $\nu_2(\dot{\gamma}) \equiv 0$ in the present problem, and the condition (4.36) in [BÖH], p. 158, is clearly satisfied. Therefore, the considered flow is possible. Concerning stability of such a flow, we refer to [B-A-H], p. 177.

For the present problem, introduce a cartesian coordinate system $x_1 x_2 x_3$ having the x_1-axis parallel to the pipe. Then Ω is a bounded region in the $x_2 x_3$-plane. The velocity field is of the form $\mathbf{V} = u(x_2, x_3)e_1$, and the velocity gradient is given by

$$\nabla\mathbf{V} = \begin{pmatrix} 0 & \frac{\partial u}{\partial x_2} & \frac{\partial u}{\partial x_3} \\ 0 & 0 & 0 \\ 0 & 0 & 0 \end{pmatrix}$$

The rate of strain tensor is

$$\mathbf{D} = \frac{1}{2}(\nabla\mathbf{V} + (\nabla\mathbf{V})^\top) = \frac{1}{2}\begin{pmatrix} 0 & \frac{\partial u}{\partial x_2} & \frac{\partial u}{\partial x_3} \\ \frac{\partial u}{\partial x_2} & 0 & 0 \\ \frac{\partial u}{\partial x_3} & 0 & 0 \end{pmatrix}$$

and one trivially finds that $\operatorname{tr}\mathbf{D}^2 = \mathbf{D}:\mathbf{D} = \frac{1}{2}|\nabla u|^2$. The constitutive law of Ostwald-de-Waele now gives

$$\mathbf{r}' = 2K|\nabla u|^{n-1}\mathbf{D}.$$

Consider then the equation of motion

$$\rho\frac{D\mathbf{V}}{Dt} = -\nabla P + \operatorname{div}\mathbf{r}' + \rho\mathbf{g},$$

where ρ = density, P = pressure, $\mathbf{g}$ = gravity. See [A-M], p. 34, or [BÖH], p. 37. In the present problem $\frac{D\mathbf{V}}{Dt} = 0$, and so

$$\operatorname{div}\mathbf{r}' = \nabla P - \rho\mathbf{g}.$$

Choose scalars α, β, γ so that $\rho\mathbf{g} = \nabla\varphi$, where $\varphi = \alpha x_1 + \beta x_2 + \gamma x_3$. Thus,

$$2K\operatorname{div}(|\nabla u|^{n-1}\mathbf{D}) = \nabla(P - \varphi),$$

and $\mathbf{D}$ is given above. In component form, this is (since $u = u(x_2, x_3)$):

$$
\begin{cases}
K \sum_{i=2}^{3} \dfrac{\partial}{\partial x_i}\left(|\nabla u|^{n-1}\dfrac{\partial u}{\partial x_i}\right) = \dfrac{\partial}{\partial x_1}(P - \varphi) & (1) \\[2ex]
0 = \dfrac{\partial}{\partial x_2}(P - \varphi) = \dfrac{\partial}{\partial x_3}(P - \varphi). & (2,3)
\end{cases}
$$

Eq. (2,3) imply that $(P - \varphi)$ and the RHS of (1) only depend on x_1. But the LHS of (1) is independent of x_1. Thus, $\frac{\partial}{\partial x_1}(P - \varphi) = M = $ constant.

Eq. (1) takes the form $K \operatorname{div}(|\nabla u|^{n-1}\nabla u) = M$ and we impose the "no slip" condition $u = 0$ at the boundary $\partial\Omega$. (Concerning this condition, see [B-A-H], p. 247.) The quantity $P - \varphi$ is called *effective pressure*, cf. [SCH], p. 74. Assume that $M < 0$, so that $u > 0$ in Ω.

Put $p = n + 1$, giving $p > 1$. A trivial rescaling leads to the boundary-value problem

$$
\begin{cases}
\operatorname{div}(|\nabla u|^{p-2}\nabla u) = -1 & \text{in } \ \Omega \\
u = 0 & \text{on } \ \partial\Omega.
\end{cases}
$$

This differential equation is sometimes called the *p-Poisson equation*.

4. On the numerical solution. Experiences. Conclusions.

For engineering purposes it is often required to know the total flow of liquid through the pipe as a function of the pressure drop. High accuracy is usually not needed. For simple geometries a satisfactory answer is obtained from "standard" formulas, more or less accurate, as for instance in [SKE], chapter 4. We also refer to [SCHE].

It is mathematically more satisfactory to have a method for numerical solution of the above boundary-value problem on an arbitrary bounded domain. A new finite element method for this and some related problems was suggested 1974–75 by R. Glowinski and A. Marrocco; see [G-M1] and [G-M2]. More material on the algorithm is found in [F-G] and [GLOW].

The method uses an augmented Lagrange function, obtained by *simultaneous* penalisation and dualisation for a function to be minimised. We refer to [G-M2] for a full description of the method.

A mathematical and numerical study of the motion of a glacier was given by M-C. Pelissier [PEL]. The ice was assumed to obey a constitutive law of power-law type. Some simplifying assumptions lead to a problem, similar to the present, for the p-Poisson equation, but with different boundary conditions. Pelissier used the numerical method of Glowinski and Marrocco, with satisfactory results. Experiences concerning the choice of numerical parameters were reported.

Our experience in Linköping is that the numerical method has good convergence properties and is also otherwise satisfactory. In most cases only a few (3-5) iterations were needed in the outer loop. It must be mentioned, however, that extra care is needed for p close to 1 (say $1 < p < 1,1$) and for big p (say $p > 7$). In these cases, convergence is not so easily achieved, and [G-M1] recommend calculation in double precision.

The programming and computer work was cleverly done by the Linköping students Lennart Pettersson and Torbjörn Andersson. Thanks to them!

[A-M] G.Astarita, G.Marrucci, Principles of Non-Newtonian Fluid Mechanics, Mc Graw-Hill, 1974.

[B-A-H] R.B.Bird, R.C.Armstrong, O.Hassager, Dynamics of Polymeric Liquids, Vol 1. Wiley, New York 1977.

[BÖH] G.Böhme, Non-newtonian fluid mechanics, North-Holland, 1987.

[F-G] M.Fortin, R.Glowinski, Augmented Lagrangian Methods. Applications to the Numerical Solution of Boundary-Value Problems, North-Holland, 1983.

[GLOW] R.Glowinski, Numerical Methods for Non-linear Variational Problems, Springer 1984.

[G-M1] R.Glowinski, A.Marrocco, Sur l'approximation par elements finis..., IRIA Rapport no.115, Avril 1975.

[G-M2] Authors and title as the preceding ref., R.A.I.R.O R2, Aout 1975, 41-76.

[PEL] M-C.Pelissier, Sur quelques problèmes non linéares en glaciologie, Thèse, Orsay 1975.

[SCHE] R.S.Schechter, On the Steady Flow of a Non-Newtonian Fluid..., A.I.Ch.E. Journal, Vol 7:3, 1961, 445-448.

[SCH] W.R.Schowalter, Mechanics of non-newtonian fluids, Pergamon Press, Oxford 1978.

[SKE] A.H.P.Skelland, Non-Newtonian Flow and Heat Transfer, Wiley 1967.

[T-G] Z.Tadmor, C.G.Gogos, Principles of Polymer Processing, Wiley 1979.

Gunnar Aronsson
Department of Mathematics
Linköping University
S-581 83 Linköping
SWEDEN

DYNAMIC SIMULATION OF PLATE DISTILLATION COLUMNS

Hj. Bart, T. Kronberger, L. Peer, Hj. Wacker

Abstract - The mathematical model for dynamic plate distillation columns includes two kinds of equations. The principle of mass and energy balance leads to differential equations with respect to time. On each plate we have balance volumes (holdup volumes of liquid and vapor phase). The mass transfer between the two phases is described by means of Withman's Two Film Theory. The other kind of equations are algebraic ones which derive e.g. from the pressure drop relationship and the thermodynamic equilibium condition at the interface. A high dimensional semi-explicit differential algebraic system of equations has to be solved. The numerical treatment of this system is discussed and an example is presented.

1. Introduction: Distillation is a popular process to separate mixtures of liquids with significant differences in their boiling points. It is based on repeated vaporization and condensation. Therefore we have a process with a high energy demand. To optimize these operation costs we have to be able to simulate a real life distillation column.

A lot of work has been done on steady state distillation towers. Most of these models assume thermodynamic equilibrium at the stages. To apply the results to real life plants, one has just to consider the tray efficiencies. This does not hold any longer in the dynamic case because these efficiencies may change during the time period considered. Therefore we used the film theory model by Withman for the mass transfer.

2. The Mathematical Model: In Fig. 1 we give a sketch of a distillation column, consisting of a boiler, a condenser and several separation plates.

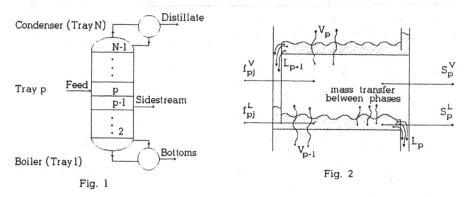

Fig. 1

Fig. 2

For ease of presentation we give the equations for the separation trays only. The equations for the boiler and the condenser are slightly modified and can be found in [3].

Balancing the input and output fluxes of a tray (see Fig. 2) we get an expression for the change of holdup with respect to time. This leads to the following

M. Heiliö (ed.), Proceedings of the Fifth European Conference on Mathematics in Industry, 139–142.

differential equations:

$$\frac{dh^L_{pj}}{dt} = f^L_{pj} - S^L_p x_{pj} + L_{p+1} x_{p+1\,j} - L_p x_{pj} + \beta^L A_p (x^x_{pj} - x_{pj})$$

$$\frac{dh^V_{pj}}{dt} = f^V_{pj} - S^V_p x_{pj} + V_{p-1} y_{p-1\,j} - V_p y_{pj} + \beta^V A_p (y^x_{pj} - y_{p-1\,j})$$

As mentioned above, the mass transfer term derives from Withman's Two Film Theory (e.g. see [4]). It is assumed that there are two turbulent bulks with laminar layers on both sides of the contact surface. In these layers we have a molecular diffusion process. The film is a third auxilary phase between the gas and liquid where thermodynamic equilibrium is assumed. As it is infinitely thin there cannot be any accumulation of mass.

Under these assumptions mass transfer is proportional to the contact surface times the gradient of concentration. Usually the contact surface is estimated by multiplying the holdup volume by the specific surface.

There is a heat stream in connection with every mass stream. We give the energy balance equations:

$$\frac{dE^L_p}{dt} = F^L_p e^{fL}_p - S^L_p e^L_p + L_{p+1} e^L_{p+1} - L_p e^L_p - \Delta E_p + Q^L_p$$

$$\frac{dE^V_p}{dt} = F^V_p e^{fV}_p - S^V_p e^V_p + V_{p-1} e^V_{p-1} - V_p e^V_p + \Delta E_p + Q^V_p$$

Here ΔE denotes the energy transfer due to the mass transfer between the liquid and the vapor phase.

To evaluate the right hand side of the differential equations we need the following set of variables on each tray: h^L, h^V, E^L, E^V, L, V, x^x, y^x, T^L, T^x, T^V, and Π. So, we have a set of $4M+8$ variables on each tray. Up to now we have derived $2M+2$ differential equations. The remaining equations are algebraic.

Given the enthalpies and the holdups, we can compute the temperatures:

$$E^L_p = H^L_p e^L_p (T^L_p, \Pi_p, x_p) \qquad\qquad E^V_p = H^V_p e^V_p (T^V_p, \Pi_p, y_p).$$

From the film theory model and the definiton of mole fractions it follows:

$$y^x_{pj} = k(y^x_p, x^x_p, T^x_p, \Pi_p) \cdot x^x_{pj} \qquad\qquad \text{(thermodynamic equilibrium)}$$

$$\beta^L A(x^x_{pj} - x_{pj}) + \beta^V A(y^x_{pj} - y_{p-1\,j}) = 0 \qquad \text{(no mass accumulation in the film)}$$

$$\sum_{i=1}^{M} y^x_{pi} = 1$$

The liquid and vapor streams depend on the geometry of the considered tray and on the other hand on the actual holdup:

$$L_p = \Phi(H^L_p, \text{plate geometry}) \qquad\qquad V_p = \Psi(H^V_p, \text{plate geomerty})$$

The last missing equation is that for the pressure. The user may define the pressure in the reboiler. For each tray we can compute the pressure drop. Therefore the pressure on each tray is defined. We have

$$\Pi_p = \Pi_{p-1} - \Delta\Pi_p \qquad \Pi_1 = \Pi_{user}$$

The pressure drop consists of two different parts: the dry and the wet pressure drop. The first one is due to the plate internals and the second one is the hydrostatical height of the liquid holdup.

Appropriate relations for the pressure drop and the mass transfer coefficients can be found e.g. for bubble cap trays in [3]. The special form of the equations for the specific enthalpy and the equilibrium constants depends upon the thermodynamical model being used e.g. UNIFAC [2].

3. Numerical Solution Technique: We have developed a mathematical model for a dynamic distillation column which consists of a $N(4M+8)$ dimensional differential algebraic equation (DAE) system. The index of our system is two. An interesting point is, that we have a special form of DAE (semi-explicit), because all the differential equations are explicit ones. We have

$$\frac{dx}{dt} = f(x,y,t) \qquad x(0) = x_O$$
$$0 = g(x,y,t) .$$

We solve this initial value problem by discretization of the differential equations (trapezoidal rule). The resulting nonlinear system at each time step is treated by Newton's method. A tray by tray arrangement of the variables (L, h^L, E^L, T^L, x^x, T^x, y^x, Π, T^V, E^V, h^V, V) and equations results in a block-tridiagonal Jacobian, where the off-diagonal blocks are less than half filled [3]. Details about DAE's can be found e.g. in [1].

4. Example: We simulate a separation column consisting of 15 trays. At the initial time we have a steady state condition. On stage 6 there is a liquid feed at 357 K and 1.5 atm. It consists of 130 mol/s methanol and 130 mol/s water. After 50 seconds a linear change in the feed composition starts. It ends after further 100 seconds at 90 mol/s methanol and 170 mol/s water. There is no further change in any input parameters afterwards.

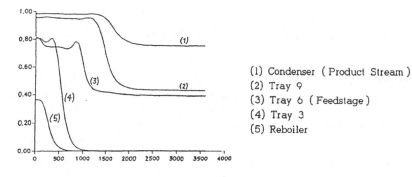

(1) Condenser (Product Stream)
(2) Tray 9
(3) Tray 6 (Feedstage)
(4) Tray 3
(5) Reboiler

Fig. 3

In Fig. 3 we plotted the methanol concentration in the gas phase of some trays. It is easy to see there, that after some 50 minutes the column attains a new steady state condition. Further graphics (liquid concentration and temperatures) and more details about the plate geometry can be found in [3].

Nomenclature

A	Contact surface between phases, m^2	x	Mole fraction (liquid)
E	Enthalpy, kJ	y	Mole fraction (vapor)
e	Specific enthalpy, kJ/mol	β	Mass transfer coefficient, $mol/(m^2 s)$
F	Total(××) feedstream, mol/s	Π	Pressure, atm
f	Feedstream, mol/s		
H	Total(××) holdup, mol	Subscripts	
h	Holdup, mol	j	Component number
k	Equilibrium constant	p	Tray number
L	Liquid stream, mol/s		
M	Number of components of the mixture	Superscripts	
N	Number of trays	f	Feed
Q	Heat duty, kJ/s	L	Liquid
S	Sidestream, mol/s	V	Vapor
T	Temperature, K	×	Interface (Film)
V	Vapor stream, mol/s		

(××) total means: sum over all components

References

[1] K.E. Brenan, S.L. Campbell, L.R. Petzold, *Numerical Solution of Initial-Value Problems in Differential-Algebraic Equations*, Elsevier Science Publishing Co., Inc. (1989)

[2] A. Fredenslund, J. Gmehling, P. Rasmussen, *Vapour-liquid equilibrium using UNIFAC*, Elsevier Scientific Publishing Co., Amsterdam (1977)

[3] T. Kronberger, *Numerical Simulation of Steady State and Nonsteady State Distillation Columns with Nontheoretical Trays*, Diploma Thesis, University Linz, (1990)

[4] K. Sattler, *Thermische Trennverfahren*, Springer Verlag, (in German), (1977)

Acknowledgement

This work has been supported by the 'Bundesministerium für Wissenschaft und Forschung' (BMfWF) and the 'Fonds zur Förderung der wissenschaftlichen Forschung' (FWF) P 6674.

Address

Department of Mathematics, University Linz
A-4040 Linz, Altenbergerstr. 69, Austria

OUTGASSING AND MASS EXCHANGE BETWEEN TWO SURFACES

A. BELLENI-MORANTE, A. FASANO, M. PRIMICERIO
Università di Firenze
Viale Morgagni 67/A
50134 Firenze
Italia

ABSTRACT. We present a mathematical model describing a process in which particles are produced on a surface and then emitted (outgassing) and collected by another surface (contamination). All particles absorbed on either surface can be re-emitted, travelling back and forth between the two surfaces. The model uses the collisionless Boltzmann equation and introduces a series of simplifications, retaining the main physical and mathematical features. The asymptotic analysis for $t \to \infty$ is performed.

1. THE MATHEMATICAL SCHEME

In some recent papers [1], [2] a mathematical model has been studied for a phenomenon (outgassing) which occurs when volatile particles are created (e. g. under the action of UV rays) in a thin layer (typically a paint coating), then reach the outer surface and are possibly emitted. The emitted particles can be collected by another surface, producing contamination. The process could be crucial e. g. for the lifetime of optical devices on satellites, and is therefore important for the aerospace industry.

In [1], [2] diffusion was taken as the dominant mass transfer mechanism, coupled with diffusion and particle production in the emitting layer in a one-dimensional geometry.

Our scheme differs mainly in the fact that collisionless Boltzmann equation is used for the particle motion.

Emission and absorption are allowed both on the contaminating and on the contaminated surface. The principal scope of our analysis is to determine the asymptotic behaviour of the particle densities on the two surfaces.

The following simplifying assumptions will be introduced.
(a) All the particles are emitted with the same velocity v, irrespective of their direction; (b) no reflection occurs.

For simplicity we suppose that the emission velocity is the same for the two surfaces, although this is not crucial. Removing assumption (b) causes a considerable complication. The problem with reflections will be treated in a forthcoming paper. The theory presented here will also be extended to include a more complicated geometry and a general spectrum of emission velocity.

M. Heiliö (ed.), Proceedings of the Fifth European Conference on Mathematics in Industry, 143–146.

For simplicity here we also assume that the thickness of the diffusing layer is negligible. As a consequence we suppose that the particle production takes place on the emitting surface itself, although the model could be modified in order to describe more general situations. The advantage of the presentation adopted here is to outline the relevant features of the phenomenon, simplifying the mathematics without destroying the physical significance.

Denoting by x the space coordinate, the "active" surface is the plane $x = 0$ and the contaminated surface is the plane $x = b > 0$. By $\theta(t)$ and $\phi(t)$ we denote the particle concentrations on the two respective planes $x = 0$, $x = b$, and by $F(x, \mu, t)$ we mean the particle distribution function for $0 \leq x \leq b$, $-1 \leq \mu \leq 1$ (μ specifies the velocity direction).

If $h(t)$ is the production rate at $x = 0$, we can write the following mass balance equations:
at $x = 0$

$$\dot{\theta}(t) = h(t) - v \int_{-1}^{1} \mu \, F(0,\mu,t) \, d\mu, \tag{1}$$

at $x = b$

$$\dot{\phi}(t) = v \int_{-1}^{1} \mu \, F(b,\mu,t) \, d\mu, \tag{2}$$

and the collisionless transport equation

$$\partial F/\partial t + \mu \, v \, \partial F/\partial x = 0, \qquad 0 < x < b, \; -1 \leq \mu \leq 1, \; t > 0, \tag{3}$$

e. g. with the initial condition

$$F(x,\mu,0) = 0. \tag{4}$$

Since we are assuming that all the incoming particles are absorbed, we only have to describe the emission dynamics:

$$v \, \mu \, F(0,\mu,t) \, = p_{+}(t,\mu) \, \theta(t), \qquad 0 \leq \mu < 1, \tag{5}$$

$$-v \, \mu \, F(b,\mu,t) = p_{-}(t,\mu) \, \phi(t), \qquad -1 < \mu \leq 0, \tag{6}$$

where p_{+}, p_{-} are given non-negative functions (extended as zero outside their natural range).

In the following we take nondimensional variables, rescaled so that b/v is the time unit and we keep the same symbols. Moreover we suppose that $p_{\pm}$ are independent of time and we define

$$\lambda_{\pm} = \int_{-1}^{1} p_{\pm}(\mu) d\mu.$$

2. ASYMPTOTIC ANALYSIS

If we suppose that emission occurs in the normal direction only, then p_{+}, p_{-} are δ-functions and (1), (2) reduce to

$$\dot{\theta}(t) = h(t) - \lambda_{+} \, \theta(t) + \lambda_{-} \, \phi(t-1), \tag{7}$$

$$\dot{\phi}(t) = - \lambda_{-} \, \phi(t) + \lambda_{+} \, \theta(t-1), \tag{8}$$

e. g. with "initial" conditions

$$\phi(t) = \theta(t) = 0, \; t < 0, \tag{9}$$

$$\phi(0) = \phi_0, \quad \theta(0) = \theta_0. \tag{10}$$

The above system is of retarded type and the theory of [3] about series representation of the solution is applicable.

In the particular case $\lambda_+ = \lambda_- = 1$, $h = 0$, setting $\sigma = \theta + \phi$, $\delta = \theta - \phi$, we obtain uncoupled equations

$$\dot{\sigma}(t) + \sigma(t) - \sigma(t-1) = 0, \tag{11}$$

$$\dot{\phi}(t) + \phi(t) + \phi(t-1) = 0, \tag{12}$$

whose eigenvalues are the roots of the respective equations

$$s + 1 - e^{-s} = 0, \tag{13}$$

$$s + 1 + e^{-s} = 0, \tag{14}$$

in the complex plane. It can be seen that all the eigenvalues are simple, that $s = 0$ solves (13) and that all other eigenvalues have negative real parts. For large n the roots of (13) are asymptotic to $- \log [2/(3+4n)\pi] \pm i \; \pi(3+4n)/2$, and the roots of (14) to $- \log [2/(1+4n)\pi] \pm i \; \pi(1+4n)/2$.

Thus it is immediately seen that σ, δ have finite limits for $t \to \infty$, leading to the conclusion that both θ and ϕ tend to $(\phi_0+\theta_0)/4$ as $t \to \infty$.

The analysis is easily extended to more general $h(t)$, linking the asymptotic behaviour of θ, ϕ to the integral of the source term $h(t)$ and to the initial data.

For the case of arbitrary λ_+, λ_- the eigenvalues are the roots of

$$s^2 + (\lambda_+ + \lambda_-) s + \lambda_+\lambda_- - \lambda_+\lambda_- e^{-2s} = 0 \tag{15}$$

and again $s = 0$ is a solution and still it can be seen that all other eigenvalues have negative real parts for any positive value of λ_+, λ_- (for a general analysis of the roots of equations of this type see [4]).

If e. g. $h(t) \geq 1$ and its growth is not exponential, for large t we have that θ/λ_- and ϕ/λ_+ both approach $c \int_0^t h(r) \, dr$, where

$$c = [\lambda_+ + \lambda_- + 2 \lambda_+\lambda_-]^{-1}.$$

Also the lower order terms can be estimated as functions of the initial data and of λ_+, λ_-.

If emission occurs in all directions equations (7), (8) are replaced by

$$\dot{\theta}(t) = h(t) - \lambda_+ \, \theta(t) + \lambda_- \int_{-1}^{0} p_-(\mu) \, \phi(t-1/|\mu|) \, d\mu, \tag{16}$$

$$\dot{\phi}(t) = - \lambda_- \, \phi(t) + \lambda_+ \int_{0}^{1} p_+(\mu) \, \theta(t-1/\mu) \, d\mu, \tag{17}$$

with the same conditions for $t \leq 0$.

For instance, taking $p_{\pm}(\mu) = \lambda_{\pm} \nu_{\pm} \mu^{-2} \exp\left[(1-1/|\mu|) \, \nu_{\pm}\right]$ and applying the Laplace trasform, we get an eigenvalue problem similar to the one described above and parallel conclusions can be drawn about the asymptotics.

Acknowledgement. The outgassing-contamination problem was brought to authors' attention by prof. Ellis Cumberbatch.

REFERENCES

[1]   S. Busenberg, W. Fang, M. Shillor. Analysis of the outgassing and contamination model. To appear on Applicable Analysis.

[2]   W. Fang et al. A mathematical model for outgassing and contamination. Preprint 1988.

[3]   R. Bellman, K.L. Cooke. Differential-Difference Equations. Academic Press, Math. Sci Engr. 6 (1963).

[4]   W. Huang. Global geometry of the stable regions for two delay differential equations, Georgia Tech. Preprint (1990).

NUMERICAL SIMULATION OF DYNAMIC SALES RESPONSE MODELS

Martina Bloß and Ronald H. W. Hoppe

Summary We consider optimal advertising in stochastic dynamic sales response models of Vidale-Wolfe type under both infinite and finite horizon utility maximization. By Bellman's principle of dynamic programming the maximum utility can be shown to satisfy a nonlinear second order partial differential equation of Hamilton-Jacobi-Bellman (HJB) type which is elliptic in the infinite horizon and parabolic in the finite horizon case. The HJB-equations are numerically solved by multi-grid techniques using finite element discretizations in the space variables with respect to a hierarchy of triangulations. In the parabolic case we compare a linearization based on the so-called Nisio formula with a more direct nonlinear approach based on implicit time discretization. The performance of the multi-grid schemes is illustrated by numerical examples.

1. The Dynamic Sales Response Models. We consider optimal advertising based on stochastic optimal control of a dynamic sales response model developed by Tapiero et al. [4] as the stochastic version of the classical deterministic sales response model due to Vidale and Wolfe [5].

In particular, we assume that a firm is selling n products P_i at prices p_i with sales $s_i(t)$ and market shares $y_i(t) = s_i(t)/M_i \in [0,1]$, $1 \leq i \leq n$, $t \in [0,T]$, where M_i denotes the market potential and $T = \infty$ (infinite horizon) or $T < \infty$ (finite horizon). We further assume that we are given m different advertising strategies $a^\mu = (a_1^\mu, ..., a_n^\mu)$, $1 \leq \mu \leq m$ with a_i^μ, $1 \leq i \leq n$, indicating the advertising costs for product P_i. The interaction of sales and advertising can be described by the market potentials M_i, the sales decay rates λ_i and the response constants r_i. Market uncertainties will be taken into account by a diffusion process involving a Wiener process $w = w(t)$ and variances $\sigma^\mu = \left(\sigma_{ij}^\mu\right)_{i,j=1}^n$, $1 \leq \mu \leq m$. Moreover, if the trajectory of the stochastic process leaves the feasible domain $\overline{\Omega} = [0,1]^n$, we decide to restore the process by reflection at the boundary Γ which can be modelled by a continuous, decreasing, adapted process $\xi = \xi(t)$. Then, starting from $y_x(0) = x$, the trajectory $y_x(t)$ of the sales process satisfies the following stochastic system:

$$(1.1) \qquad dy_x(t) = b^\mu(y_x(t)) \, dt + \sigma^\mu(y_x(t)) \, dw(t) - \chi_\Gamma(y_x(t)) \, \gamma^\mu(y_x(t)) \, d\xi(t)$$

where the drift term $b^\mu = (b_1^\mu, ..., b_n^\mu)^T$ is given by $b_i^\mu(y) = -\lambda_i \, y_i + q_i \, a_i^\mu (1 - y_i)$, $q_i = r_i/M_i$, χ_Γ is the characteristic function of Γ and γ^μ stands for a function such that $\gamma^\mu(z)$, $z \in \Gamma$, is pointing towards the exterior of $\overline{\Omega}$. The control objective is to choose the advertising strategies in such a way that the expected, discounted total profit is

M. Heiliö (ed.), Proceedings of the Fifth European Conference on Mathematics in Industry, 147–150.
© 1991 B. G. Teubner Stuttgart and Kluwer Academic Publishers. Printed in the Netherlands.

maximized

$$(1.2) \qquad u(x) = \max_{1 \leq \mu \leq m} E\left(\int_0^T \exp(-c^\mu t)\, U(\Pi^\mu(y_x(t)))\, dt\right)$$

where $U = U(\Pi^\mu)$ stands for the utility of profits, $\Pi^\mu(y) = \sum_{i=1}^n \left(p_i\, M_i\, y_i - a_i^\mu\right)$ denotes the instantaneous profit and the $c^{\mu'}$s with $c^\mu \geq c_0 > 0$, $1 \leq \mu \leq m$ are the discount factors.

By Bellman's principle of dynamic programming (cf. e.g. [1]) the optimal profit $u = u(x)$ turns out to be the solution of a strongly nonlinear p.d.e. with oblique derivative boundary conditions which is called a Hamilton-Jacobi-Bellman (HJB-) equation. In the finite horizon case $T < \infty$ we get the parabolic problem

$$(1.3) \quad u_t - \max_{1 \leq \mu \leq m}\left(A^\mu u + U(\Pi^\mu)\right) = 0 \quad \text{in } \Omega \times (0,T), \quad \max_{1 \leq \mu \leq m}\left(\gamma^\mu \cdot \nabla u\right) = 0 \text{ on } \Gamma \times (0,T)$$

with some appropriate initial condition where

$$A^\mu = \sum_{i,j=1}^n a_{ij}^\mu\, \partial x_i\, \partial x_j + \sum_{i=1}^n b_i^\mu\, \partial x_i - c^\mu$$

and the a_{ij}^μ's, $1 \leq i,j \leq n$, are the matrix elements of $a^\mu = \sigma^\mu(\sigma^\mu)^T/2$, $1 \leq \mu \leq m$.
Note that the coefficients in A^μ may explicitly depend on the time variable t.
However, in the infinite horizon case $T = \infty$ which can be viewed as the stationary state of (1.3) for $t \to \infty$, the coefficients are independent of t and we end up with a corresponding elliptic problem ($u_t = 0$ in (1.3)).

Under suitable assumptions on the data of the problem existence, uniqueness and regularity of solutions to HJB-equations of the above type have been investigated by P.L. Lions and N.S. Trudinger [3].

We further remark that (1.3) does not only provide the maximum profit but also indicates how to choose the optimal strategy. For that purpose let us denote by $\Omega^\nu(t)$, $1 \leq \nu \leq m$, the set of all states $x \in \Omega$ such that in (1.3) the maximum is attained for $\mu = \nu$. Then, if for some time interval $t_s - \varepsilon < t < t_s$, $\varepsilon > 0$ we do have $x \in \Omega^{\nu_1}(t)$ while $x \in \Omega^{\nu_2}(t)$ for $t_s < t < t_s + \varepsilon$, this means that at time instant t_s the optimal strategy requires to switch from a^{ν_1} to a^{ν_2}. In this context (1.3) may be viewed as a moving boundary problem with moving boundary $\Gamma(t) = \bigcup_{\nu \neq \mu} \Gamma^{\nu\mu}(t)$ where $\Gamma^{\nu\mu}$ represents the free boundary separating Ω^ν from Ω^μ, $1 \leq \nu \neq \mu \leq m$.

2. Multi-Grid Solution of the HJB-Equation. As far as the numerical solution of the parabolic HJB-equation is concerned, one way is to use a linearization technique based on Nisio's formula [1]. In practice, given a uniform partition of the time interval [0,T] into subintervals $[t_j, t_{j+1}]$, $t_j = j \Delta t$, $0 \leq j \leq M-1$, $\Delta t = T/M$, and discretizing implicitly in time, this technique amounts to the simultaneous solution of m linear elliptic p.d.e.'s in each subinterval $[t_j, t_{j+1}]$, $0 \leq j \leq M-1$.

2.1a) $\qquad u_{j+1}^{\mu} - \Delta t \left(A^{\mu} u_{j+1}^{\mu} + U(\Pi^{\mu})_{j+1} \right) = u_j$ in Ω , $\gamma^{\mu} \cdot \nabla u_{j+1} = 0$ on Γ

where u_j is given by the pointwise maximum of the solutions u_j^{μ} with respect to the previous time interval

(2.1b) $\qquad\qquad\qquad u_j = \max_{1 \leq \mu \leq m} u_j^{\mu}$

A more direct approach is to discretize the nonlinear parabolic HJB-equation implicitly in time by the backward Euler scheme which means that in each subinterval $[t_j, t_{j+1}]$ we have to solve a nonlinear elliptic p.d.e. of HJB-type

(2.2) $\max_{1 \leq \mu \leq m} \left((\Delta t \, A^{\mu} - I) u_{j+1} + \Delta t U(\Pi^{\mu}) + u_j \right) = 0$ in Ω , $\max_{1 \leq \mu \leq m} \left(\gamma^{\mu} \cdot \nabla u_{j+1} \right) = 0$ on Γ

Discretizing in space by continuous, piecewise linear finite elements or finite diffe- rences, respectively, the linearization technique requires the simultaneous solution of m linear algebraic systems while using the second scheme we have to solve a nonlinear algebraic system. To speed up convergence we have used standard linear multi-grid methods in the first case and a nonlinear multi-grid technique of FAS-type in the second case. The main characteristics of the nonlinear method are the use of the nonlinear Gauss-Seidel-iteration as a smoother and an adaptive local choice of restrictions and prolongations in the fine-to-coarse and coarse-to-fine-transfers of the multi-grid cycles. The reader is referred to [2] where the elliptic problem is treated in detail.

3. Numerical results. The performance of the numerical schemes can be measured in terms of the approximation of the continuous free boundaries by their discrete coun- terparts. For some model problems where the exact free boundary is known we found that the nonlinear approach gave very accurate approximations whereas the linearization did not unless the time step size was reduced to much smaller values thus leading to considerably higher CPU-times.

As an example let us consider the case of a firm selling two products under the four advertising strategies $a_1=(a/2, a/2)$, $a_2=(a, 0)$, $a_3=(0, a)$ and $a_4=(0, 0)$ with $a=1000$. Market uncertainties are modelled by the variances $\sigma^\mu = \mathrm{diag}\big(\sigma_{ii}^\mu\big)_{i=1}^2$, with $\big(\sigma_{ii}^\mu(y)\big)^2 = \big(\lambda_i\, y_i + q_i\, a_i^\mu\, (1-y_i)\big)^{1/2}$, $1\leq i\leq 2$, $1\leq\mu\leq 4$ and $\gamma^\mu(z)$, $z\in\Gamma$, $1\leq\mu\leq 4$, is chosen as the exterior normal to Γ. The utility of profits is given by $U(\Pi^\mu)=\Pi^\mu$.

Figures 1–3 display the temporal evolution of the free boundaries at some selected time instants in case of time varying market potentials $M_1(t)=60\,000\,\exp(\max\{2-2*t/5, 0\})$ and $M_2(t) =120000$, but otherwise identical parameters. The regions marked 1, 2, 3 and 4 represent the sets Ω^μ (t) associated with the strategy a^μ, $1\leq\mu\leq 4$.

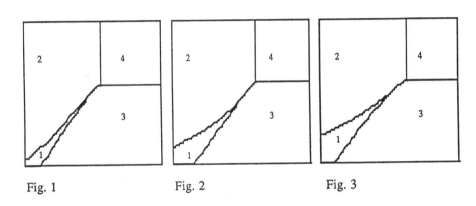

Fig. 1   Fig. 2   Fig. 3

References

[1] Bensoussan, A. and Lions, J.L.: Impulse control and quasivariational inequalities. Paris: Gauthier-Villars 1984
[2] Bloß, M. and Hoppe, R.H.W.: Numerical computation of the value function of optimally controlled stochastic switching processes by multi-grid techniques. Numer. Funct. Anal. and Optim. 10 (1989) 275–304
[3] Lions, P.L. and Trudinger N.S.: Linear oblique derivative problems for the uniformly elliptic Hamilton-Jacobi-Bellman equations. Math. Z. 191 (1986) 1–15
[4] Tapiero, C., Eliashberg, J. and Wind Y.: Risk behaviour and optimum advertising with a stochastic dynamic sales response. Optim. Control Appl. Methods 8 (1987) 299–304
[5] Vidale, M.L. and Wolfe, H.B.: An operations research study of sales response to advertising. Oper. Res. 5 (1957) 370–381

M. Bloß, FB Mathematik, TU Berlin, Straße des 17. Juni 135, D–1000 Berlin 12

R.H.W. Hoppe, Math. Inst., TU München, Arcisstr. 21, D–8000 München 2

TUBES

J.J.A.M. BRANDS
Eindhoven University of Technology
Department of Mathematics and Computing Science,
P.O. Box 513, 5600 MB Eindhoven, The Netherlands

ABSTRACT. Tubes, stable or unstable, provide bounds for solutions of differential equations. Two applications are presented. One application proves that the separatrix of a system describing an infectious disease intersects the coordinate axes. The second application shows the utility of tubes in the study of the asymptotic behaviour of solutions of differential equations.

1. Introduction

We consider first order scalar differential equations

$$(1) \qquad \dot{x} = f(t,x),$$

were $f \in C(D \to \mathbb{R})$, $D \subset [0,\infty) \times \mathbb{R}$, is such that initial value problems have unique solutions. We will formulate the definitions for $D = [0,\infty) \times \mathbb{R}$.

Let $\phi : [t_0,\infty) \to \mathbb{R}$ (with $t_0 \geq 0$) be differentiable. If $\dot{\phi}(t) \leq f(t,\phi(t))$ $(t \geq t_0)$ then ϕ is called a lower fence (of (1)) on $[t_0,\infty)$; if $\dot{\phi}(t) \geq f(t,\phi(t))$ $(t \geq t_0)$ then ϕ is called an upper fence on $[t_0,\infty)$. Let x be a solution of (1). The domain of x is denoted by $DOMx$.

If ϕ is a lower fence on $[t_0,\infty)$ and $x(t_0) \geq \phi(t_0)$ then $x(t) \geq \phi(t)$ $(t \in [t_0,\infty) \cap DOMx)$. If ϕ is an upper fence and $x(t_0) \leq \phi(t_0)$ then $x(t) \leq \phi(t)$ $(t \in [t_0,\infty) \cap DOMx)$.

Let ϕ be an upper fence and ψ a lower fence both on $[t_0,\infty)$. If $\phi(t_0) \geq \psi(t_0)$ then clearly $\phi(t) \geq \psi(t)$ $(t \geq t_0)$. The pair (ψ,ϕ) is called a stable tube on $[t_0,\infty)$; if $\phi(t) < \psi(t)$ on $[t_0,\infty)$ then the pair (ϕ,ψ) is called an unstable tube.

The existence of a stable tube (ψ,ϕ) on $[t_0,\infty)$ guarantees the existence of (often and infinity of) solutions x of (1) with $\psi(t) \leq x(t) \leq \phi(t)$ $(t \geq t_0)$. An unstable tube (ϕ,ψ) on $[t_0,\infty)$ guarantees the existence of at least one solution x with $\phi(t) \leq x(t) \leq \psi(t)$ $(t \geq t_0)$.

(Proof: Define $I_n := \{x(t_0) : \phi(t_0+n) \leq x(t_0+n) \leq \psi(t_0+n)\}$ $(n \in \mathbb{N})$. The intervals I_n are closed, non-empty and bounded; moreover $I_{n+1} \subset I_n$ $(n \in \mathbb{N})$. Hence $\bigcup\limits_{n \in \mathbb{N}} I_n \neq \varnothing$.)

REMARK. Fences and tubes can be defined on finite intervals in an obvious way. Also generalizations to higher dimensions are possible. Furthermore, fences and tubes on $[a, \infty)$ can be defined for equations of the kind

151

M. Heiliö (ed.), Proceedings of the Fifth European Conference on Mathematics in Industry, 151–154.

$$\dot{x}(t) = f(t,x(t)) + \int_a^t g(t,s,x(s)) \, ds \, ,$$

where $g(u,v,w)$ is continuous and increasing in w.

The notions of fence and tube can be very useful in the approximation of solutions of (1), especially in studying the asymptotic behaviour. N.G. de Bruijn applies these ideas in the study of the asymptotics of Riccati equations. (See [1]). In this paper we will give two applications: The first one is about the separatrix of a system which is one of the simplest models describing an endemic disease with feedback. ([2], also for further references.) A second application shows how tubes can be used to study the asymptotics of a nonlinear differential equation.

2. The Separatrix

The system reads (after rescaling)

(2)
$$\dot{x} = -x + y$$
$$\dot{y} = -\alpha(1+\beta)^{-1} y + g(x)$$

where α and β are positive constants, $\beta < 1$ and $g(x) = \alpha x^2 (1+\beta x^2)^{-1}$. (In a practical case [3] $\alpha \approx 0.12$, $\beta \approx 0.25$.) There are three equilibrium points in the first quadrant: two stable nodes in $(0,0)$ and (β^{-1},β^{-1}) and one saddlepoint in $(1,1)$. We want to know if the separatrix through $(1,1)$, which separates the two domains of attraction, intersects the axes. Linearizing around $(1,1)$ we get

(3)
$$\dot{u} = -u + v$$
$$\dot{v} = 2\alpha(1+\beta)^{-2} u - \alpha(1+\beta)^{-1} v \, .$$

The characteristic equation of (3) is

(4)
$$\lambda^2 + (1+\alpha(1+\beta)^{-1})\lambda - \alpha(1-\beta)(1+\beta)^{-2} = 0 \, .$$

Denoting the negative eigenvalue by $-\gamma$ we can show that

(5)
$$1 + \alpha(1+\beta)^{-1} < \gamma < 1 + 2\alpha(1+\beta)^{-1} \, .$$

An eigenvector for the eigenvalue $-\gamma$ is $[1, 1-\gamma]^T$, a tangent of the separatrix at $(1,1)$. By consideration of the direction field of (2) we easily see that the separatrix can be described by a function $y = y(x)$ with $y'(x) < 0$ for all $x \geq 0$ for which $y(x)$ is defined. From (2) it follows

(6)
$$(-x+y)y' = -\alpha(1+\beta)^{-1} y + g(x), \quad y(1)=1, \quad y'(1)=1-\gamma \, .$$

We will study the two parts (left and right of $(1,1)$) separately.

THE LEFT PART. Putting $x = 1 - \xi$, $y = w(\xi) + x$ in (6) we get

$$w \frac{dw}{d\xi} = (1+\alpha(1+\beta)^{-1})w + \alpha(1+\beta)^{-1}(1-\xi) - g(1-\xi),$$

$$w(0) = 0, \quad \frac{dw}{d\xi}(0) = \gamma.$$

The pair $(1+\alpha(1+\beta)^{-1})\xi$, $(1+2\alpha(1+\beta)^{-1})\xi$ forms a stable tube on $(0,1]$. (That means: a stable tube on every interval $[\delta,1]$ with $0 < \delta < 1$.) Moreover $w(\xi) \sim \gamma\xi$ $(\xi \downarrow 0)$. Whence, using (5), we infer that

$$1 + \alpha(1+\beta)^{-1}(1-x) \le y(x) \le 1 + 2\alpha(1+\beta)^{-1}(1-x) \quad (0 \le x \le 1).$$

THE RIGHT PART. Putting $x = 1 + \xi$, $y = x - z(\xi)$ in (6) we find

(7) $$z \frac{dz}{d\xi} = (1+\alpha(1+\beta)^{-1})z + g(1+\xi) - \alpha(1+\beta)^{-1}(1+\xi)$$

$$z(0) = 0, \quad \frac{dz}{d\xi}(0) = \gamma.$$

It is easy to verify that the pair ξ, $c\xi$ is a stable tube on $(0,\infty)$ if c is sufficiently large. Whence (using (5)) $\xi \le z \le c\xi$ $(\xi \ge 0)$. Using these inequalities in (7) we find that $\frac{dz}{d\xi} \ge 1 + c^{-1}(g(1+\xi) - g(1))\xi^{-1}$. It follows that $z(\xi) \ge \xi + b \log \xi$ for ξ sufficiently large where $b > 0$. Guided by this result we try to find a lower fence of the form $\phi(\xi) = \xi + d \log(1+\xi)$. Indeed, if $d = \alpha(1+\beta)^{-1}$ then ϕ is a lower fence on $(0,\infty)$. It follows (again using (5)) that $y(x) \le 1 - \alpha(1+\beta)^{-1} \log x \quad (x \ge 1)$.

We can conclude that the separatrix intersects the axes.

3. Asymptotics

Tubes can be a very useful tool in studying the asymptotics of a large class of differential equations. We give an example.

(8) $$t^2(t^2 + x^2)\dot{x} = 2x^2(t^2 - x).$$

After some heuristics we conjecture the following.

(9) For every $c \in \mathbb{R}$ there exists precisely one solution x with $x(t) \sim c$ $(t \to \infty)$; this unique solution has an asymptotic expansion $x(t) \approx c + \sum_{k=1}^{\infty} a_k t^{-k}$ the coefficients a_k of which can be calculated by formal substitution.

(10) There is precisely one solution x with $x(t) = t + o(t \log^{-1} t)$ $(t \to \infty)$; this unique solution has an asymptotic expansion $x(t) \approx t + \sum_{k=0}^{\infty} b_k t^{-k}$ $(t \to \infty)$ the coefficients b_k of which can be calculated by formal substitution.

(11) For every $c \in \mathbb{R}$ there exists precisely one solution x with $x(t) = t + 2t \log^{-1} t + ct \log^{-2} t + o(t \log^{-2} t)$ $(t \to \infty)$; this unique solution x has the asymptotic behaviour $x(t) = t + tu + O(1)$ $(t \to \infty)$, where u is the (unique) solution of the equation $$(2 + 2u + u^2)\dot{u} = -t^{-1}(u^2 + u^3)$$ with $u(t) = 2\log^{-1} t + c \log^{-2} t + O(\log^{-3} t)$ $(t \to \infty)$. This function u has an asymptotic expansion in powers of $\log^{-1} t$.

The existence follows from the existence of unstable tubes on $[t_0, \infty)$, t_0 sufficiently large,

for (9) by $c - (2c^2 + 1)t^{-1}$, $c - (2c^2 - 1)t^{-1}$,

for (10) by $t + 1$, $t + 1 + 2t^{-1}$,

for (11) by $t(1 + 2\log^{-1} t + c \log^{-2} t \pm A \log^{-3} t)$, A sufficiently large.

Uniqueness can be proved by showing that the difference of two different solutions (of the same kind) is too large. As an example we will prove uniqueness in case (10). Suppose that there are two different solutions x_1, x_2 with $v := x_1 - x_2 > 0$ and $x_i(t) = t + o(t \log^{-1} t)$ $(t \to \infty)$. Then, using the mean value theorem, we get $\dot{v} = t^{-1}(1 + o(\log^{-1} t))v$. Whence for t sufficiently large, say $t \geq t_0$, $\dot{v} > t^{-1}(1 - \frac{1}{2} \log^{-1} t)v$. It follows that $v(t) \geq v(t_0) t \log^{-\frac{1}{2}} t$ $(t \geq t_0)$, a contradiction.

Proofs of the existence of asymptotic expansions in (9) and (10) involve induction. Some possibilities can be found in [1]. Since (in both cases (9) and (10)) an asymptotic expansion for x implies an asymptotic expansion of the same kind for $\dot{x}$ we conclude that coefficients can be calculated by formal substitution.

References

[1]   N.G. de Bruijn; Asymptotic Methods in Analysis, 1981 Dover Publications, Inc. New York.

[2]   S.J.L. van Eijndhoven and A.G. de Waal; Mathematical models for the spread of infectious diseases, ECMI Report 89-07, Eindhoven University of Technology, The Netherlands.

[3]   M.A. Hooykaas, H.A.F. Leermakers and T.J.G. Zwartkruis; A mathematical model for the spread of infectious diseases based on a system of ordinary differential equations, ECMI Student Report 90-07, Eindhoven University of Technology, The Netherlands.

A Linear Programming Approach for Scheduling an S.M.T. Assembly Line

Enzo Brembilla and Giancarlo Poli

University of Milan

Via Cicognara, 7

20129 - MILANO - ITALY

Abstract

In this paper the authors describe a linear programming approach to the problem of optimal scheduling of electronic cards production process with S.M.T. (Surface Mounting Technology). First they briefly describe the SMT production process, then they define the linear programming model used to compute the optimal production schedule on the one week time basis; finally some computational result are provided.

1. Introduction

The main differences between the SMT technology and the old traditional Pin in Hole technology are:

- minor components volume;
- minor components weight;
- minor cards area;
- more components density;
- possible use of both sides of the cards.

The process consists of consecutive processing of cards by a sequence of different operations and not completly automatized. The fig. 1 shows the complete production process of this type of electronic cards; our attention is concerned about the Pick and Places machines, because of theyr high costs and require, consequently, the study of scheduling procedures for their optimal utilization.

Production line consists of a number of machines falling into different types, machines belonging to each type are identical.

There are several types of cards to be produced within planning period. The volume

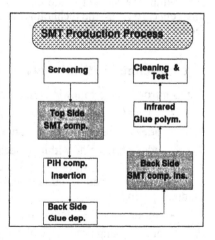

Fig. 1

M. Heiliö (ed.), Proceedings of the Fifth European Conference on Mathematics in Industry, 155–159.
© 1991 *B. G. Teubner Stuttgart and Kluwer Academic Publishers. Printed in the Netherlands.*

of production for each type is known prior to the beginning of a planning period.

Each card of each type should be processed by exactly one machine of each type. At the beginning of the production period machines are set up for production of specific types of cards, each machine could be set up for production of several types of cards.

The critical element of the SMT cards production process is the very high setup time which is necessary for the placement of the components on the machine feeders, when the change of production from one card to another is performed.

The objective of the proposed scheduling is to minimize setup repetition during planning period in order to reduce the idle time; the developed solution consists in getting the Pick and Place machines ready to process several different cards exploiting the flexibility of the feeders, available of hold components which belong to different cards in order to obtain a zero setup production line (fig. 2).

In other words, it is necessary to define a subset of available machines which would not be set up again during the planning period and maximize the total value of cards produced on these machines; the remaining cards will be produced in another line called Mix line.

The aim of this paper is to describe the Linear Mixed-Integer Programming Model which has been developed to face

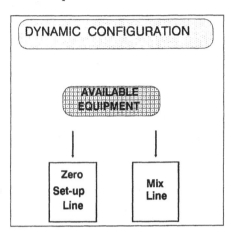

Fig. 2

the SMT machines scheduling problem; the objective function of our model is the maximization of the economic value of the cards, which turn out to be located on the zero-setup line, subject to the follow constraints: the weekly production plan of the different card types, the technological characteristics of the machines, the component feeders capacity and flexibility. This model is formalizes as a follow.

2. Description of the model.

Let us introduce the following notations:

i - index of machine type, $i = 1,...,M$;

j - index of the card type (part number), $j = 1,...,N$;

l - index of machine within a machine type, $l = 1,...,n_i$

Q_j - number of cards of the type j which is necessary to produce during planning period. This is defined by demand and becomes known prior to the beginning of the planning period.

t_{ij} - amount of time needed to the machine of type i to process the card of type j.

d_{il} - total working time available for the machine number l of the type i

v_j - value of one card of the type j

c_{ij} - capacity of feeder of a machine of the type i used by the card type j

C_i - feeder capacity of a machine of the type i.

The following quantities are **variables** which optimal values should be defined :

x_{ijl} - amount of cards of the type j allocated to the machine number l of the type i

z_{il} - Boolean variable which equals 1 if the machine number l of the type i is allocated to the zero set up (high volume) production line (i.e. this machine will not be set up anew in the middle of the planning period), otherwise it equals 0.

w_j - Boolean variable which equals 1 if cards of the type j are allocated to high volume production, otherwise it equals 0. It is assumed that either all cards of the type j are allocated to high volume production or none of them.

y_{ijl} - Boolean variable which equals 1 if part number j is allocated to the high volume production on the machine number l of type i, otherwise it equals 0.

The objective function would be to find the values of x_{ijl}, z_{il}, w_j, y_{ijl} which would maximize total value of the high volume production:

$$(1) \qquad \max \sum_{j=1}^{N} w_j v_j Q_j$$

subject to the following **constraints**:

$$(2) \qquad \sum_{l=1}^{n_i} x_{ijl} = w_j Q_j , \quad i=1,..,M, \quad j=1,...,N$$

i.e. sum of the high volume production on machines of each type should equal the total volume of a given part number.

$$(3) \qquad \sum_{j=1}^{N} y_{ijl} c_{ij} \leq C_i , \quad i=1,..,M, \quad l=1,...,n_i$$

i. e. for each machine required feeder capacity should not exceed the total ,capacity of the feeder.

$$(4) \qquad \sum_{j=1}^{N} x_{ijl} t_{ij} \leq d_{il} z_{il} , \quad i=1,..,M, \quad l=1,...,n_i$$

i.e. the total production time for each machine should not exceed the total available production time

$$(5) \qquad \sum_{j=1}^{N} Q_j t_{ij} (1-w_j) \leq \sum_{l=1}^{n_i} (1-z_{il}) d_{il} , \quad i=1,..,M$$

i.e. the production time on machines which would not be allocated to high volume production should be sufficient to cope with the rest of production.

$$(6) \qquad y_{ijl} \leq z_{il}, \quad i=1,...,M, \quad j=1,...,N, \quad l=1,...,n_i$$

i.e. part number j can be allocated to a particular machine in a zero set up production line only if this machine is set up for the high volume production.

$$(7) \qquad y_{ijl} \leq w_j, \quad i=1,...,M, \quad j=1,...,N, \quad l=1,...,n_i$$

i.e. part number j can be allocated to a particular machine in a zero set up production line only if it is allocated to the high volume production.

(8)
$$\sum_{l=1}^{n_i} y_{ijl} \geq w_j, \quad i=1,...,M, \quad j=1,...,N,$$

i.e. if part number j is allocated to the high volume production then it should be processed at least on one machine of each type.

(9)
$$x_{ijl} \leq K y_{ijl}, \quad i=1,...,M, \quad j=1,...,N, \quad l=1,...,n_i, \quad K=10000$$

This is technical constraint introduced to bound the volume of production from above by large number.

To solve this model an IBM software product has been used, MPSX/370 V2 (Mathematical Programming Sistem EXtended, which is available in the VM Operating System; MPSX uses the revised simplex method for the resolution of linear models and the branch and bound technique for the integer solution research.

3. An example of application

The input data for the problem are an average of the typical production plan of Vimercate plant of IBM Italy in which we have tested our model.

To assembly the components we have 3 machines for the top side of the card and 6 machines for the back sides.The weekly production plan consists of 12 different cars, some high volume and some high cost.

The number of constraints, in this case is 392, but the number increases rapidly with more different cards or with more available equipments (fig. 3).

The results of the model are shown in fig.4, in which the 90 % of the production plan is scheduled in a zero setup line, and the remaining production is allocated in the mix line in which setup is necessary.

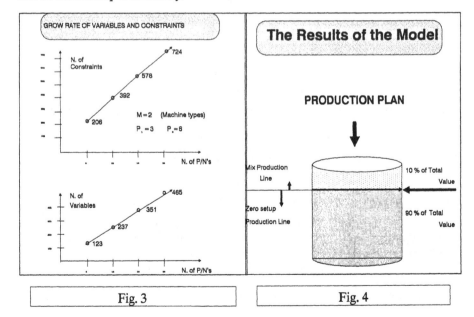

Fig. 3   Fig. 4

References

[1] R. J. Wittrock, "*A Linear Programming Approach to Mix Allocation for Flexible Assembly Line*", IBM T.J. Watson Research Center, N.Y., 1984

[2] "*MPSX Reference Guide*", IBM Corporation, 1989

[3] Archetti et al., "*Metodi della Ricerca Operativa*", Giappichelli Editore, Torino, 1989

[4] J. Pasquier, "*High level Simulation of Flexible Card Assembly Lines* ", IBM T.J. Watson Research Center, N.Y., !984

[5] R. S. Garfinkel, "*Integer Programming*", Addison Wesley, 1975

[6] M. S. Bazaraa - J. J. Jarvis, "*Linear Programming and Network Flow*", Wiley, 1977

[7] R. Fletcher, "*Practical Methods of Optimization*" Vol. 2 : "*Constrained Optimization*", Wiley, 1981

NONLINEAR ANALYSIS OF A COMPRESSOR WITH ACTIVE STABILIZATION

Morten Brøns and Peter Gross

1. INTRODUCTION

Turbo-compressors find a wide range of applications: industrial plants, turbochargers, aircraft propulsion, etc. In many cases, the stability of the flow is an important limiting factor. In particular, *surge* is an unwanted resonance phenomenon involving both the compressor and ambient piping and volumes. Recently, a simple technique for suppression of surge has been proposed, based on active control of the resonance volume [2]. Experiments indicate that the region of stable operation is greatly enhanced.

Here we propose a simple nonlinear mathematical model for a controlled compressor. Physically, the model is essentially a simplification of a linear model [2]. Earlier studies of uncontrolled systems show that they can be modelled successfully by relatively simple systems of nonlinear ordinary differential equations [1]. Following this approach, the model is analyzed in terms of bifurcation theory, and we show that the loss of stability occurs in the model as a Hopf bifurcation. A bistable regime is associated with this bifurcation, and we discuss its importance.

2. THE MODEL

Following [2], we consider the physical system shown schematically in fig. 1, a test stand for a compressor. The controller measures the difference between the actual pressure p in the compressor and the desired pressure C_{SS}. The controller exerts a force f on the surface A_3, in addition to a force from the spring. The displacement ξ of the surface produces a change of the plenum volume of size ξA_3.

If the mean flow velocity in the compressor is denoted u, the balance of momentum of the fluid in the compressor is

$$\frac{d}{dt}(\rho_0 A_C L_C u) = A_C(C - p). \tag{1}$$

Here ρ_0 is the density of the air, which we assume to be constant in the compressor. The compressor length is L_C, and the cross-section area is A_C. The pressure rise delivered by the compressor is C. This is denoted the characteristic, and is assumed to be an experimentally determined function of u.

The balance of mass for the plenum is

$$\frac{d}{dt}(\rho V) = \rho_0 A_C u - M, \tag{2}$$

where V is the volume of the plenum, and M denotes the mass flow through the throttle valve. Ignoring inertia in the throttle, we assume that M is a function of the plenum pressure p.

161

M. Heiliö (ed.), Proceedings of the Fifth European Conference on Mathematics in Industry, 161–165.
© 1991 B. G. Teubner Stuttgart and Kluwer Academic Publishers. Printed in the Netherlands.

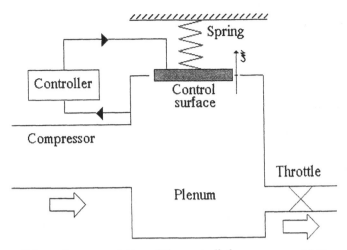

Figure 1. Schematic representation of the controlled compressor system.

The final dynamic equation is the balance of momentum for the control surface

$$m\frac{d^2\xi}{dt^2} = -K\xi + A_3 p - f, \tag{3}$$

where m is the mass of the control surface, and K is the spring constant.

To simplify these equations, we first assume that the density in the plenum changes adiabatically around ρ_0, such that $dp/dt = a^2 d\rho/dt$, where a is the speed of sound and since $V = V_0 + A_3\xi$, where $A_3\xi$ is a small perturbation, we get the approximation

$$\frac{d}{dt}(\rho V) \approx \frac{V_0}{a^2}\frac{dp}{dt} + A_3\rho_0\frac{d\xi}{dt}. \tag{4}$$

If the mass m of the control surface is ignored, we get from (3)

$$\xi = \frac{1}{K}(A_3 p - f). \tag{5}$$

Differentiating (5) with respect to time, and substituting in (2) with (4), we get

$$\left(\frac{V_0}{a^2} + \frac{A_3^2\rho_0}{K}\right)\frac{dp}{dt} = \rho_0 A_C u - M + \frac{A_3\rho_0}{K}\frac{df}{dt}. \tag{6}$$

In modelling the controller force f, we assume a linear stationary response from the controller unit. That is, if the plenum pressure varies sinusoidally around the desired value,

$$p - C_{SS} = \delta p \cos\omega t, \tag{7}$$

the response df/dt is also sinusoidal with a gain g and a phase shift ϕ,

$$\frac{df}{dt} = g\,\delta p \cos(\omega t + \phi). \tag{8}$$

Using (7), this can be rewritten as

$$\frac{df}{dt} = g\left((p - C_{SS})\cos\phi + \frac{1}{\omega}\frac{dp}{dt}\sin\phi\right). \tag{9}$$

When (9) is substituted in (6) we have, together with (1) a two-dimensional dynamical system to determine the plenum pressure and the compressor flow. Following standard conventions, we introduce dimensionless variables

$$u = U\tilde{u}, \ p = \frac{1}{2}\rho U^2\tilde{p}, \ t = \frac{1}{\omega_H}\tilde{t}, \ C = \frac{1}{2}\rho U^2\tilde{C}, \ g = A_3\omega_H\tilde{g}, \tag{10}$$

where U is a characteristic rotation speed of the compressor and ω_H is the Helmholtz frequency of the compressor-plenum system, $\omega_H = a\sqrt{A_C/L_C V_0}$, which is the frequency of oscillations in the system when $C = 0$. We assume for simplicity that $\omega = \omega_H$, and with the two dimensionless parameters

$$B = \frac{U}{2L_C\omega_H}, \ B_C = \frac{U A_3^2\rho_0\omega_H}{2K A_C}, \tag{11}$$

we finally get — dropping the ~'s again for convenience — the following system

$$\frac{du}{dt} = B(C(u) - p), \tag{12}$$

$$(B + B_C(1 - g\sin\phi))\frac{dp}{dt} = u - M(p) - B_C g\cos\phi(p - C_{SS}). \tag{13}$$

3. LINEAR ANALYSIS

To analyze the equations (12)-(13), the system characteristics and parameters must be specified. We will use a third-order approximation to the characteristic obtained in [4],

$$C(u) = 0.85 + 36.6u^2 - 127u^3, \tag{14}$$

and for the throttle characteristic we make the standard choice, $M(p) = \sqrt{p/S}$, appropriate for low-speed turbulent resistance. Here, S is a parameter reflecting the position of the throttle.

When $B_C = 0$, no control surface is present, and the system is the reduced Greitzer model, studied in detail in [1]. For all values of S, the system has a steady state at the point where the two characteristics C and M intersect, and the compressor delivers a pressure rise $C_{SS} = C_{SS}(S)$.

To analyze the stability properties of this state in the controlled system, the eigenvalues of the jacobian matrix, evaluated at the steady state, must be calculated. Based on experimental data from [2], we choose $B = 0.5$ and $B_C = 0.3$. For fixed values of the gain g the value of S where the stability changes can be calculated as a function of the phase shift ϕ. This gives rise to neutral stability curves as shown in fig. 2.

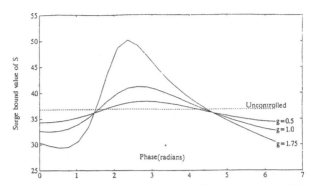

Figure 2. Stability limit of throttle parameter S as a function of the phase ϕ. The uncontrolled system loses stability at $S = 36.8$.

4. NONLINEAR ANALYSIS

To further reduce the number of independent parameters, we eliminate the phase ϕ by choosing, for a fixed value of S, the value which gives the largest stability region. This leaves two independent parameters, S and g.

At the loss of stability, limit cycles are born in a Hopf bifurcation [1], and the stability of the limit cycles can be calculated from a stability number a_3.

For each value of S, a Hopf bifurcation value g can be calculated. This gives rise to a curve of Hopf bifurcations in the (S, g) parameter plane, as shown in fig. 3. Along the Hopf curve, the stability number a_3 must be calculated to determine the type of bifurcation. Proceeding as in [1], we find the formula

$$16a_3 = B^2 C''' - H^2 M''' - \frac{BC'}{\omega_0^2}\left(\frac{(H^2 M'')^2}{H} - \frac{(B^2 C'')^2}{B}\right), \tag{15}$$

where

$$H = \frac{1}{B + B_C(1 - g\sin\phi)}. \tag{16}$$

Using (15), we find that a_3 changes sign along the Hopf curve, from negative for small values of g to positive for larger values of g. From the general Hopf theory, it follows that a curve of *saddle-node bifurcations of limit cycles* must emerge from the point on the Hopf curve where $a_3 = 0$, denoted (S_D, g_D). In a saddle-node bifurcation of limit cycles, two limit cycles, one stable and one unstable, merge and disappear. Using simulations, the start of this curve has been located, as shown in fig. 3.

5. DISCUSSION

The addition of a controller can make a significant increase in the region of local stabillity of a compressor. However, the region of global stability is smaller than region predicted by linear theory. This is due to the presence of saddle-node bifurcations, which result in a range of bistability. In this region, between the Hopf and

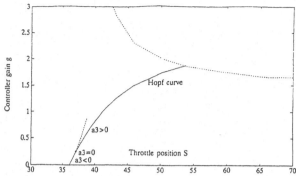

Figure 3. Partial bifurcation diagram for the controlled system. The full line is Hopf bifurcation, which meets a curve of stationary bifurcations, shown dotted. From the point on the Hopf curve where $a_3 = 0$, a curve of saddle-node bifurcations emerge, shown dotted.

saddle-node curves, the system may jump from the stable operation to surge, if an external perturbation is applied.

If the parameters are chosen such that $a_3 < 0$, the controller will be efficient even in deep surge. One simply has to turn on the controller with a value of g larger than g_H, and the steady state will have gained stability. However, if $a_3 > 0$, the gain must also be larger than g_{SN} to kill the oscillations. Consequently, one should aim at designing the compressor system such that the bifurcations are supercritical.

Finally, it should be remarked that additional steady states exist in the model for $g > 0.87$. A detailed analysis of these are, however, outside the scope of this preliminary study.

REFERENCES

[1] Brøns, M.: Bifurcations and instabilities in the Greitzer model for compressor system surge. Math. Engng. Ind. **2** (1988) 51-63.

[2] Ffowcs Williams, J. E. and X. Y. Huang: Active stabilization of compressor surge. J. Fluid Mech. **204** (1989) 245-262.

[3] Greitzer, E. M.: Surge and rotating stall in axial flow compressors, I, II. J. Engng. Power **98** (1976) 190-217.

[4] Hansen, K. E., P. Jørgensen and P. S. Larsen: Experimental and theoretical Study of surge in a small centrifugal compressor. J. Fluids Engng. **103** (1981) 391-395.

Mathematical Institute
The Technical University of Denmark
DK-2800 Lyngby
Denmark

A TIME-DEPENDENT MODEL FOR THE SPACE-CHARGE DISTRIBUTION IN SOLID DIELECTRICS UNDER HIGH ELECTRIC FIELD

CHRIS BUDD   RICHARD HARE

ABSTRACT. High-field electrical breakdown of solid dielectrics is of particular interest to the electricity generation and transmission industries, where life expectancy of insulation is often uncertain. In order to study and explain the phenomenon a collaborative project has been set up involving National Power, National Grid, King's College London, City of London Polytechnic, Manchester University, Glasgow College of Technology and Bristol University. The current paper proposes a time-dependent, theoretical model of charge injection into a solid dielectric. The geometry used is a hyperboloidal, high d.c. voltage needle adjacent to an earthed plane, similar to a common experimental situation. Mobility is taken to be low for low fields and some orders of magnitude higher for high fields, varying rapidly between them at a threshold field. It is found that a quasi-stable space charge cloud forms around the needle in the order of 10^{-4} or 10^{-3} s. The cloud then dissipates relatively slowly to form a true steady-state distribution of charge throughout the dielectric.

1 Introduction

When a solid dielectric is subjected over a long period of time to electric fields in excess of 1 or 0.1 $MVcm^{-1}$, a visible tree-like structure begins to grow. When the tree grows right through the dielectric, connecting the electrodes, electrical breakdown quickly occurs and major physical damage is caused to the formerly insulating material.

This phenomenon is of concern to the electricity generation and transmission industries, being a potential cause of insulation failure in Gas Insulated Substations, switchgear and electrical machines. A collaborative research programme involving National Power, National Grid, King's College London, City of London Polytechnic, Manchester University, Glasgow College of Technology and Bristol University has been set up to try to understand and explain breakdown in solid dielectrics. The ultimate aim is to develop a model which predicts the life expectancy of a dielectric once the defect likely to cause failure is known. To this end, a study of space-charge injection and distribution prior to tree inception is needed.

From a mathematical point of view, the problem is similar to the drift of charges in a gas from a corona boundary to a collector electrode. Much of the literature on gas discharges, therefore, is applicable to the problem of charge motion in solid dielectrics. For an excellent review of the subject see Sigmond (1986).

Sigmond's paper points out the disadvantages of purely numerical solutions, particularly their expense, slowness and inherent loss of generality. Reasonable approximations are thus valid: what they lose in accuracy they make up for in clarity. One widely-used approximation is named after Deutsch (1933). Here the Laplacian, space-charge-free problem is solved and field lines identified. Then injected charge is artificially constrained to flow along these field lines. The equipotential surfaces remain the same shapes as in the Laplacian solution: only the potential on them is allowed to change.

A less restrictive form of Deutsch's assumption underlies the present paper. Here we consider that field and current densities vary much more rapidly on moving away from the injecting electrode than moving parallel to it. This allows the same simplification of the equations without constraining the equipotentials to their Laplacian positions.

M. Heiliö (ed.), Proceedings of the Fifth European Conference on Mathematics in Industry, 167–176.

When the electrodes are parallel planes, concentric cylinders or concentric spheres, Deutsch's assumption is equivalent to an invocation of symmetry and is thus exact. These geometries have been treated in detail by Mott and Gurney (1948), Meltzer (1960), Smith (1987) and Budd and Wheeler (1988). A more common experimental setup is a high voltage, sharp steel needle (of tip radius typically 5 microns), embedded in dielectric material, close to an earthed plane electrode. In theoretical models the needle is usually represented by a hyperboloid. The problem can then be neatly expressed in prolate spheroidal coordinates (u, v, θ) with each electrode being a surface of constant v. The use of Deutsch's assumption in this needle-plane situation is not exact, as demonstrated by Sigmond (1986) and Wintle (1987), and even the less restrictive "relaxed" form is still an over-simplification. However, a straightforward needle-plane solution using the latter approximation has been produced (Hare and Hill (1991)) and its conceptual development from the steady-state to the time-dependent case is the subject of the present paper. Work on quantifying the error involved is currently in hand.

A major concern in this work is the question of mobility. For solid dielectrics the existence of charge traps must be taken into account. Hibma and Zeller (1986) proposed a field-limited space-charge (FLSC) model for this phenomenon. This is characterised by a critical field E_1 above which the mobility rises by several orders of magnitude. As charge is injected into the dielectric with high local field and high mobility, the field is reduced. The FLSC model sets the low mobility to zero, so that charges stop moving when their local fields drop below E_1. The result is a static space charge cloud around the needle.

The present paper uses a non-zero low mobility and traces the evolution of space charge in the needle-plane region from an initial Laplacian situation, through a quasi-stable Hibma-and-Zeller-like charge cloud to a steady-state Hare-Hill solution.

2   Theory

The governing equations for time-dependent space-charge flow for a single charge carrier in a dielectric are as follows:

$$\text{Continuity} \qquad \underline{\nabla}.\underline{j} + \frac{\partial \rho}{\partial t} = 0 \qquad\qquad (1)$$

$$\text{Transport} \qquad \underline{j} = \mu \rho \underline{E} \qquad\qquad (2)$$

$$\text{Poisson} \qquad \underline{\nabla}.\underline{E} = \frac{\rho}{\epsilon \epsilon_0} \qquad\qquad (3)$$

In almost all practical cases diffusion effects are very small and have been neglected in the above formulae (Sigmond, 1986). The symbols have their usual meanings which are detailed in Table 1. Typical magnitudes for each of the quantities are given in Section 4. The field-dependent mobility μ is assumed to rise sharply from a small value $\delta\mu_1$ ($\delta \ll 1$) to a high value μ_1 when the local field exceeds some constant E_1. A convenient continuous formulation of the mobility which satisfies these conditions is:

$$\mu = \mu_1 \left\{ \delta + \tfrac{1}{2}(1 - \delta)[1 + \tanh \gamma (E/E_1 - 1)] \right\} \qquad\qquad (4)$$

where the width of the rise $\Delta E \propto 1/\gamma$ for fixed E_1.

The prolate spheroidal coordinate system (u, v, θ) is related to Cartesian coordinates as follows:

$$\left. \begin{array}{rcl} x &=& av \left(u^2 + 1\right)^{\frac{1}{2}} \\ y &=& au \left(1 - v^2\right)^{\frac{1}{2}} \cos \theta \\ z &=& au \left(1 - v^2\right)^{\frac{1}{2}} \sin \theta \end{array} \right\} \qquad\qquad (5)$$

It is well suited to the problem since Laplace's equation is then separable, giving equipotentials which are hyperboloids, surfaces of constant v. In particular, the charged needle

Table 1: Description of Variables

Parameter	Meaning	Equivalent in terms of the associated non-dimensional parameter
μ	Mobility	$\mu_1 \hat{\mu}$
t	Time	$\mu_1^{-1} E_1^{-1} a \hat{t}$
ρ	Space charge density	$\epsilon \epsilon_0 a^{-1} E_1 \hat{\rho}$
E	Field	$E_1 \hat{E}$
ϕ	Potential	$E_1 a \hat{\phi}$
ϕ_0	Potential at needle	$E_1 a \hat{\phi}_0$

and earthed plane boundaries are defined simply by $v =$ constant (Fig. 1). In transforming coordinates, the x-axis becomes the axis of symmetry onto which the ellipsoids and hyperboloids vanish as $u \to 0$ and $v \to 1$. An appeal to rotational symmetry about this axis sets all $\partial/\partial\theta$ terms to zero.

The problem is expressed in (u, v, θ) coordinates in dimensionless form (see Table 1). Non-dimensionalisation is based on $\hat{E}_1 = 1$, being the critical field at which mobility rises rapidly. Combining (1) and (2) to eliminate the current density $\underline{j}$ means the governing equations become:

$$\frac{(u^2 + 1)^{\frac{1}{2}}}{u(1 - v^2 + u^2)} \left\{ \frac{\partial}{\partial u} \left[u(1 - v^2 + u^2)^{\frac{1}{2}} \hat{\mu} \hat{\rho} \hat{E}_u \right] + \right.$$
$$\left. \frac{\partial}{\partial v} \left[u(1 - v^2 + u^2)^{\frac{1}{2}} (1 - v^2)^{\frac{1}{2}} \hat{\mu} \hat{\rho} \hat{E}_v \right] \right\} + \frac{\partial \hat{\rho}}{\partial \hat{t}} = 0 \tag{6}$$

$$\frac{(u^2 + 1)^{\frac{1}{2}}}{u(1 - v^2 + u^2)} \left\{ \frac{\partial}{\partial u} \left[u(1 - v^2 + u^2)^{\frac{1}{2}} \hat{E}_u \right] + \right.$$
$$\left. \frac{\partial}{\partial v} \left[u(1 - v^2 + u^2)^{\frac{1}{2}} (1 - v^2)^{\frac{1}{2}} \hat{E}_v \right] \right\} = \hat{\rho} \tag{7}$$

where the field $\hat{E} = (\hat{E}_u, \hat{E}_v, 0)$. The relaxed form of Deutsch's assumption in the needle-plane geometry is that $\partial/\partial u$ terms are negligible when compared with $\partial/\partial v$ terms.

With $\hat{E}_v$ written simply as $\hat{E}$, the governing equations now become:

$$\frac{1}{(1 - v^2 + u^2)} \frac{\partial}{\partial v} \left[(1 - v^2 + u^2)^{\frac{1}{2}} (1 - v^2)^{\frac{1}{2}} \hat{\mu} \hat{\rho} \hat{E} \right] + \frac{\partial \hat{\rho}}{\partial \hat{t}} = 0 \tag{8}$$

$$\frac{1}{(1 - v^2 + u^2)} \frac{\partial}{\partial v} \left[(1 - v^2 + u^2)^{\frac{1}{2}} (1 - v^2)^{\frac{1}{2}} \hat{E} \right] = \hat{\rho} \tag{9}$$

To define the problem completely, an initial condition and three boundary conditions are needed. The voltage $\hat{\phi}_0$ is first applied to the needle at time $\hat{t} = 0$ so at that instant

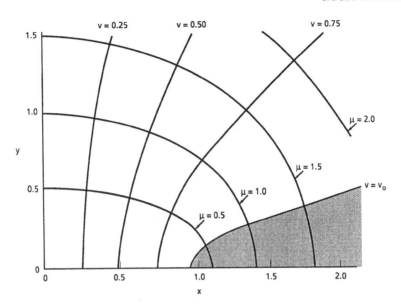

Figure 1 The prolate spheroidal coordinate system. The system is cylindrically
symmetrical about the x axis and v = v₀ defines the surface of the metal
pin contact at potential φ. The plane x = 0 defines the earth plane.

there is no space charge within the region of interest. The field distribution is thus given
by the special case of equation (3).

One boundary condition is that the potential on the needle (at $v = v_0$) is fixed at $\hat{\phi}_0$
for all time. Another is that the plane electrode is earthed: its potential is zero. The
third boundary condition governs charge injection at the needle and is less obvious. Its
form has been derived for the related corona problem by Budd (1991). Hare and Hill
(1991) point out that for solid dielectrics there is some justification for each of a number
of alternatives. For the present exploratory work a constant space charge density at the
needle $\hat{\rho} = \hat{\rho}_0$ has been chosen. The derivation of a physically realistic condition based on
known injection processes is the subject of ongoing research.

Because of the field-dependent mobility and the strong non-linearity in the field, the
region near the needle is of most interest. In order to bias the solution grid towards this
area, a change of variable is required. Setting:

$$w = \frac{e^q - e^{qv/v_0}}{e^q - 1} \tag{10}$$

with constant q means that $w = 0$ at the needle, $w = 1$ at the earthed plane and w varies
fastest when v is close to v_0. Equations (8) and (9) now become:

$$\frac{1}{(1 - v(w)^2 + u^2)} \frac{dw}{dv} \frac{\partial}{\partial w} \left[(1 - v(w)^2 + u^2)^{\frac{1}{2}} (1 - v(w)^2)^{\frac{1}{2}} \hat{\mu} \hat{\rho} \hat{E} \right] + \frac{\partial \hat{\rho}}{\partial \hat{t}} = 0 \; (11)$$

$$\frac{1}{(1 - v(w)^2 + u^2)} \frac{dw}{dv} \frac{\partial}{\partial w} \left[(1 - v(w)^2 + u^2)^{\frac{1}{2}} (1 - v(w)^2)^{\frac{1}{2}} \hat{E} \right] = \hat{\rho} \; \tag{12}$$

The solution method for equations (11) and (12) together with the boundary and initial
conditions is described in the next section.

3 Numerical Algorithm

The Two-Step Lax-Wendroff numerical scheme, described in Morrow (1981) and Press, Flannery, Teukolsky and Vetterling (1989), is a second-order-in-time method that avoids large numerical dissipation and mesh drifting. It often introduces spurious oscillations to the solutions but, as long as these are small, the disadvantages are outweighed by the transparency of the method. To make use of it, equation (11) is first written in the form:

$$\frac{\partial G}{\partial \hat{t}} = -\frac{\partial F(G)}{\partial w} \tag{13}$$

Equation (12) becomes:

$$\frac{\partial}{\partial w} \left[(u^2 - v^2 + 1)^{\frac{1}{2}} (1 - v^2)^{\frac{1}{2}} \hat{E} \right] = G \tag{14}$$

Because of Deutsch's assumption, each field line can be treated independently and the two independent variables are w and t. A considerable simplification of the numerical scheme has therefore been achieved. The algorithm works on a two-dimensional grid, a general point on which is called (j, n). The interval between rows j and $j+1$ is the distance Δw, whereas the interval between columns n and $n+1$ is the time $\Delta \hat{t}$. Given the solution $\hat{\rho}_j^n$ and $\hat{E}_j^n$ at all points j at time index n, the solution at time index $n + 1$ is obtained using the Lax-Wendroff scheme on equation (13). The field $\hat{E}$ is updated using equation (14) each time a new set of G's or $\hat{\rho}$'s is calculated.

The time-step $\Delta \hat{t}$ is chosen to be as large as stability considerations will allow. At regular intervals throughout the run, it is increased to take account of improving numerical stability as space charge is dissipated in the dielectric.

4 Results and Discussion

Fig. 2a–d show the evolution of space charge density and field along the axis of symmetry $u = 0$. The needle shaft asymptotes to an angle of $15°$ from the axis, giving $v_0 = 0.9659$. This corresponds to a rather thick needle, whose use as a baseline model greatly speeds up the numerical calculations. The voltage at the needle (on the left-hand side of each graph) is $\hat{\phi}_0 = 0.2$ and the space-charge density there is fixed at $\hat{\rho}_0 = 5$. The "distance" axis w is related to v as in equation (10), with the bias parameter $q = 5$. This means that the midpoint between needle and plane $v = \frac{1}{2} v_0$ corresponds to $w = 0.92$, very near the right-hand edge of the grid.

The mobility/field dependence in equation (4) has steepness coefficient $\gamma = 10$ and the low:high mobility ratio is $\delta = 0.01$. The time-step in each graph is constant but it increases between consecutive pictures. The simulation begins at time $\hat{t} = 0$ with no internal space-charge and Laplacian field (Fig. 2a). The high local field means that charge penetrates the dielectric very rapidly, but soon loses mobility as the field drops below the threshold $\hat{E}_1 = 1$ (at around $\hat{t} = 0.02$). As expected, a quasi-steady state evolves, with constant near-threshold field inside a charge "cloud" close to the needle, and zero charge outside the cloud. This is the situation described by Hibma and Zeller (1986).

Over much longer time-scales (Fig. 2b-d) the low mobility is the dominant parameter. Field is now below the threshold everywhere and the charge cloud gradually dissipates throughout the dielectric. The slight, grid-related oscillations following the leading peak are spurious effects caused by the numerical scheme.

Other runs have been carried out for different geometries and different forms for the mobility. The results are summarised in Table 2. Field lines close to the axis of symmetry $(u = 0)$ are very similar, but at $u = 0.5$ the Laplacian field is no longer large enough to produce any high mobility effects and no space charge cloud occurs. A sharper needle is

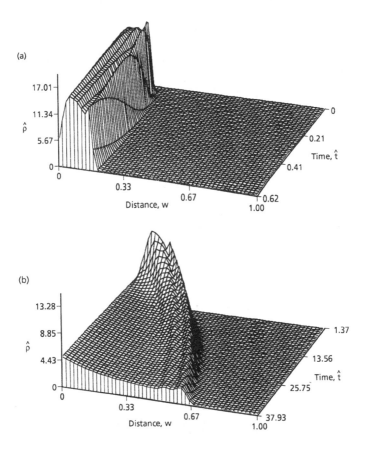

Figure 2 Space charge density on the axis of symmetry

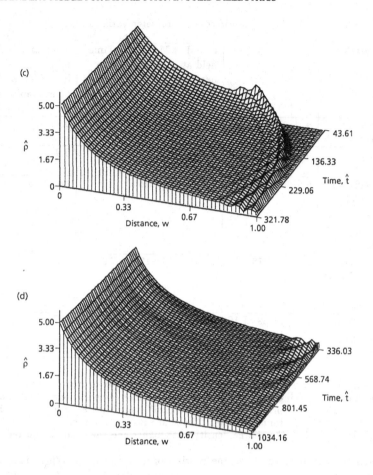

Figure 2 Space charge density on the axis of symmetry

Table 2: The Effects of Parameter Variation

Variation	u	v_0	γ	Initial field at needle $\hat{E}$	Max. ρ in cloud	Time $\hat{t}$ to steady state	Approximate v/v_0 at boundary of quasi-stable charge cloud
Standard model	0	.9659	10	1.47	17	10^3	0.97
Field line $u = 0.5$	0.1	.9659	10	1.37	16	10^3	0.97
Field line $u = 0.5$ (field below threshold)	0.5	.9659	10	0.68	no cloud	10^3	no cloud
Standard model with $2.86°$ needle	0	.99875	10	21.7	80–120	10^3	0.94
Mobility change is less steep	0	.9659	10	1.47	14	10^3	0.96

closer to experimental reality: the value of $v_0 = 0.99875$ is chosen to give a $5\mu m$ tip radius when the tip-plane spacing is 2 mm. It does require more computational effort, however, to ensure convergence, which is why the "standard" model has $v_0 = 0.9659$ (with a tip radius of 139 μm for the same spacing). The sharp needle produces similar results, but the initial field is higher, the charge densities are higher and the cloud is injected further into the dielectric.

The simulations are sensitive to the precise form of the mobility/field relationship. Halving the steepness parameter γ means that low mobility effects occur sooner: the cloud is broader and less pronounced. There is very little sensitivity, however, to the needle boundary condition: its effect does not penetrate far into the dielectric.

Restoring the dimensions to the parameters in the standard model gives some guide as to physical interpretation of the results. The use of a relatively low v gives a rather fat needle which, in consequence, needs a very high voltage to produce above-threshold fields. If the threshold E_1 is 3 $MVcm^{-1}$ and the tip-plane spacing av_0 is 2 mm the parameters used above require the needle voltage to be $\phi_0 = 120$ kV. The maximum charge density in the cloud is $(2.55 \times 10^{12}\ Vm^{-2})\epsilon\epsilon_0$ and its width is 0.06 mm. Mobilities in dielectrics can range from $10^{-12}\ m^2V^{-1}s^{-1}$ (good insulator) up to $10^{-8}\ m^2V^{-1}s^{-1}$ (bad insulator). As $\delta = 0.01$ in the standard model, a value of $\mu_1 = 10^{-9}\ m^2V^{-1}s^{-1}$ seems reasonable; the low mobility is then $\delta\mu_1 = 10^{-11}\ m^2V^{-1}s^{-1}$. In this case, $t = 6.7 \times 10^{-3}\ \hat{t}$, the cloud evolves in just 10^{-4} s and the steady state is reached within 10 s. Scaling down the ratio δ or the high mobility μ_1 would increase proportionally the overall time for charge to dissipate.

The nature of the problem makes it difficult to define precisely when the charge cloud appears and when the steady-state is reached. In practice, the times are judged from the graphs in Fig. 2 and should be taken as characteristic rather than exact.

5 Conclusions

The major application of the foregoing work is to the study of high-field dielectric break-down. It is generally thought that one of the contributors to breakdown involves the motion of excited, high-mobility charges. The process has been modelled here by using a strongly field-dependent mobility, which leads to the following conclusions.

When the local field near the injecting needle electrode is above the high-mobility threshold E_1, a space-charge cloud forms around the needle in the order of 10^{-4} or 10^{-3} s. Within the cloud the field is constant and near to the threshold E_1. The boundary condition at the needle dominates only a small part of the dielectric.

The cloud is the one reported by Hibma and Zeller (1986) but in the present model its stability is short-lived. The lower mobility gains precedence and sweeps the cloud through the dielectric. Eventually the true steady-state is reached and the field and charge distributions are those calculated by Hare and Hill (1991).

For a dielectric thickness of 2 mm, an applied voltage of 120 kV and a low-field mobility of 10^{-11} $m^2V^{-1}s^{-1}$ the steady-state is attained in about 10 s.

This exploratory model may bear some relation to impulse breakdown. Here a short burst of charge is injected into the dielectric and breakdown occurs either immediately or not at all. The model would support the view that the phenomenon is due to high-mobility charge motion. The conditions for high mobility are present only in the first fraction of a second from switch-on: injection of space-charge very quickly reduces all fields to a threshold below which high mobility charge motion cannot be sustained. If high-mobility effects *are* important, however, the limited penetration of the charge cloud into the dielectric means that additional, material-related effects not modelled here must contribute significantly towards breakdown.

Theoretical and experimental validation of the model, including space-charge measure-ments, is the subject of ongoing research at National Power TEC and Bristol University in conjunction with the other establishments participating in the collaborative project.

6 Acknowledgements

The second author is employed by National Power plc, who have supported this work and granted permission for its publication. Both authors would like to thank Dr J H Pickles and Prof R M Hill for valuable discussions and Miss J E Seal for help and advice on the graphical output.

7 References

Budd, C.J. and Wheeler, A.A. (1988) 'Exact Solutions of the Space Charge Equation Using the Hodograph Method', IMA. J. Appl. Maths. **40**, 1–14.

Budd, C.J. (1991) 'Coronas and the Space Charge Problem', Eur. J. Applied Maths (to appear).

Deutsch, W. (1933) 'Über die Dichtverteilung unipolarer Ionenströme', Annalen der Phy-sik, **5**, 588–612.

Hare, R.W. and Hill, R.M. (1991) 'Space-Charge in Needle-Plane Geometry', J. Phys. D. (to appear).

Hibma, T. and Zeller, H.R. (1986) 'Direct Measurement of Space-Charge Injection from a Needle Electrode into Dielectrics', J. Appl. Phys., **59**, 1614–20.

Meltzer, B. (1960) 'Space-Charge-Limited Currents in Cylindrical and Spherical Insulator Diodes', J. Electron. and Control, **8**, 171–6.

Morrow, R. (1981) 'Numerical Solution of Hyperbolic Equations for Electron Drift in Strongly Non-Uniform Electric Fields', J. Comp. Phys. **43**, 1–15.

Mott, N.F. and Gurney, R.W. (1948) Electronic Processes in Ionic Crystals, Clarendon Press, Oxford, 2nd edition, p172.

Press, W.H., Flannery, B.P., Teukolsky, S.A. and Vetterling, W.T. (1989) Numerical Recipes, Cambridge University Press, p633.

Sigmond, R.S. (1986) 'The Unipolar Corona Space Charge Flow Problem', J. Electrostatics, **18**, 249–72.

Smith, S.A. (1987) 'Congruent Harmonic and Space Charge Electrostatic Fields', IMA J. Appl. Maths. **39**, 189–214.

Wintle, H.J. (1987) 'Space Charge Limited Current in the Needle-Plane Geometry', J. Electrostatics, **19**, 257–274.

8 Authors' Details

CHRIS BUDD   RICHARD HARE
University of Bristol   National Power
School of Mathematics   Technology and Environmental Centre
University Walk   Kelvin Avenue
Bristol BS8 1TW   Leatherhead KT22 7SE
United Kingdom   United Kingdom

CATALYTIC CONVERTERS AND POROUS MEDIUM COMBUSTION
Helen Byrne

Physical Interpretation. We study an implicit free boundary value problem
(BVP) (1)–(4) which derives from a model of porous medium combustion [3] when
a certain asymptotic limit is taken. Our model may describe the action of a catalytic converter [4], a device used in motor-vehicles to reduce pollution effects.
The dependent variables are temperature, u, and gaseous concentration, g, whilst
r denotes the reaction rate. The step-function factors in r, see (3), signify that
the reaction can only be sustained when both the temperature and the gas concentration exceed critical values. The key parameters are λ and μ where

$$\lambda = \frac{L^2}{u_c} \text{ and } \mu = \frac{La}{vg_a},$$

L is the length of the catalyst; u_a is the critical switching temperature; a is the
ratio of gaseous to solid fuel consumption; v is the inlet gas velocity; g_a is the inlet
gas concentration.

We examine the existence and stability with respect to small time-dependent
perturbations of steady-state solutions of (1)–(4). We show that $\lambda > \mu$ is necessary
for the existence of steady solutions. In the original variables this yields a lower
bound on the length of a converter, $L > au_c/vg_a$. We conclude by showing that for
two asymptotic limits stable and unstable solutions to the problem exist. The stable solutions correspond to reactions terminated by gas exhaustion (G-solutions),
whilst the unstable ones represent reactions terminated by the temperature falling
below the critical value (U-solutions).

The Model. We study the mixed hyperbolic-parabolic system

$$\frac{\partial u}{\partial t} = \frac{\partial^2 u}{\partial x^2} + \lambda r, \tag{1}$$

$$\frac{\partial g}{\partial x} = -\mu r, \tag{2}$$

$$r = \mathcal{H}(u - 1)\, \mathcal{H}(g)\, f(u, g). \tag{3}$$

Here $\mathcal{H}(.)$ is the Heaviside step-function and λ and μ are positive parameters. The
following initial and boundary conditions are imposed on the system:

$$u_x(0, t) = 0 = u(1, t), g(0, t) = 1, u(x, 0) = u_0(x) \text{ a prescribed function.} \tag{4}$$

We seek non-trivial solutions of equations (1)–(4) for which $\exists\, s(t) \in (0, 1)$ such
that $r > 0\ \forall\ x \in (0, s)$ and $r = 0\ \forall\ x \in (s, 1)$. The variables u, u_x and g are
assumed continuous across $x = s(t)$. The form of r enables us to deduce that at
$x = s(t)$ either a *U-solution* occurs, with $(u = 1, g > 0)$, or a *G-solution* occurs,
with $(u > 1, g = 0)$, or a *degenerate U/G-solution* occurs, with $(u = 1, g = 0)$.

M. Heiliö (ed.), Proceedings of the Fifth European Conference on Mathematics in Industry, 177–180.
© 1991 *B. G. Teubner Stuttgart and Kluwer Academic Publishers. Printed in the Netherlands.*

Steady-State Solutions. Setting $\partial/\partial t = 0$ in equations (1)–(4) leads to the steady-state problem defined for $x \in (0, s)$

$$u_x = -\frac{\lambda}{\mu}(1 - g), g_x = -\mu f(u, g) \Rightarrow \frac{dg}{du} = \frac{\mu^2}{\lambda} \frac{f(u, g)}{1 - g},$$

$$u_x(0) = 0, g(0) = 1, u_x(s) = -\frac{u(s)}{1 - s} = -\frac{\lambda}{\mu}(1 - g(s)),$$

and one of the matching conditions stated above ($u(s), g(s)$ are constants to be determined as part of the solution). The above relationship between $u(s)$ and $g(s)$, obtained from continuity of u_x across $x = s(t)$ and solving explicitly for u and g in $s < x < 1$, enables us to deduce the following:

Lemma 1 *A necessary condition for the existence of steady-state solutions is* $\lambda > \mu$.

By restricting attention to the class of functions $f(u, g) = u^m, m \geq 1$, we can exploit the form of the problem above to obtain a series of algebraic equations which determine the unknowns $u(0), u(s), g(s)$ and s. Defining

$$M = u(0), R = \frac{u(s)}{u(0)}, G = 1 - g(s),$$

these identities become

$$G^2 = \frac{2\mu^2}{\lambda(m + 1)} M^{m+1} (1 - R^{m+1}),$$

$$s = \left[\frac{M^{1-m}(m + 1)}{2\lambda}\right]^{1/2} \int_R^1 dv(1 - v^{m+1})^{-1/2},$$

$$\frac{MR}{G} = \frac{\lambda}{\mu}(1 - s),$$

with either $\{R = M^{-1}, 0 < G < 1\}$, for a U-solution, or $\{r > M^{-1}, G = 1\}$, for a G-solution. Asymptotic analysis of these relations is possible for the limiting cases $(0 < s \ll 1)$ and $(0 < 1 - s \ll 1)$. For example, we can deduce the existence of U-solutions having

$$0 < s \ll 1, \lambda \sim \frac{1}{s}, 0 < \mu < \frac{1}{s}, u(0) \sim 1 + \frac{s}{2}.$$

Moreover, it can be shown that all valid solutions having $0 < s \ll 1$ satisfy

$$\lambda \geq \frac{2 + (1 - m)s}{2s} \sim \frac{1}{s}, \tag{5}$$

provided $0 < |s(1 - m)| \ll 1$. The above analysis was used to construct figure 1 which describes the existence regimes for steady U/G-solutions in (λ, μ) space. The remaining analysis concentrates on the specific case $f(u, g) = u$, chosen for its simplicity.

Numerical Results. The Keller Box Scheme and Newton iteration were used to solve the time-dependent equations (1)–(4). A more detailed account of the numerical methods employed can be found in [1]. Figure 2 describes results obtained for the case ($\lambda = 25, \mu = 12$), chosen for comparison with our asymptotic analysis – $\lambda = 25 \Rightarrow s(0) \sim 0.04 \sim 1/\lambda$. The curves $G1(t), U1(t), G2(t)$ and $U2(t)$ in figure 2 trace free boundaries whose initial profiles satisfy $0 < s_{G1}(0) = s_{U1}(0) < s(0) < s_{G2}(0) = s_{U2}(0)$. The divergence of $U1$ and $U2$ together with the (minimal) convergence of $G1$ and $G2$ suggest that G-solutions are stable and U-solutions unstable.

Stability Analysis, We seek small time-dependent perturbations from the steady-state solutions of the form

$$u(x,t) = u(x) + \delta e^{\rho t} \phi(x). \qquad (6)$$

Eigenvalue problems for U- and G-solutions are obtained by linearising the perturbed equations. In each case the problem reduces to the solution of a dispersion relation for ρ in terms of λ (and μ, when $m > 1$). Using standard techniques [2], the stability of U- and G-solutions for the limits $\underline{A}$: ($0 < s \ll 1, \lambda \sim 1/s$) and $\underline{B}$: ($0 < 1 - s \ll 1, \lambda \sim \pi^2/4$) are examined. The results are summarised thus:

Theorem 1 *(i) Steady-state G-solutions of (1)–(4 for the asymptotic cases $\underline{A}$ and $\underline{B}$ are stable with respect to small time-dependent perturbations of the form (6).
(ii) Steady-state U-solutions of (1)–(4) for the same limits are unstable.*

Notes. The author wishes to thank the Science and Engineering Research Council for funding both her D.Phil. and her attendance at the ECMI90 and Free Boundary Conferences. The work reported here forms part of her D.Phil. thesis which is to be submitted for examination at Oxford University in 1991.

References

[1] Byrne, H. and Norbury, J., *Catalytic Converters and Porous Medium Combustion*, Proceedings of the Fifth International Colloquia on Free Boundary Problems, Montreal (1990), to appear.

[2] Drazin, P.G. and Reid, W.H., *Hydrodynamic Stability*, Cambridge University Press, 1981.

[3] Norbury, J. and Stuart, A.M., *A Model for Porous-Medium Combustion*, Quart. J. Mech. and Appl. Math., 42 (1989), pp. 159-178.

[4] Wade, W.R., White, J.E. and Florek, J.J., *Diesel Particulate Trap Regeneration Techniques*, SAE Technical Paper Series, Number 810118, 1981.

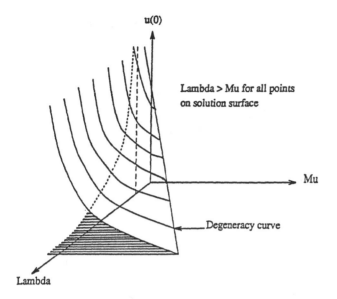

Figure 1: Steady-State Solution Surface; $f(u,g) = u$: the upper, stable G-solution surface meets the lower, unstable U-solution surface on the degeneracy curve.

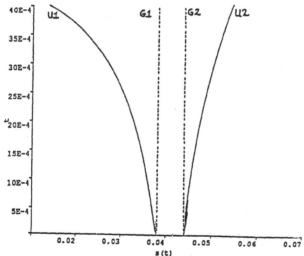

Figure 2: Evolution of Free Boundary; $\lambda = 25, \mu = 12$: the dashed curves G1(t) and G2(t) converge towards the steady state, whilst the continuous curves U1(t) and U2(t) diverge away from the steady state.

Oxford University Computing Laboratory,
8-11 Keble Road,
Oxford, England, OX1 3QD.

UNCERTAINTY QUANTIFICATION OF STRUCTURAL RESPONSE BY HIGHER MOMENTS ANALYSIS

C.Carino

Abstract
A maximum-entropy model for describing non-gaussian probability distributions is considered in this paper. Based on moments information, the technique is investigated in order to establish uncertainty associated with response predictions in structural systems with imperfect knowledge.

1. INTRODUCTION

Uncertainties associated with loads acting upon a mechanical system and/or with material and geometrical properties introduce random variability in response characteristics, such as internal actions, stresses or displacements. For example, the influence of uncertainty in aerodynamic loading and structural characteristics of wind-excited systems, due to lack of data or control, is propagated in accordance with a functional relationship (which is known in particular situations [1]) that relates these random variables to the structural response . In other circumstances, such as reliability analyses of complex offshore platforms, modeling the system behavior in terms of all sea-state, soil and structural parameters is required, as mathematical models are not available: in these cases response surface techniques may be employed as methods capable of fitting the input-output relationship [2], before uncertainty propagation can be carried out. .

In this study, the random characterization of structural response is accomplished through the analysis of the aforementioned function of random input variables, assumed it is known: once the relevant approximate moments are evaluated with minimal computational effort, the probability density function (PDF) is assessed and a comparison with simulated results is performed.

2. QUADRATIC APPROXIMATION

By assuming to operate in the uncorrelated standard gaussian space ($\underline{X}$) (otherwise suitable transformations could be adopted [3][4]), let $R(\underline{X})$ be the input-output relationship which defines the structural response as a function of n random physical parameters X_i. In general, $R(\underline{X})$ is nonlinear, hence its distribution can be only roughly estimated as gaussian, when linearization techniques are adopted. Differently, a quadratic approxi-

181

M. Heiliö (ed.), Proceedings of the Fifth European Conference on Mathematics in Industry, 181–184.

mation to the response function leads to knowledge of its higher moments, required for more accurate descriptions.

Assume that $R(\underline{X})$ is defined on an interval which contains the point $P(\overline{E}[\underline{X}]) \equiv P(0,\ldots,0)$, where $E[\underline{X}] = \{E[X_1],\ldots,E[X_n]\}$ is the vector given by the zero-mean values of the standardized variables X_i. Further, suppose that R is twice differentiable at P. By a second-order Taylor's expansion of $R(\underline{X})$ about P and through appropriate matrix transformations [5][6] $R(\underline{X})$ turns into the form

$$R(\underline{X}) = Q(\underline{X})/2 - K \tag{1}$$

in which Q represents a standard quadratic form in unit normal independent variables [7] and K is a constant depending on the problem being analyzed. In particular, $Q(\underline{X})$ is expressed by

$$Q(\underline{X}) = \Sigma_1^n \, \delta_i \, (X_i - d_i)^2 \tag{2}$$

where the notations used are successively described in the following:

- δ_i the eigenvalues of the matrix $\underline{G}$, containing the second and mixed derivatives of R in P
- $\underline{d} = -\underline{\Omega}^{-1}\underline{l}$.
- $\underline{\Omega}$ the diagonal matrix of the eigenvalues of $\underline{G}$
- $\underline{l} = \underline{T}^{-1}\underline{h}$
- $\underline{T}$ the modal matrix of $\underline{G}$
- $\underline{h}$ the vector of first-order derivatives of R in P.

By considering $R(X)$ as a single stochastic variable itself, and by introducing the corresponding standardized variate R' (having $E[R']=0$ and $Var(R')=1$), ordinary moments of R' can be obtained through the analysis of the standard quadratic form $Q(\underline{X})$, as shown in previous works [4][6] and resumed in Appendix I.

3. HIGHER MOMENTS ANALYSIS

A well known method for approximating the distribution of a random variable V, when its first ordinary moments are prescribed, is based on the maximum-entropy principle (MEP) [8]. The technique enables one to obtain the relevant PDF in the form

$$f_v(v) = \exp(-\sum_{i=0}^{N} \beta_i \, v^i) \tag{3}$$

where coefficients β_i have to be computed in such a way to satisfy moment constraints and N is chosen so as to obtain the desired accuracy in describing $f_v(v)$. Rosenblueth [9] has recently derived the following recursive relationship between model parameters and assigned moments:

$$m \, E[V^{m-1}] - \sum_{j=1}^{N} j \, \beta_j \, E[V^{m+j-1}] = 0 \tag{4}$$

where m is any integer greater or equal to 1. By varying m,

Eq.(4) can be transformed into a system of linear equations in the unknown parameters β_j (j=1,...,N).

With reference to the case N=4, for example, it is possible to write 4 simultaneous linear equations by changing m from 1 to 4 and hence evaluate $\beta_1,...,\beta_4$, if 7 moments of V are known. After which, β_0 can be computed as a function of the coefficients β_j (j=1,...,4) through the following relationship

$$\beta_0 = \log_e \left[\int_a^b \exp\left(- \sum_{i=1}^{4} \beta_i v^i \right) dv \right] \tag{5}$$

obtained by imposing that the area under $f_v(v)$, which is assumed to be defined on [a,b], is equal to unity.

Closed-form expressions for β_i (i=1,..,4) are reported in [10], as derived from solution of Eqs.(4).

4. APPLICATIONS - CONCLUDING REMARKS

Applicability of model (3) is being investigated in parallel works for dynamic response analyses. For the sake of brevity, the elementary example illustrated in Fig.1 is considered in this paper. The random characterization of beam vertical displacement R in point M has been carried out by assessing the PDF $f_{R'}(r')$ of R' (see Fig.2). Load L, inertia J and modulus of elasticity E have been regarded as random gaussian variables, with characteristic parameters also reported in Fig.1. Particular attention has been devoted to the choice of a and b in Eq.(5), as some conditions discussed in detail in [10] should be fulfilled. Results exhibit good agreement with the numerically-simulated response, while the comparison with the standard normal density $\phi(r')$ emphasizes the error that one would have by considering linearization techniques in performing exceedance probabilities calculations.

ACKNOWLEDGEMENT - Support by the National Research Council is gratefully acknowledged.

APPENDIX I - MOMENTS EVALUATION - By properties of expectation algebra, it is seen from eq.(1) that $E[R'^k] = E[Q'^k]$ (k=1,2,3,4,..), where Q' indicates standardization for Q. Therefore, the problem of evaluating moments of R' can be tackled by calculus of $E[Q'^k]$. For this aim, define the quantity $A=Q/[Var(Q)]^{1/2}$. Again for properties of expectation, one has

$$E[Q'^k] = \mu_{k,A} = \sum_{j=0}^{k} (-1)^j \binom{k}{j} m_{k-j,A} \ m_{1,A}^j \tag{6}$$

where notations μ and m indicate central and ordinary moment, respectively. With reference to [11] and [7] for a quadratic form of the kind (2), Eq.(6) can be solved by successively considering the following relations:

$$m_{s,A} = m_{s,Q} \left\{ \sum_1^n 2\delta_i^2(2d_i^2 + 1) \right\}^{-s/2} \quad (s=1,...,k) \tag{7}$$

$m_{o,A}=1$

$$m_{s,Q} = \sum_{j=1}^{s} \binom{s-1}{j-1} 2^{j-1} (j-1)!\; m_{s-j,Q} \left\{ \sum_{i=1}^{n} \delta_i^j (jd_i^2 + 1) \right\}$$ (8)

with $m_{o,Q}=1$ as the starting value for recursion in Eq.(8).

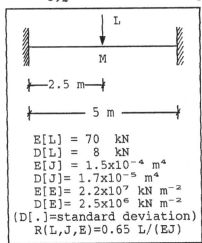

E[L] = 70 kN
D[L] = 8 kN
E[J] = 1.5x10⁻⁴ m⁴
D[J]= 1.7x10⁻⁵ m⁴
E[E]= 2.2x10⁷ kN m⁻²
D[E]= 2.5x10⁶ kN m⁻²
(D[.]=standard deviation)
R(L,J,E)=0.65 L/(EJ)

Fig.1 - The example considered

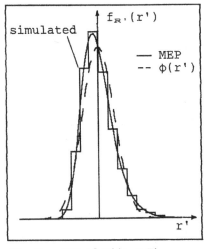

Fig.2 - PDF of R'(L,J,E)

REFERENCES
[1] Kareem A.,"Aerodynamic Response of Structures with Parametric Uncertainties", Structural Safety, 5, 1988
[2] Chiesa G.,Van Dick J., Zuccarelli F.,"Response Surface Fitting in Offshore Structural Design", Proc.ICOSSAR'89, San Francisco, 1989
[3] Madsen H.O.,Krenk S.,Lind N.C.,"Methods of Structural Safety", Prent.Hall, 1986
[4] Carino C.,"An Application of Asymptotic Polynomial Expressions in the Reliability Analysis of Structural Components", Proc.Symp. on Reliability based Design, Lausanne, 1988
[5] Fiessler B.,Neumann H.J.,Rackwitz R.,"Quadratic Limit States in Structural Reliability", J.Eng.Mech.Div.,ASCE, Vol.105, No.4,1979
[6] Carino C.,Carli F.,"Higher Moments Descriptions of Uncertain Quantities in the Analysis of Structural Components",Proc. ICOSSAR'89, Intl.Conf.on Struct.Safety and Reliability, San Francisco, 1989
[7] Johnson N.L.,Kotz S.,"Distributions in Statistics",Wiley & S.,1976
[8] Jaynes E.,"Probab.Theory in Science and Engrg.",McGraw-Hill,1961
[9] Rosenblueth E.,Karmeshu,Hong H.P.,"Maximum Entropy and Discretization of Probability Distributions",Prob.Engrg.Mechanics,2,1987
[10] Carino C.,"Uncertainty Analysis of Random Structural Behaviour by Maximum-entropy Model", Proc.Intl.Symp. on Uncertainty Modeling and Analysis, College Park, MD, U.S.A., 1990 (to be published)
[11] Kendall M.G.,Stuart A.,"Adv. Theory of Statistics",C.Griffin, 1979

Claudio Carino - Dept. of Structural Mechanics - University of Pavia
Via Abbiategrasso 211 - 27100 PAVIA - Italy

RELIABILITY OF TIME–SERIES FORECASTS

D. ČEPAR, B. MOTNIKAR

ABSTRACT. The Box–Jenkins method allows us to compute time series forecasts and confidence intervals. Confidence level $1-\epsilon$ is the probability that a single future value will be between the computed confidence limits. However, a single confidence level is not the probability that two, three or more future values will be inside the corresponding $1-\epsilon$ confidence intervals. Previously, we have developed an algorithm for calculating this probability for stationary models. The confidence region was restricted to a set of confidence intervals, all of them with the confidence level $1-\epsilon$. The algorithm is described in Bohte, Čepar, Motnikar (1989).

The subject of the present paper are two generalizations which are of particular importance in practice; (i) the algorithm has been generalized to a time–series with trend and (ii), the confidence region has been generalized to an arbitrary set of confidence intervals. Lower and upper limits of the intervals can now be determined arbitrarily by the user. A program for computing this probability is available on an IBM/PC.

1. Introduction and motivation

At the beginning we introduce some basic elements of the Box–Jenkins time–series forecasting method and the main results from Bohte, Čepar, Motnikar (1989). For a stationary ARMA model, each time–series value z_t can be expressed as an infinite linear combination of residuals a_i: $z_t = \sum_{j=0}^{\infty} \psi_j a_{t-j}$, where $\psi_0 = 1$ and coefficients ψ_i are model parameters (a method for calculating the parameters is described in Box, Jenkins (1970)).

The minimum mean square error forecast of the value z_{t+l} is $\hat{z}_t(l) = \psi_l a_t + \psi_{l+1} a_{t-1} + \ldots$. We will analyse time series forecasts for different lead times from the same fixed origin t, where time series values $z_1, z_2, \ldots, z_t$ and therefore also $a_1, a_2, \ldots, a_t$ are considered to be known constants.

By $z_{t+l}(\pm) = \hat{z}_t(l) \pm u_{\epsilon/2}\sigma_a(1 + \psi_1^2 + \ldots + \psi_{l-1}^2)^{1/2}$ we denote $1-\epsilon$ confidence limits for z_{t+l}, where $u_{\epsilon/2}$ satisfies the equality $\Phi(u_{\epsilon/2}) = (1-\epsilon)/2$. This means $P[z_{t+l}(-) \leq z_{t+l} \leq z_{t+l}(+)] = 1-\epsilon$. The probability $P_n(\epsilon)$ that all future values z_{t+l}, $l = 1, \ldots, n$ will be inside the corresponding $1-\epsilon$ confidence intervals is $P_n(\epsilon) = P[z_{t+1}(-) \leq z_{t+1} \leq z_{t+1}(+), \ldots, z_{t+n}(-) \leq z_{t+n} \leq z_{t+n}(+)]$.

It was shown that $P_n(\epsilon)$ depends on parameters $\psi_1, \psi_2, \ldots, \psi_{n-1}$ and can be expressed by

$$\frac{1}{(2\pi)^{\frac{n}{2}}} \int_{-u_{\epsilon/2}}^{u_{\epsilon/2}} e^{-v_1^2/2} dv_1 \int_{-\psi_1 v_1 - u_{\epsilon/2}\sqrt{1+\psi_1^2}}^{-\psi_1 v_1 + u_{\epsilon/2}\sqrt{1+\psi_1^2}} e^{-v_2^2/2} dv_2 \ldots \int_{-\psi_1 . v_{n-1} - \ldots - \psi_{n-1} v_1 - u_{\epsilon/2}\sqrt{1+\psi_1^2+\ldots+\psi_{n-1}^2}}^{-\psi_1 v_{n-1} - \ldots - \psi_{n-1} v_1 + u_{\epsilon/2}\sqrt{1+\psi_1^2+\ldots+\psi_{n-1}^2}} e^{-v_n^2/2} dv_n.$$

$$(1)$$

We developed an algorithm for numerical integration of this n–fold integral. The probability $P_n(\epsilon)$ has been derived and computed for stationary time series models. The confidence region was restricted to a set of confidence intervals, all of them with the confidence level $1-\epsilon$. In the following we generalize the algorithm so that $P_n(\epsilon)$ can also

185

M. Heiliö (ed.), *Proceedings of the Fifth European Conference on Mathematics in Industry*, 185–188.
© 1991 *B. G. Teubner Stuttgart and Kluwer Academic Publishers. Printed in the Netherlands.*

be computed for (seasonal) differenced time series and to have the possibility of specifying arbitrary confidence region.

2. Generalization for nonstationary time series

Some time series with trend (z_t) will become stationary (w_t) by nonseasonal differencing: $w_t = z_t - z_{t-1}$, $\forall t = 2, 3, \ldots$. A model and the forecasts are developed for stationary (differenced) time series: $w_t = \sum_{j=0}^{\infty} \psi_j a_{t-j}$. Future values and their forecasts for original time series can be expressed by inverse differencing:

$$z_{t+i} = w_{t+i} + z_{t+i-1} = w_{t+i} + w_{t+i-1} + \ldots + w_{t+1} + z_t$$

and

$$\hat{z}_t(i) = \hat{w}_t(i) + \hat{z}_t(i-1) = \hat{w}_t(i) + \hat{w}_t(i-1) + \ldots + \hat{w}_t(1) + z_t .$$

Forecast error can be written as a linear combination of residuals:

$$z_{t+i} - \hat{z}_t(i) = a_{t+i} + (1 + \psi_1)a_{t+i-1} + \ldots + (1 + \psi_1 + \ldots + \psi_{i-1})a_{t+1} .$$

From this expression parameters ψ_i' of the original model (belonging to the time series z_t) can be read:

$$\psi_j' = 1 + \psi_1 + \ldots + \psi_j = \psi_{j-1}' + \psi_j \quad \text{for} \quad j = 1, 2, \ldots, i-1 , \qquad \text{where} \quad \psi_0' = \psi_0 = 1 .$$

In the algorithm for computing the probability $P_n(\epsilon)$ the same formula (1) is used as in the stationary models, but the parameters ψ_i must be replaced by the parameters ψ_i'.

Another type of nonstationarity arises from seasonality. Some seasonal time series (z_t) with length of season s will become stationary (w_t) by seasonal differencing: $w_t = z_t - z_{t-s}$, $\forall t = s + 1, s + 2, \ldots$.
The same procedure as in the former case gives us the parameters ψ_i'' (belonging to the original time series) expressed with the parameters ψ_i (belonging to the seasonal differenced time series):

$$\psi_j'' = \psi_j \quad \text{and} \quad \psi_{ks+j}'' = \psi_{(k-1)s+j}'' + \psi_{ks+j} ; \quad j = 1, 2, \ldots, s-1 ; \quad k = 1, 2, \ldots .$$

When computing probability $P_n(\epsilon)$ the parameters ψ_i must be replaced by the parameters ψ_i''. It is possible (and sometimes necessary) to difference and/or seasonal difference time series more than once. For each differencing we just have to recalculate the parameters ψ_i and then use the formula (1).

3. Generalization of the confidence region

Often we want to know the probability that a set of future values will be in a confidence region, specified by upper and lower limits of intervals for single forecasts. In the generalized version of the algorithm these limits are arbitrary. We denote them by ul_i (upper limits) and ll_i (lower limits). The probability is denoted by $P_n(limits)$ to emphasise the dependance on $ll_1, ll_2, \ldots, ll_n$ and $ul_1, ul_2, \ldots, ul_n$.

The procedure for developing an expression for $P_n(limits)$ which is suitable for numerical integration is the same as for probability $P_n(\epsilon)$. So let us consider the final result:

$$P_n(limits) = \frac{1}{(2\pi)^{\frac{n}{2}}} \int_{\frac{ll_1 - \hat{z}_t(1)}{\sigma_a}}^{\frac{ul_1 - \hat{z}_t(1)}{\sigma_a}} e^{-v_1^2/2} dv_1 \int_{\frac{ll_2 - \hat{z}_t(2)}{\sigma_a} - v_1 \psi_1}^{\frac{ul_2 - \hat{z}_t(2)}{\sigma_a} - v_1 \psi_1} e^{-v_2^2/2} dv_2 \ldots \int_{\frac{ll_n - \hat{z}_t(n)}{\sigma_a} - \sum_{j=1}^{n-1} v_j \psi_{n-j}}^{\frac{ul_n - \hat{z}_t(n)}{\sigma_a} - \sum_{j=1}^{n-1} v_j \psi_{n-j}} e^{-v_n^2/2} dv_n.$$

Probability $P_n(limits)$ depends also on parameters ψ_i, forecasts $\hat{z}_t(i)$, correlations between forecasts and the standard deviation of residuals σ_a.

In practice the true model parameters $\psi_1, \psi_2, \ldots$ and σ_a are not always known. In the case they are not known we use their estimates (as suggested by Box and Jenkins (1970) in the case of estimation of forecast error variance). So the results are not the true values of $P_n(\epsilon)$ and $P_n(limits)$ but their estimates.

A program using the generalized algorithm for computing probability $P_n(limits)$ and $P_n(\epsilon)$ and is available on an IBM/PC.

4. An example

Let us consider time series z_t of length 272 representing the electrical energy load. Data were collected for each hour from 6 a.m. to 22 p.m. from the 1st to the 26th of January.

Time series z_t has a nonstationary model, but time series $w_t = z_t - z_{t-17}$ (length of the season $s = 17$ represents the number of hours) is stationary. The identified model for w_t is ARMA(1,1) with $\psi_1 = 0.50$ and $\psi_2 = 0.38$. By the Box–Jenkins method forecasts and confidence limits can be computed. Probabilities calculated by our algorithm at a confidence level of 0.90 are $P_1(0.10) = 0.90$, $P_2(0.10) = 0.82$ and $P_3(0.10) = 0.76$. Futhermore, we can answer the following questions:

1. What is the joint probability that the loads at 6, 7 and 8 a.m. on the 27th of January will differ from the respective loads on the 26th (denoted by $z_{26.1.,6}$, $z_{26.1.,7}$ and $z_{26.1.,8}$) for at most 5%?

   As input data we must determine the following lower and upper limits of intervals using the known values:

   $ll_1 = z_{26.1.,6} * 0.95 = 952.85$, $ul_1 = z_{26.1.,6} * 1.05 = 1053.15$,

   $ll_2 = z_{26.1.,7} * 0.95 = 1157.10$, $ul_2 = z_{26.1.,7} * 1.05 = 1278.90$,

   $ll_3 = z_{26.1.,8} * 0.95 = 1245.45$, $ul_3 = z_{26.1.,8} * 1.05 = 1376.55$.

   Calculated probability $P_3(limits)$ is 0.80.

2. What is the joint probability that the loads at 6, 7 and 8 a.m. on the 27th of January will be less than 1300 kW? All lower limits are 0 and all upper limits are 1300. Calculated probability $P_3(limits)$ is 0.40.

Conclusions

In the business world many questions about the probability of future events arise every day and encourage research towards making improvements in statistical forecasting methods. In the present paper we analysed the reliability of forecasts obtained by the Box–Jenkins method. Using the proposed method we are able to answer such questions as:

- What is the probability that hotel rooms will be at least 75% occupied in the next three months?

- What is the probability that by the end of the year we are able to sell at least 2000 pieces of our product monthly (so the warehouse is not needed) but that customer demand will not exceed 3000 pieces monthly (so there is no need to employ extra workers)?

- In which limits will the exchange rate of DEM be next week at 5% risk?

The probability that a single future value will be between the upper and lower confidence limit at a certain moment is given by the confidence level $1 - \epsilon$. Of course, the probability of the product of two or more events is smaller than the probability of a single event. So someone can make too optimistic conclusions in interpreting confidence level $1 - \epsilon$ as probability $P_n(\epsilon)$ when $n > 1$. Especially in long–term decision making, great mistakes can be made. Using the results of this paper we can calculate probability $P_n(\epsilon)$ of avoiding the mistakes described above.

We would like to point out new possibilities given by the improvement in the forecasting method. Using the Box–Jenkins method we can choose the **probability** (confidence level $1 - \epsilon$) and then calculate the corresponding confidence **region** (lower and upper limits of $1 - \epsilon$ confidence intervals). By our algorithm we can choose an arbitrary **region** (lower and upper limits of intervals) and then calculate the corresponding **probability** ($P_n(limits)$).

Acknowledgment

We would like to thank Prof. Peter Žunko for his helpful comments and Peter Pehani and Boštjan Vovk for their generous support of coding the computer program.

References

G. E. P. Box, G. M. Jenkins (1970), Time Series Analysis Forecasting and Control, Holden-Day, San Francisco.

Z. Bohte, D. Čepar, B. Motnikar (1989): Confidence level for the region of single time series forecasts with different lead times; Proc. 14. Symposium on Operations Research, Ulm, Methods of operation research 62, p. 503-509.

Drago Čepar, Barbara Motnikar, "Jožef Stefan" Institute,
Jamova 39, 61111 Ljubljana, Slovenia, Yugoslavia.

ARMA MODELS OF TIME SERIES WITH MISSING DATA

D. ČEPAR, Z. RADALJ

ABSTRACT. A method for estimating ARMA models of time series with missing observations is considered. The minimum variance estimator of missing observations, based on ARMA models, is used in an iterative algorithm for identification and estimation of parsimonious ARMA models. Performance of the method is illustrated by a case study.

1. Introduction

In the analysis of time series one often has a situation where some observations are defective or even missing. Due to human or technical errors some observations are not appropriate and cannot be used in analysis. In this paper we are concerned with the minimum variance estimation of the missing observations and with ARMA model identification and parameter estimation for time series with missing observations. The method is described and its performance is illustrated by a case study.

2. Method Description

We are given a time series z_i of length N for which we want to identify and estimate a parsimonious ARMA model. For the sake of simplicity we discuss the case of a non-seasonal stationary model with the constant equal to zero:

$$\Phi(B)z_i = \Theta(B)a_i , \qquad (1)$$

where B is the backward shift operator defined by $Bz_i = z_{i-1}$, and the operators $\Phi(B)$ and $\Theta(B)$ are polynomials of B with the unknown parameters: $\Phi(B) = 1 - \phi_1 B - \ldots - \phi_p B^p$ and $\Theta(B) = 1 - \theta_1 B - \ldots - \theta_q B^q$; the a_i are a sequence of independent random variables distributed according to $N(0, \sigma_a^2)$, and z_i ; $i = 1, \ldots, N$, are time series values. We assume that some values z_i, given by (1), $z_{t+1}, z_{t+2}, \ldots, z_{t+m-1}$ $(t + m < N)$ are missing.

We shall identify and estimate the model, given by (1), with an iterative procedure which consists of four steps: (i) initialization, (ii) identification and estimation of the parsimonious ARMA model, (iii) estimation of the missing observations and (iv) testing the convergence. Step (i) is performed once and afterwards steps (ii), (iii) and (iv) are performed at each iteration. The iteration is stopped when one of the convergence criteria is satisfied.

The procedure allows an estimation of time series, with missing values distributed in an arbitrary way accross the time series, and not only as $m - 1$ successive missing values.

2.1. Initialization. To start the iterative procedure we must have the complete time series. We calculate the average value $\bar{z}$ of all existing observations ($\bar{z} = \frac{1}{t} \sum_{i=1}^{t} z_i + \frac{1}{N-(t+m-1)} \sum_{i=t+m-1}^{N} z_i$) and set all missing values to $\bar{z}$: $z_j = \bar{z}$, $j = t+1, \ldots, t+m-1$.

M. Heiliö (ed.), Proceedings of the Fifth European Conference on Mathematics in Industry, 189–192.
© 1991 *B. G. Teubner Stuttgart and Kluwer Academic Publishers. Printed in the Netherlands.*

2.2. Identification and estimation of the parsimonious ARMA model. In each $K - th$ iteration we use the method for estimation of the ARMAG models introduced by Čepar and Dekleva (1986) to identify a parsimonious ARMAG (p,q) model and estimate parameters $\phi_i^{(K)}$, $i = 1, \ldots, p$, and $\theta_j^{(K)}$, $j = 1, \ldots, q$.

2.3. Estimation of the missing observations. With the estimated ARMA model parameters we construct an estimate $\tilde{z}_{t+l}$ of the missing value z_{t+l}.

$$\tilde{z}_{t+l} = x_0(l)\hat{z}_t(l) + y_0(l)\check{z}_{t+m}(k) \qquad ; l = 1, 2, \ldots, m - 1 ,$$

where $l + k = m$, and $\hat{z}_t(l)$ is the "forward" forecast $(\hat{z}_t(l) = \psi_l a_t + \psi_{l+1}a_{t-1} + \cdots$, $l = 1, 2, \ldots, m-1)$ and $\check{z}_{t+m}(k)$ is the "backward" forecast $(\check{z}_{t+m}(k) = \psi_k e_{t+m} + \psi_{k+1}e_{t+m+1} + \cdots$, $k = 1, 2, \ldots, m-1)$. The values of ψ_j can be obtained from the equation $\Phi(B)\Psi(B) = \Theta(B)$, where $\Psi(B) = 1 + \psi_1 B + \ldots$ (see Box and Jenkins (1970), p.134) and the e_i are a sequence of independent random variables with the same distribution as a_i. This procedure is valid when the zeros of the polynomials $\Phi(B)$ and $\Theta(B)$ lie outside the unit circle.

Here we list only the basic formulas used in the procedure. All details and mathematical background is described in Bole, Čepar and Radalj (1989). The weights $x_0(l)$ and $y_0(l)$ are determined, so as to minimize the variance of the error $z_{t+l} - \tilde{z}_{t+l}$:

$$x_0(l) = \frac{e(l)(d(l) - c(l))}{d^2(l) - e(l)c(l)}, \qquad y_0(l) = \frac{c(l)(d(l) - e(l))}{d^2(l) - e(l)c(l)}, \qquad (2)$$

where

$$c(l) = \sum_{i=l}^{\infty} \psi_i^2, \qquad e(l) = \sum_{i=k}^{\infty} \psi_i^2,$$

$$d(l) = \sum_{i=l}^{\infty} \psi_i^2 - \sum_{i=0}^{k-1} \psi_i^2 + \frac{1}{\sigma_a^2}K(\sum_{i=0}^{l-1} \psi_i a_{t+l-i}, \sum_{i=0}^{k-1} \psi_i e_{t+l+i}),$$

for $l = 1, 2, \ldots, m - 1$. The covariance between forward and backward forecast errors is

$$K(\sum_{i=0}^{l-1} \psi_i a_{t+l-i}, \sum_{i=0}^{k-1} \psi_i e_{t+l+i}) = \sigma_a^2 \sum_{j=0}^{k-1}\sum_{i=0}^{l-1} \psi_j \psi_i \psi_{i+j}^* ,$$

where

$$\psi*_n = \sum_{i=0}^{\infty} \pi_i \psi_{n+i},$$

where $n = -\infty, \ldots, \infty$ and $\psi_i = 0$ for $i < 0$. The values of π_i can be obtained from the equation $\Pi(B)\Theta(B) = \Phi(B)$, where $\Pi(B) = 1 + \pi_1 B + \pi_2 B^2 + \ldots$.

Calculation of (2) is possible when $d^2(l) - e(l)c(l) \neq 0$. When $d^2(l) - e(l)c(l) = 0$, we have the following possibilities:

(i) $c \neq 0$ and $e = 0 \Longrightarrow x = 1$, $y = 0$;   (ii) $c = 0$ and $e \neq 0 \Longrightarrow x = 0$, $y = 1$;

(iii) $c \neq 0$ and $e \neq 0$ or $c = 0$ and $e = 0 \Longrightarrow x = y = 0.5$.

2.4. Testing the convergence. At each step of the iterative procedure we obtain new estimates of the model parameters and new values of missing observations. To obtain a suitable criterium for stopping the iterative procedure we define the distance between two subsequent iterations in three different ways:

$$d_1(K-1, K) = \sqrt{\frac{1}{m-1} \sum_{l=1}^{m-1} (z_{t+l}^{(K-1)} - z_{t+l}^{(K)})^2}$$

$$d_2(K-1, K) = \max_{l=1,\ldots,m-1} | z_{t+l}^{(K-1)} - z_{t+l}^{(K)} |$$

$$d_3(K-1, K) = \sqrt{\frac{1}{p+q} [\sum_{i=1}^{p} (\phi_i^{(K-1)} - \phi_i^{(K)})^2 + \sum_{i=1}^{q} (\theta_i^{(K-1)} - \theta_i^{(K)})^2]}$$

The proposed iterative procedure is repeated until at least one of the defined distances d_i falls under the corresponding value ε_i chosen in advance. We stop the procedure also when we exceed the maximal prescribed number of iterations.

3. Illustration of the method performance

We obtain an idea on the performance of the method as follows. We analyze a given time series of a given length. We identify and estimate the parameters of the ARMA model for this time series. After that we omit some observations to obtain time series with missing data. Then we analyze the time series with missing data. We identify a new ARMA model and estimate ARMA model parameters for this series using the algorithm described in this paper. The smaller the differences between the estimates obtained on the complete time series and the estimates obtained on the time series with missing data, the more consistent we consider the proposed method.

We report the results for two time series with 200 observations (obtained by simulation). The values of the estimated model parameters on complete time series are:

TS 1: $z_t - 0.4139 z_{t-1} = a_t$, TS 2: $z_t = a_t - 0.5788 a_{t-1}$.

     (± 0.0643)   (± 0.0565)

For both time series we omit some observations to obtain time series with the missing observations. We omit 1,5,10,and 20 observations "in the middle" and at both "ends" of the time series. So we obtain 12 time series with missing data from each original time series. For all time series the searching procedure has reproduced the original ARMA model as the optimal model. In Table 1, the values of the estimated parameters of time series with missing observations and the differences from estimates obtained on complete time series are given. From Table 1 we can see that estimates obtained on time series with missing data are contained in 66 % confidence intervals of estimates on the complete time series. The differences vary for AR(1) from -0.0227 to 0.0174, and for MA(1) from -0.0382 to 0.0079.

4. Conclusion

A method for estimating ARMA models of time series with missing observations is considered. The minimum variance estimator of the missing observations, based on the ARMA models, is used in an iterative algorithm for the identification and estimation of the parsimonious ARMA model. Performance of the method is illustrated on two cases. For both cases the ARMA model of the original time series is reproduced by our procedure on time series with missing data. The results of our performance test suggest that the parameters estimates obtained on time series with missing data are stable: the difference between

Number of missing observations	Position of missing observations	TS 1 AR (1) ϕ_1	$diff$	TS 2 MA (1) θ_1	$diff$
1	6	0.4224	-0.0031	0.5805	-0.0017
5	6 - 10	0.4019	0.0174	0.5745	0.0043
10	6 - 15	0.4047	0.0146	0.5744	0.0044
20	6 - 25	0.4131	0.0062	0.5709	0.0079
1	100	0.4197	-0.0004	0.5834	-0.0046
5	98 - 102	0.4279	-0.0086	0.5936	-0.0148
10	96 - 105	0.4173	0.0020	0.5954	-0.0166
20	91 - 110	0.4193	0.0000	0.5840	-0.0052
1	195	0.4193	0.0000	0.5787	0.0001
5	191 - 195	0.4208	-0.0015	0.5725	0.0063
10	186 - 195	0.4420	-0.0227	0.5933	-0.0145
20	176 - 195	0.4274	-0.0081	0.6170	-0.0382

Table 1: Parameter estimates (ϕ_1 and θ_1) of time series with missing values and their differences from estimates on the complete time series.

these estimates and the estimates obtained on the complete time series vary only in the second decimal place. The values of estimated parameters on time series with missing data are in the limits of the standard deviation for the parameter estimates obtained on complete time series.

Our method is consistent even for the analysis of a time series with 10% observations missing, what is considered as a large number of missing data according to examples reported by other authors, for example Hunt and Triggs (1989), who consider 9 missing observations of time series of length 80 as a large number of missing observations. The computer program based on our method is a useful tool for analysts working with time series with missing data. By entering the existing time series observations into the computer, ARMA model parameters and missing values estimates are obtained. The users need no extra knowledge to run this program. All these features advocate the use of the program in practice.

Acknowledgment
We thank Boštjan Vovk for developing an interactive computer program in Pascal on IBM PC.

References
Box G.E.P., Jenkins G.M. (1970), Time Series Analysis Forecasting and Control, Holden–Day, San Francisco.

Bole V.,Čepar D. and Radalj Z. (1989), Estimating missing values in time series, Methods of Operations Research Vol. 62, p.511 - 516, A. Hain Verlag 1990.

Čepar D. and Dekleva J. (1986), On the Generalized Univariate ARMA Models and their Identification, Methods of Operations Research, 54, 41 - 50, A. Hain Verlag 1986.

Hunt D.N. and Triggs C.M. (1989), Iterative Missing Value Estimation, Appl. Statist., 38, No. 2, 293 - 300, Royal Statistical Society 1989.

D. Čepar, Z. Radalj, "Jožef Stefan" Institute, Jamova 39, 61111 Ljubljana, Yugoslavia

NONLINEAR OSCILLATIONS OF MINIATURE SYNCHRONOUS MOTORS

W D COLLINS

ABSTRACT. Mathematical models for single and double stator synchronous motors are set up and possible resonant oscillations analysed.

A synchronous motor is a rotational motor used for the steady transfer of electrical to mechanical energy. Under ideal conditions it operates at a constant angular speed for a given supply frequency. The term synchronous refers to the relationship between the angular speed ω_m of the motor and the angular supply frequency, this being $\omega_m = \omega_s/p$ radians s^{-1} for p pole pairs. Miniature permanent magnet synchronous motors are manufactured in millions each year and widely used in both industrial and domestic appliances, for example, motor driven valves, switches, central heating timers, hair dryers, record players, lemon squeezers and electric carving knifes. An average size is 3cm and they operate at a variety of speeds, typically in the range 250 to 750 rpm. Despite their mass production there is still considerable scope for improving their design to make them more efficient. In operation they are prone to relatively large amplitude oscillations about their synchronous speeds which it is the aim of this work to model and analyse as part of a programme on these motors in the Department of Electronic and Electrical Engineering at the University of Sheffield. The essential components of a miniature synchronous motor are an outer stationary element, the stator, and an inner element, the rotor, mounted on bearings fixed to the stator. The stator comprises two interleaved crowns of steel pole fingers issuing from two end plates with a single coil positioned round the poles and connected to the supply. The rotor is a permanent magnet, a ferrite cylinder with a regular pattern of north and south poles magnetically imprinted around its circumference with as many poles on the rotor as on the stator. For this single stator motor the magnetic flux can be represented by two rotating fields with opposite senses of rotation, the torque due to the coil supply arising from the interaction of these rotating fields with the field of the magnet. When synchronized the magnet field follows one of the rotating fields, the other field however generating an alternating torque which may cause substantial oscillations on the mean speed at low values of damping and inertia.

When a motor is required to rotate unidirectionally but with the option of either direction, a second set of stator poles displaced spatially from the first set but joined to the same rotor can be used to establish a rotating field, giving a double stator motor. Even then the backward rotating field can never be completely suppressed and large oscillations can still result at synchronous speed.

To analyse these oscillations we model the motion of the rotor by treating it as a rigid cylinder and using $d - q$ (direct-quadrature) theory to obtain the electromagnetic torque on it. The $d - q$ hypothesis is that a rotating two-pole magnet can be replaced by two

M. Heiliö (ed.), Proceedings of the Fifth European Conference on Mathematics in Industry, 193–196.

identical coils positioned on perpendicular $d-$ and $q-$axes and producing the same field components along these axes as the magnet. The coils are thus fixed but have rotational voltages induced in them. When the rotor magnet has p pole pairs, the representation of the two-pole magnet is repeated p times. The $d-$axis is taken to coincide with the stator coil axis, so that the basic model for a single stator configuration consists of two coils on the $d-$axis, the magnet equivalent $d-$coil and the stator coil, and one coil on the $q-$axis, the magnet equivalent $q-$coil.

The double stator motor is constructed by joining together two identical stator structures with a spatial angular displacement of $\pi/2$, the total electromagnetic torque being the sum of the torques from the two halves. With θ the angular displacement of the rotor and i_{c1}, i_{c2}, the currents in the two stator coils the total torque is

$$T = -pMI_m i_{c1} \sin p\theta - pMI_m i_{c2} \cos p\theta,$$

where M is the mutual inductance between the magnet $d-$coil and the stator coil. The equation of motion of the rotor is then

$$J\ddot{\theta} = total\ torque,$$

where J is the moment of inertia of the rotor and its shaft. The total torque is made up of the electromagnetic torque, a torque due to viscous friction assumed proportional to $\dot{\theta}$, a friction torque T_f from the bearings assumed constant and an external load torque T_L applied at the shaft of the rotor arising, for instance, from the mechanical system being driven and taken as constant.

In the current fed case the stators are fed by alternating current supplies so that $i_{c1} = I_{c1} \cos\omega t, i_{c2} = I_{c2} \cos(\omega t + \gamma), \omega$ the supply frequency. In the voltage fed case the supply is an alternating voltage source and the equation of motion is now coupled with two circuit equations. Since we live in a voltage fed world, this latter is the practically important case. However, the current fed case being simpler to analyse and the results for the voltage fed case broadly comparable, only this case is considered here. The equation of motion can be expressed as (Yonnet, 1982)

$$\frac{d^2\theta}{dt^2} + D\frac{d\theta}{dt} + T_f + T_L = T^+ \cos(p\theta - \omega t) + T^- \cos(p\theta + \omega t),$$

where D is a damping coefficient and T^+, T^-, are peak torques due to the forward and backward components of flux given in terms of I_{c1}, I_{c2} and γ. For the double stator motor in forward running one component of flux is significantly less than the other, so the ratio $\mu = T^-/T^+$ is small. Setting $p\theta - \omega t = \pi/2 + \psi$, we obtain the equation for ψ as

$$\frac{d^2\psi}{dt^2} + \Omega^2 \sin\psi = -\mu\Omega^2 \sin(2\omega t + \psi) - 2b\Omega\frac{d\psi}{dt} - \lambda\Omega^2 - 2b\Omega\omega,$$

where $\Omega^2 = p\frac{T^+}{J}, 2b\Omega = \frac{D}{J}, \lambda = (T_L + T_f)/T^+$. We assume b small.

This equation is the same as that for a simple pendulum under constant torque whose point of support undergoes small horizontal and vertical oscillations. When the constant torque term $\lambda\Omega^2$ is small, the exact solution of the unperturbed equation is known in terms of elliptic functions and can be used as the basis of an averaging method. When as for the motor the torque is not small, the exact solution of the unperturbed problem is known only in terms of quadratures. However we can set $\psi = -\alpha + \chi$, where $\lambda = sin\alpha, \alpha$ the load angle, and assume χ is small, this being so in practice. The method of multiple scales can now be used to investigate the primary resonance and the second and third subharmonics, corresponding to $2\nu \simeq 1, 2, 3$, where $\nu = \frac{\omega}{\omega_o}, \omega_o^2 = \Omega^2 cos\alpha$. The leading term in the expansion of χ given by $\chi = \chi_o + \epsilon\chi_1 + ...$, is of the form $\chi_o = Rcos(2\nu t/n + \phi), n = 1, 2, 3$, where $\epsilon = \mu^{1/3}, \mu^{1/2}, \mu$, for $2\nu \simeq 1, 2, 3$, and R and ϕ are functions of the slow time variables. Steady states correspond to R and ϕ constant.

For the primary resonance either one stable steady state or three steady states, two stable and one unstable, are possible, all the states being non-zero. Hysterisis effects are possible for fixed α when 2ν varies through unity and also for fixed ν when the load angle α varies. For $2\nu \simeq 2$ either the zero steady state is stable or there are three steady states, two non-zero stable ones and the unstable zero state, or there are five steady states, the zero and two non-zero states being stable and two other non-zero states unstable. Useful information on the domains of attraction of these states and the variation of their amplitudes with the motor parameters can be obtained.

Experimentally all these resonances as well as higher ones arise. To investigate the fourth and fifth subharmonics for which $2\nu \simeq 4, 5$ and also the subharmonic with $2\nu \simeq 3/2$ simplified equations can first be considered and the analysis extended to the motor equations. For $2\nu \simeq 4$ the simplified equation is

$$\ddot{u} + u + 2\epsilon^2 b\dot{u} + \epsilon\alpha u^3 + \epsilon^2\beta u^4 = fcos[(4 + \epsilon\sigma)t],$$

ϵ the small parameter and σ a detuning parameter. We set

$$u = Rcos\left(\left(1 + \frac{1}{4}\epsilon\sigma\right)t + \phi\right) - \lambda cos[(4 + \epsilon\sigma)t], \lambda = \frac{f}{15},$$

where R and ϕ are slowly varying functions of t. The usual averaging procedure then gives equations for R and ϕ to which we apply a near identify transformation

$$R = R_o + \epsilon R_1 + \epsilon^2 R_2 + .., \phi = \phi_o + \epsilon\phi_1 + \epsilon^2\phi_2 + ...$$

with the aim of choosing R_1, R_2, ϕ_1, ϕ_2, so as to obtain equations for R_o and ϕ_o as simple as possible. Away from the resonance manifold, given by $R_o = \overline{R}_o$, where $\frac{3\alpha}{2}\left(\overline{R}_o^2 + 2\lambda^2\right) = \sigma$, we obtain outer expansion equations

$$\dot{R}_o = -\epsilon^2\mu R_o, \dot{\phi}_o = \frac{\epsilon}{4}\left(\frac{3\alpha}{2}(R_o^2 + 2\lambda^2) - \sigma\right),$$

whilst near the resonance manifold $R_o = \overline{R}_o + \epsilon^{1/2}P$ and ϕ_o satisfies the inner expansion equation

$$\frac{d^2\phi_o}{d\tau^2} + 2\epsilon^{\frac{1}{2}}\mu\frac{d\phi_o}{d\tau} + \frac{3\alpha\beta\lambda}{16}\overline{R}_o^4 sin4\phi_o = -3\alpha\mu\frac{\overline{R}_o^2}{4},$$

where $\tau = \epsilon^{\frac{3}{2}}t - \tau_o(\epsilon), \tau_o$ being determined in the subsequent matching. This is the equation of a damped simple pendulum under constant torque, part of whose phase plane along with that of the undamped case is shown in the case when $4\mu < |\beta\lambda| \overline{R}^2$. The effect of the damping is to open up the separatrices into the saddle, giving two branches, one

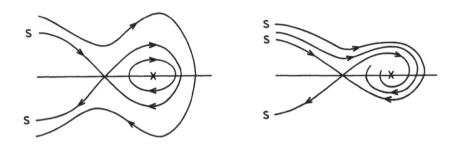

s = separatrix

straight to the saddle and the other to the saddle after a turn around the attractor. By matching the inner expansion solution for large negative τ with the outer expansion solution it can be shown that there are solutions whose trajectories lie between these separatrices and so go to the attractor, thus tending to a steady state subharmonic, whilst other solutions remain near the attractor for some time but then continue to zero. This indicates that, whilst subharmonics of order four can occur, they may not be too easily set up, this being borne out experimentally .

References

Yonnet, C (1982) 'Speed oscillations of miniature synchronous motors'. PhD thesis, University of Sheffield, England.

Department of Applied & Computational Mathematics,

University of Sheffield, Sheffield, S10 2TN, England, UK.

WAVELET METHOD FOR THE STEFAN PROBLEM

E. CORTINA *   S. GOMES **
* Servicio Naval de Investigación y Desarrollo - Avda. Libertador 327 - 1638
Vicente Lopez - Argentina
** Instituto de Pesquisas Espaciais - Caixa Postal 515 - 12201 Sao Jose dos
Campos SP - Brasil

I - Introduction

A wavelet based numerical scheme to solve the Stefan problem is proposed. In the enthalpy formulation of the problem, the space approximations are performed by expanding the solution in the Daubechies' bases of compactly supported scaling functions [1]. These scaling functions arise in the multiscale analysis for the construction of special orthonormal bases of $L^2(R)$, the wavelets. Wavelets (and scaling functions) are generated from a single function called basic wavelet (basic scaling function) by translations and dilations.

In representing functions using scaling functions and wavelets, good results are obtained both in space and frequency localisation. It seems likely that wavelet based schemes may combine the advantages of spectral and finite elements (differences) techniques thus providing powerful methods for solving partial differential equations with irregular solutions. Results in this direction are in [2] and [3]. A detailed discussion of the results presented here is in [4].

II - The one phase Stefan problem

We consider the enthalpy formulation of the problem:

$$\partial H/\partial t = \partial^2 u/\partial x^2$$

and use a regularised version:

$$\widetilde{H}'(u)\partial u/\partial t = \partial^2 u/\partial x^2$$

where

$$\widetilde{H}'(u) = \begin{cases} 1 & \text{if} \quad u < -\Delta u \\ 1 + (u + \Delta u)/(\Delta u)^2 & \text{if} \quad -\Delta u \leq u < 0 \\ 1 + (\Delta u - u) + (\Delta u)^2 & \text{if} \quad 0 \leq u \leq \Delta u \\ 1 & \text{if} \quad u > \Delta u \end{cases}$$

M. Heiliö (ed.), Proceedings of the Fifth European Conference on Mathematics in Industry, 197–200.

We assume that all the physical parameters are equal to one.
The fully implicit time discretisation scheme:

$$\widetilde{H}'[u^{n+1} - u^n] = \Delta t \, \partial^2 \, u^{n+1}/\partial x^2$$

is solved iteratively by expanding u^{n+1} in the orthonormal basis of the scaling functions
and their integar translates

$$u^{n+1}(x) = \sum_{k=k_{\min}}^{k_{\max}} c_k \, {}_N\Phi \left(2^{-m}x - k\right) \quad ; \quad m, k \in Z$$

where m is the dilation parameter, N the regularity parameter of the scaling functions,
$k_{\min} = 2 - 2N$ and $k_{\max} = 2^{-m} - 1$.

III - Numerical results

EXAMPLE 1

A problem of ice melting with Neumann boundary conditions in $[0,1]$ is considered. The
initial temperature is:

$$u_0(x) = \begin{cases} 0 & \text{if} \quad 0 \le x < 0.25 \\ 2x - 0.5 & \text{if} \quad 0.25 \le x < 0.75 \\ 1 & \text{if} \quad 0.75 \le x \le 1 \end{cases}$$

Figures 1 and 2 show the numerical solutions of the approximate Galerkin-wavelet formulation. We chose ${}_6\Phi$ as basic scaling function and $m = -7$ is the dilation rate, which
involves 137 unknowns. The regularisation parameter is $\Delta u = 0.05$ and the time step
$\Delta t = 0.005$. At $t = 0.15$ only a small part of ice is not completely melted, and at $t = 0.4$
the constant steady state is almost reached.

EXAMPLE 2

We checked the proposed method by comparing numerical results for the problem of ice
formation in stagnant water with the analytical solution of Lamé and Clapeyron. In our
calculations $u(0,t) = -2$ and we started from the analytical solution at $t = 0.005$. In
Figs. 3 and 4, the computed temperature distibution, for different values of the dilation
parameter m and the regularity parameter N of the scaling function, are compared with
the analytical solution. As it was expected, a better approximation is obtained by using a
more regular wavelet. The most important parameter is m. As m decreases finer approximations are achieved. High frequencies (as in the presence of discontinuities) can only be
captured with very small m. The influence of this parameter is illustrated in Fig. 5, where
the difference between the numerical results and the exact solution at time $t = 0.0125$ is
plotted for $m = -5$ and $m = -7$. Δu is the regularisation parameter for the enthalpy
function H. Fig. 6 shows the difference between the computed results and the exact solution at $t = 0.0125$ for $\Delta u = 0.01$. An improvement in the approximation near the free
boundary is observed for the smaller value of Δu.

REMARK

The scaling functions we have used in this work were calculated by the iterative method described in [1]. However, we must remark that after presenting this paper we have obtained a more accurate evaluation of the scaling functions and their derivatives by using results from [5].

References

1. Daubechies, I. (1988) 'Orthonormal bases of compactly supported wavelets', Comm. in Pure and Appl. Math., XL1 (7), 909-996.
2. Glowinski, R.; Lawton, W. and Ravachol, M. (1989) 'Wavelet solution of linear elliptic, parabolic and hyperbolic problems in one space dimension', Preprint, Cambridge, MA, One Cambridge Center.
3. Latto, A and Tenenbaum, E. (1990) 'Les ondelettes à support compact et la solution numerique de l'équation de Burgers', Preprint, AWARE, Inc., Cambridge, MA.
4. Cortina, E. and Gomes, S. (1990) 'A wavelet based numerical method applied to free boundary problems', Report INPE- 5076/623.
5. Strang, G. (1989) 'Wavelets and dilations: a brief introduction', SIAM Review 31 (4), 614-627.

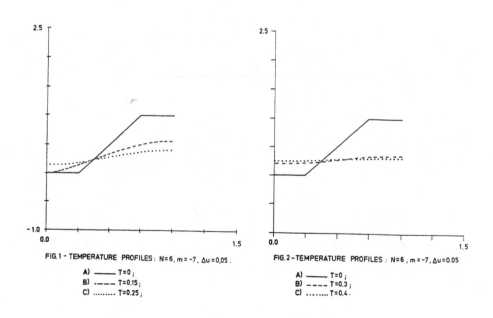

FIG.1 - TEMPERATURE PROFILES: N=6, m=-7, Δu=0,05.

A) ———— T=0 ;
B) — — — T=0.15 ;
C) .... T=0.25 ;

FIG.2 - TEMPERATURE PROFILES : N=6 , m=-7, Δu=0.05

A) ———— T=0 ;
B) — — — T=0.3 ;
C) .... T=0.4 .

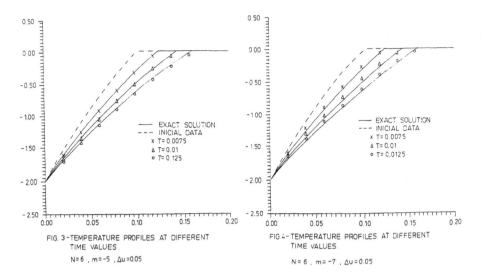

FIG. 3 - TEMPERATURE PROFILES AT DIFFERENT
TIME VALUES.
N = 6 , m = -5 , Δu = 0.05

FIG. 4 - TEMPERATURE PROFILES AT DIFFERENT
TIME VALUES.
N = 6 , m = -7 , Δu = 0.05

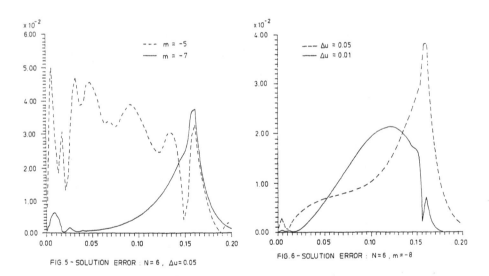

FIG. 5 - SOLUTION ERROR : N = 6 , Δu = 0.05

FIG. 6 - SOLUTION ERROR : N = 6 , m = -8

WORKING GROUP HEADER: THRUST BEARING

André J. Damslora
Norwegian Inst. of Technology
Div. of Mathematical Sciences
Trondheim, Norway

Jo P.M. Laumen
Eindhoven Univ. of Technology
Dep. of Mathematics
Eindhoven, The Netherlands

ABSTRACT. The article summarizes a preliminary analysis of the stability of a thrust bearing.

1. Introduction – problem description

A thrust bearing is a basic part of a jet engine and may be regarded as a half open cylinder with a disc rotating near its bottom as shown in figure 1 a). The figure is not drawn to scale. The motion is lubricated by highly compressed air and the lubrication zone is extremely narrow compared to the other dimensions of the problem. In contemporary systems the rotation axis is physically fixed but a new principle has been introduced where the disc is driven by air jets tangential to its periphery so that the rotation axis is no longer fixed. An important question is whether the new principle leads to a construction which is stable with respect to small perturbations of the disc motion. An instability may destroy the jet engine.

This text is a summary of a 4 days work carried out by 5 students and, as supervisor, Dr. Alistair Fitt from Southampton University, during a student workshop arranged by ECMI, Kaiserslautern. The group wants to thank Dr. Alistair Fitt for his cooperation and, especially, Dr. H. Neunzert from Kaiserslautern for his kind support. The goal of the work was, by means of using mathematical modelling, to analyse the following question: If the disc is tilted or its center of mass is moved away from the center axis of the thrust bearing cylinder, *will this cause the motion to become unstable?*

The disc is attached to the wall by means of a mechanism which we assume may be modeled as springs. The mass, radius and thickness of the disc are 15 kg, 0.5 m and a few millimeters, respectively. The disc rotates about a moving virtual axis, its angular velocity being of order 10000 RPM. The disc should be able to sustain a pressure load of order 1 MPa. The pressure forces are parallel to the center axis of the thrust bearing. Of particular interest is the thickness of the lubrication zone which is only 20 μ. The air leakage from the lubrication zone to its surroundings is assumed to be negligible.

Assuming that the disc obeys the laws of Newtonian mechanics a stability analysis involves a study of the equations:

$$\sum_j F_j = m\ddot{r} \qquad (1)$$

M. Heiliö (ed.), *Proceedings of the Fifth European Conference on Mathematics in Industry*, 201–204.
© 1991 *B. G. Teubner Stuttgart and Kluwer Academic Publishers. Printed in the Netherlands.*

and

$$\sum_j \tau_j = \dot{L} = I\dot{\omega}, \qquad (2)$$

where F_j and τ_j are the forces and torques acting on the disc. Its mass and inertia matrix are denoted by m and I. The vectors $\dot{L}$ and $\dot{\omega}$ are the material derivatives of the angular momentum and angular velocity of the disc. Finally r is the position vector of the center of mass of the disc.

2.   Translational motion

Consider the case when the disc is not tilted and where the motion of the disc is in a plane perpendicular to the center axis of the thrust bearing. The translational forces is assumed to be dominated by elastic and frictional forces from the springs. The elastic force is assumed to vary linearly with the radial displacement of the spring and to act perpendicular to the disc. The frictional force is assumed to vary linearly with the elastic force and to act tangentially to the disc. Then equation 1 takes the form:

$$m\ddot{r} = m(\ddot{x}i + \ddot{y}j) = \int_0^{2\pi} [R(\theta) + F(\theta)] \, d\theta$$

$$= \int_0^{2\pi} \lambda \frac{b - \mu(\theta)}{b} \{(-\cos\theta i - \sin\theta j) + \mu'(\sin\theta i - \cos\theta j)\} d\theta, \qquad (3)$$

where $b, \mu(\theta), \lambda$ is the natural length, instantaneous length and elastic coefficient of the springs, respectively. μ' is the frictional coefficient. R and F are the elastic and the frictional forces, respectively. The coordinate system used for analysis of translational motion is shown in figure 1 b).   Using the geometry of the figure it is possible to express $\mu(\theta)$ in terms of x, y, θ and known constants. It can be shown that the system reduces to:

$$\frac{d^2}{dt^2} \begin{bmatrix} x \\ y \end{bmatrix} = kA \begin{bmatrix} x \\ y \end{bmatrix} = \frac{\lambda\pi}{mb} \begin{bmatrix} -1 & \mu' \\ -\mu' & -1 \end{bmatrix} \begin{bmatrix} x \\ y \end{bmatrix}. \qquad (4)$$

The square root of the eigenvalues of the matrix, A, may be shown to have nonzero real part and therefore the system is unstable. Thus, according to a linear analysis, the translational part of the motion may be unstable.

3.   Rotational motion

Consider the case when the center of mass of the disc is fixed at the center axis of the thrust bearing. Then equation 1 holds. The pressure forces from the lubricant are assumed to be the main contribution to the torque on the left hand side of equation 2. The lubricator is assumed to behave like a Newtonian fluid, i.e. the pressure obeys the Navier Stokes equations. Using the geometry of the system to scale the Navier Stokes and continuity equations the resulting equation, which is known from lubrication theory [Ba], is

$$\begin{array}{ccccc} u_r + r^{-1}v_\theta + w_z + r^{-1}u & = & 0 & , & p_r = \mu u_{zz}, \\ p_\theta & = & \mu r v_{zz} & , & p_z = 0, \end{array} \qquad (5)$$

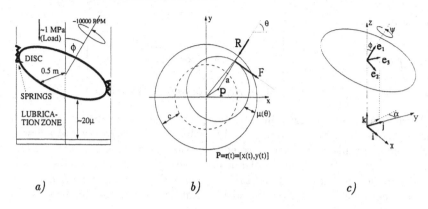

Figure 1: *a) A sketch of the thrust bearing. b) Coordinate system for the analysis of translational motion. The radius of the disc is denoted by a and $\mu(\theta)$ is the instantaneous length of the springs. P, R, F is the center of the disc and the elastic and frictional forces from the springs. c) Coordinate system for the analysis of rotational motion. ϕ, α, ψ are called the tilt, precession and pure rotation angles, respectively. The basis $\mathbf{i}, \mathbf{j}, \mathbf{k}$ is fixed in space, whereas $\mathbf{e}_1$ is fixed normal to the disc and $\mathbf{e}_2$ and $\mathbf{e}_3$ lie in the plane containing the disc, but do not rotate with the disc.*

with boundary conditions[1] given by $(u, v, w)^T = (0, 0, 0)^T$ at $z = 0$, $(u, v, w)^T = (u_d, v_d, w_d)^T$ at $z = h(r, \theta, t)$ and $p = 0$ at $r = a$, where (u, v, w) is the fluid velocity in cylindrical coordinates r, θ, z, and where p, a, μ denotes the pressure in the lubricator, the radius of the thrust bearing cylinder and the dynamic viscosity in the lubricator, respectively. The $()_d$-subscript denotes the velocity of the disc surface. Integrating equation 5 and using the boundary conditions it is possible to reduce the system to one single PDE:

$$\frac{\partial}{\partial r}\left[\frac{rp_r h^3}{6\mu}\right] + \frac{\partial}{\partial \theta}\left[\frac{p_\theta h^3}{6\mu r}\right] = \frac{\partial}{\partial r}\left[u_d rh\right] + \frac{\partial}{\partial \theta}\left[v_d h\right] + 2rh_t + 2ru_d h_r + 2v_d h_\theta \quad (6)$$

with the boundary condition $p = 0$ at $r = a$.

The coordinate system for the analysis of rotational motion is shown in figure 1 c). The torque on the left hand side of equation 2 may be calculated from the pressure by using the relation

$$\boldsymbol{\tau} = \boldsymbol{r} \times \boldsymbol{F} = \iint_{Disc} \boldsymbol{r} \times \boldsymbol{n}p(\boldsymbol{r})dS, \quad (7)$$

that is an integral of the pressure forces times the distance from the disc center. The angular momentum on the right hand side of equation 2 follows from elementary

---
[1]The conditions on u, v, w is merely the usual no slip condition, where h describes the disc position. Moreover, the pressure under the disc is very large whereas at the disc periphery it is of atmospheric order which is negligible in this context.

classical mechanics:

$$\iint_{Disc} \boldsymbol{r} \times n p(\boldsymbol{r}) dS = \begin{bmatrix} 2A\,\overbrace{(\dot{\alpha}\cos\phi + \dot{\psi})} \\ -A\ddot{\alpha}\sin\phi + 2A\dot{\phi}\dot{\psi} \\ A\ddot{\phi} + A\dot{\alpha}^2\sin\phi\cos\phi + 2A\dot{\alpha}\dot{\psi}\sin\phi \end{bmatrix}, \qquad (8)$$

where the right hand side is given in the basis $\{e_i\}_{i=1}^3$ and where the pressure, p, is given by equation 6. This equation is hard to solve, even numerically, but a first impression may be achieved by making some approximations. Assume that the tilt angle is very small and that the precession is much slower than the pure rotation of the disc. Then we may set $u_d = 0$ and $v_d = r\Omega$ where Ω is the angular velocity of the disc. Furthermore $h_t \approx 0$. The dependent variables p and h may be expanded in the small tilt angle. The first order solution of equation 6 is then

$$p = p_0 + \phi p_1 = \phi \cdot C' \sin\theta (r^3 - ra^2). \qquad (9)$$

Because of the small tilt angle $\cos\phi \approx 1$ and $\sin\phi \approx \phi$. By substituting into equation 8 we end up with the following system:

$$\begin{aligned} 0 &= \overbrace{(\dot{\alpha} + \dot{\psi})} \\ C\phi &= -\ddot{\alpha}\phi + 2\dot{\phi}\dot{\psi} \\ 0 &= \ddot{\phi} + \dot{\alpha}^2\phi + 2\dot{\alpha}\dot{\psi}\phi, \end{aligned} \qquad (10)$$

where C is a constant. Also $\dot{\alpha} + \dot{\psi} = C_1$ is constant. If the precession is assumed to be steady, say $\dot{\alpha} = C_2 \neq C_1$ the system reduces to

$$\ddot{\phi} = \phi \frac{1}{4}\left(\frac{C}{C_1 - C_2}\right)^2, \qquad (11)$$

which is unstable. Thus the first order approximation tells that if the disc is tilted then the tilt angle may grow indefinitely as a function of time.

The disc motion is governed by a strongly coupled system of nonlinear differential equations. Our approximation indicates that the disc may not reach an equilibrium with zero tilt angle and a center of mass at rest. Of course there may exist equilibria with nonzero tilt angle due to nonlinear effects.

Reference

[Ba]   BATCHELOR, G.K., *An introduction to fluid dynamics*, Cambridge University Press, Cambridge, 1967

Self-validating Numerics
for Industrial Problems

H.-J. Dobner, W. Klein

1 Introduction

The growing and complex use of computers by non mathematicans makes it necessary to require a great reliability from the computed results. A disadvantage of contemporary numerical software is the lack of realistic and easy obtainable error bounds. Therefore in modern computations, numerical software, where a reliable error analysis is part of the output becomes more and more important. We present here a new kind of numerics - Self Validating Numerics or E-Methods - which can be characterized as follows: *"An E-Method is an algorithm, where close and mathematically guaranteed error bounds are part of the output"*.

2 Elements of inclusion theory

We give a summary of the tools necessary for E-Methods, where we distinguish between finite and infinite dimensional problems.

2.1 Tools for all kind of problems

- Fundamental for all inclusion procedures is a precise formulation of floating point arithmetic on computers (going back to Kulisch), which can be described shortly by the requirement *"that there is no machine number between the correct and the computed rounded result of a single floating point operation"*. The floating point numbers are indicated with R.
- The error tolerance is described by sets (intervals) of the form

$$[a] = [\underline{a}, \overline{a}] = \{x \in I\!R | \underline{a} \leq x \leq \overline{a}\}, \quad \underline{a}, \overline{a} \in R.$$

  For objects of this kind the elementary operations $+, -, \cdot, /, \subseteq$, are defined in terms of their bounds (see [1]). IR denotes the set of all those intervals.
- Another important tool is a programming language, where both the rounding concept of Kulisch and interval analysis are available (e.g. FORTRAN-SC).

2.2 Tools for finite dimensional problems

Consider the linear system $Ax = b$ with $A \in I\!R^{n \times n}, b \in I\!R^n$, $n \in I\!N$. Then an E-Algorithm consists of the following steps:

I) Compute an approximation Q for the inverse of A.

II) Iterate in IR^n : $X_{i+1} = Qb + (E - QA)X_i$ until $\overline{X_{i+1}} \subseteq \overset{\circ}{X_i}$, E : Identity.
Then it is proved by computational means, that A and Q are nonsingular and that there exists one and only one solution x of $Ax = b$, wich is enclosed within X_{i+1}.

2.3 Tools for infinite dimensional problems

Here the situation is more complicated (cf. [4]). Let M be a normed linear space

M. Heiliö (ed.), Proceedings of the Fifth European Conference on Mathematics in Industry, 205–208.
© 1991 B. G. Teubner Stuttgart and Kluwer Academic Publishers. Printed in the Netherlands.

and $b_i = b_i(s)$, $i \in I\!N$, a generating system of it. In the power set PM of M the operations $+, -, \cdot, /, \int$ are defined pointwise. A subset $I_n M \subseteq PM, n \in I\!N$, of the form $I_n M = \{\sum_{i=1}^n [\alpha_i] b_i(s) \mid [\alpha_i] \in I\!R\}$ is called an interval screen. The elements $[f]$ of $I_n M$ are function tubes, describing a subset of M in the graph sense. A mapping $\Diamond : PM \to I_n M$ is called directed rounding, iff

$$1)\ \Diamond b_i = b_i,\ i = 1, \ldots, n\ ,\quad 2)\ f \in \Diamond f,\ f \in M\ ,\quad 3)\ F \subseteq \Diamond F,\ F \in PM\ .$$

The operations in the screen $I_n M$ are explained according to

$$[f] \Diamond\!\!\!\!\Diamond [g] := \Diamond([f] * [g]), \quad * \in \{+, -, \cdot, /\}$$

and $(I_n M, \Diamond\!\!\!\!\Diamond)$ is called a directed functoid. With these constructs it is possible to derive E-Methods for problems of the kind $Tx = x$ where $T : M \to M$ is compact:

I) Compute an interval extension $[T]$ of T, i.e. $Tu \in [T]u$ holds for $u \in M$.

II) Iterate in $I_n M$: $[X_{k+1}] = [T][X_k]$, until $[X_{k+1}] \subseteq [X_k]$, $[X_{k+1}] \neq \emptyset$.

Then it is proved by computational means that there exists a fixed point $\hat{x}$ of T for which the inclusion (error estimation) $\hat{x} \in [X_{k+1}]$ is asserted.

3   VIB - Verified Inclusions of Critical Bending Vibrations

In a joint project of the Institutes of Applied Mathematics and Technical Mechanics of the University of Karlsruhe ([2]) a program system was developed for the calculation and animation of the bending vibrations of rotating shafts. The mechanical model of the rotor is based on the Bernoulli-Euler beam and evaluated by the use of transfer matrices.

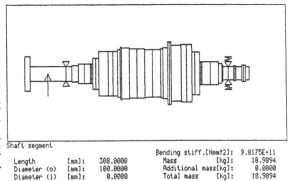

By means of verified inclusions the eigenvalues can be determined with high accuracy up to high speeds of rotation.

In the CAD-part of this VIB program the construction of a rotor configuration with different bearing types and different segments for the rotor shaft is done.

The CAE-part of VIB performs a modal analysis of the rotor system by means of transfer matrices. The calculation of bending vibrations starts with an iterative determination of the natural frequencies for a chosen range of rotation speeds and frequency steps. Extending the considered domain to higher frequencies leads to a numerical behaviour of the bending vibration contradicting the expected physical or mechanical performance. So the algorithm was carefully analyzed and improved.

1. The coefficients of each transfer matrix $U_i, i \in \{1, 2, \ldots, N\}$, depend on trigonometric and hyperbolic functions. When computing these functions using a floating point arithmetic with 13 decimal digits and using arguments $x > 30$, identical results are obtained for all of the following expressions: $(\sinh(x) + \sin(x))/2 = (\cosh(x) + \cos(x))/2 = \sinh(x)/2 = \cosh(x)/2 = \exp(x)/4$. By the use of standard functions with arbitrary precision and verified results, all coefficients of the transfer matrices U_i are included in intervals determined up to an accuracy of 52 decimal digits.

2. The boundary state of each rotor segment has to be determined by the product of its representing transfer matrix U_i and the boundary state of the preceeding segment $z_{i+1} = U_i * z_i$. But repeated matrix multiplications increases rounding errors enormously, and so the transfer matrices U_i are stored in only one large band-shaped matrix. The resulting linear system of equations is solved with the residuum technique (cf. section 2.2) and the exact scalar product.

3. In the last step the determinant of the resulting product matrix has to be computed: here the LU-factorization of a matrix with predetermined precision and verified results was used.

These improvements (bold line) produced, in contrast to the old program (tight line), the results of the following picture. With the new program VIB, the eigenvalues for high frequencies as well for lower frequencies can be determined with high accuracy. Furthermore, the results are verified to be correct.

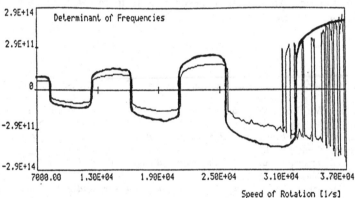

4 Horizontal forces on piles in granular soil

We consider the case of a pile embedded in sand. The movement of the sand causes a horizontal force, hence a displacement of the pile. This situation is described mathematically by the following boundary value problem (abbr. bvp)

$$y''''(x) = -x^{1-\beta} \cdot \text{sgn} \, |y(x)|^\alpha, \quad 0 < \alpha \leq 1, \ 0 \leq \beta \leq 1, \ \alpha, \beta \in \mathbb{R},$$

the parameters α and β arising from the ansatz. The second and third derivatives at the top and bottom always have the form $y''(0) = y''(1) = y'''(1) = 0, y'''(0) = Q_0$;

Q_0 denoting the horizontal force at the top of the pile. The result is shown for $\alpha = \frac{5}{8}, \beta = \frac{33}{40}, Q_0 = 10^{-5}$ and the boundary conditions $y(0) = 10^{-5}, y'(0) = y(1) = y'(1) = -10^{-4}$, in the diagram. The correct analytical solution is indicated by the continous line, whereas the dotted line represents the solution computed with a standard difference scheme of the NAG library. The error estimated by this program was claimed to be $1.5 \cdot 10^{-8}$.

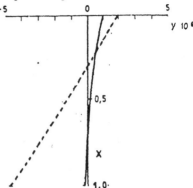

This is obviously not a mathematically guaranteed estimate and so this method is not trustworthy if employed on sensitive problems. For a treatment with an E-Method, we formulated the bvp equivalently as an integral equation. The results obtained now are guaranteed by mathematical means to be in error less than 1 %.

The results of the standard methods cannot be improved by reducing step sizes or performing the calculations with double precission -just the opposite occurs. Therefore the whole problem must be reformulated as a fixed point problem, so that the inclusion tools can be applied. Then a code must be generated using Kulisch and interval arithmetic. An E-Method needs slightly more space and execution time as conventional schemes.

References

[1] Alefeld, G., Herzberger, J.: *An Introduction to Interval Computations.* Academic Press, New York, 1983.
[2] Ams, A., Klein, W.: *VIB - Verified Inclusions of Critical Bending Vibrations;* Computing, Suppl. 6, 1988 (pp 91 - 98).
[3] Hettler, A.: *Horizontal belastete Pfähle mit nichtlinearer Bettung in körnigen Böden;* Habilitation, Karlsruhe 1985.
[4] Kaucher, E., Miranker, W. L.: *Self-Validating Numerics for Function Space Problems;* Academic Press, New York, 1984.
[5] Kulisch, U., Miranker, W. L.: *Computer Arithmetic in Theory and Practice.* Academic Press, New York, 1981.

Authors: H.- J. Dobner , W. Klein , Fakultät für Mathematik
Universität Karlsruhe, Kaiserstr. 12, 7500 Karlsruhe, Germany

TOPOLOGICAL ALGORITHMS IN PICTURE PROCESSING

Ulrich Eckhardt
Universität Hamburg
Institut für Angewandte Mathematik
Bundesstaße 55
D–2000 Hamburg 13

Gerd Maderlechner
SIEMENS AG
Corporate Research and Development
Otto–Hahn–Ring 6
D–8000 München

Abstract

For automatic analysis of texts or engineering drawings fast parallel data reduction methods are needed. Skeletonization methods have the advantage of retaining topological properties during the data reduction process. Although these methods are often used in many situations, there remain many open questions concerning their mathematical properties.

Documents and engineering drawings are usually given as black–and–white images. The information contained in such an image is completely determined by the set of all of its black points. Closed sets in the Euclidean plane can be described by means of their *skeletons*. The skeleton of a closed set is the set of all centers of maximal spheres contained in the set. It is known that the topolocical closure of the skeleton of a set is homeomorphic to the set and that it is in some sense 'thin'. When each point on the skeleton carries a label indicating its distance to the complement of the original set, the latter can be reconstructed [12, Chapter XI.C], [2,5]. It is known that the skeletonization process is 'ill–posed', i.e. the mapping of a set on its skeleton is not continuous in the Hausdorff metric for sets. For finite sets, the skeleton coincides with the Voronoi diagram.

The theoretical basis for skeletonization was laid by Motzkin [10]. Skeletons were introduced by Blum [1] in picture processing and investigation of perception processes as a means for representing the 'form' of an object.

An important application for skeletonization is the investigation of topological properties of the contents of an image. It is not possible to decide topological predicates efficiently in parallel [9]. However, the skeleton of a set usually has a simpler structure than the original set while it is homeomorphic to the latter [2,12]. When it can be determined in parallel, then the hard task of investigation of the topology of the set under consideration can be performed by a sequential method on the genenerally simpler skeleton.

There are several applications where it is adviceable to investigate the skeleton of an image instead of the whole picture. This is especially true for pictures containing lines and curves such as engineering drawings. In such applications one has a huge amount of data which is dramatically reduced by skeletonization [7]. The application of skeletons for analyzing office documents was discussed in [4]. When a more complex transformation has to be performed on an image, then it can be more efficient to perform this transformation on the skeleton and to reconstruct the

M. Heiliö (ed.), Proceedings of the Fifth European Conference on Mathematics in Industry, 209–212.
© 1991 B. G. Teubner Stuttgart and Kluwer Academic Publishers. Printed in the Netherlands.

transformed image from the transformed skeleton. There are different proposals in the literature for this approach, e.g. smoothing of digital sets [6] scaling [8] or computation of projections and moments [11].

In order to perform the skeletonization process on digitized images, the concept of skeletonization has to be formulated in discrete form. Unfortunately, however, all properties of skeletons mentioned above were proved only for smooth and strictly convex norms, even for Euclidean norms. Moreover, it is known by counterexamples that almost all of these properties are actually not true for polyhedral norms such as the L_1 or L_∞ norms. For discretized images, however, only polyhedral norms are adequate, hence any discrete formulation of the skeleton depends on how much weight is given to the properties mentioned above of the continuous skeleton. Specifically, there cannot be a 'canonical' way for translation of the continuous concept into the discrete setting. This is the reason for the enormous proliferation of skeletonization methods in the literature.

The authors proposed a discrete method for skeletonization which has the following properties

- the method is *well defined*, i.e. the result of the application of it does not depend on details of the specific implementation.

- The proposed method is *motion invariant*, i. e. it does not depend on 90^0-rotations, translations or reflections of the grid point plane.

- The method is genuinely *parallel*.

- By means of a suitable labeling scheme, it is possible to *reconstruct* the original set from its skeleton. This means that the skeletonization process causes no loss of information and preserves the 'shape' of the image.

Although these requirements seem to be quite natural, they are not met by most of the existing thinning methods. On the other hand, by the requirements a skeletonization method is (more or less) uniquely determined. The autors investigated this method both theoretically and by means of practical experiments. A detailed discussion of these results will appear in the report [3].

There are many open questions related to this topic. We only mention the following problems:

- Motzkins results should be reformulated for more general normed spaces.

- Since the skeletonization process is an ill-posed problem, regularization is necessary. There exist already interesting approaches by Klein [8] and Yu [13].

The following pictures show the result of application of the method to a document. The document was poorly scanned in order to exhibit the effects caused by inadequate scanning. The skeleton obtained by our method is compared with the skeleton obtained by application of a standard method (KONTRON IPS–System). While the standard method yields a thinner skeleton, many details essential for the recognition process were destroyed.

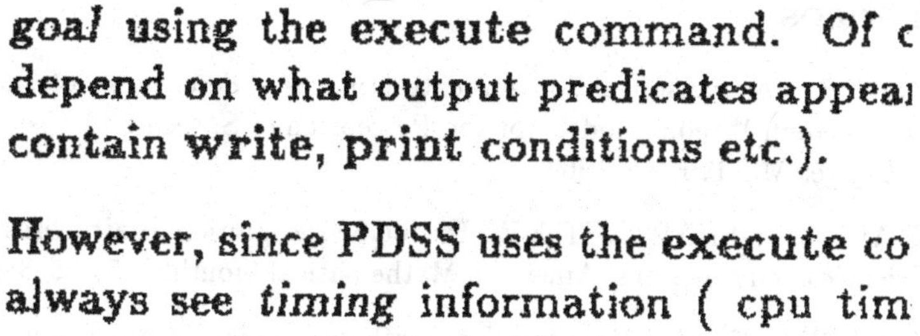

goal using the execute command. Of c
depend on what output predicates appea
contain write, print conditions etc.).

**However, since PDSS uses the execute co
always see** *timing* **information (cpu tim**

Figure 1: Original document

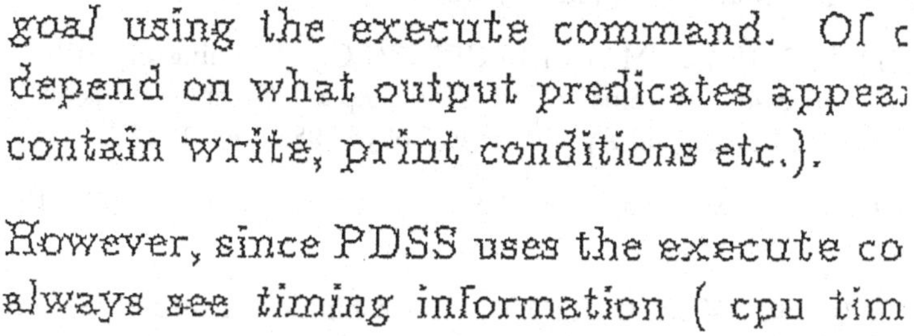

goal using the execute command. Of c
depend on what output predicates appea
contain write, print conditions etc.).

However, since PDSS uses the execute co
always see timing information (cpu tim

Figure 2: Skeleton of text document obtained by application of a labeling method yielding an invertible result

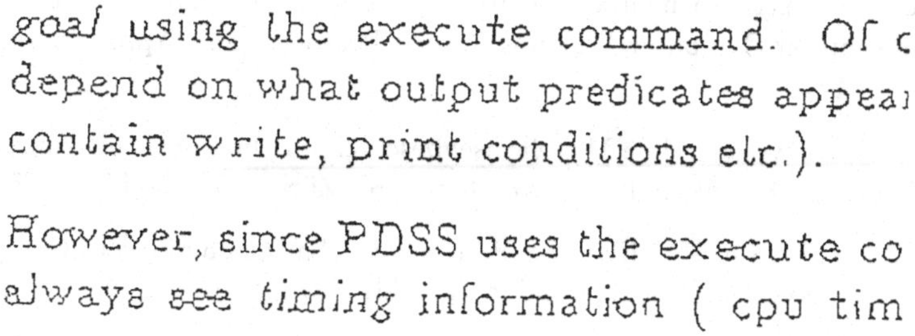

goal using the execute command. Of c
depend on what output predicates appea
contain write, print conditions etc.).

However, since PDSS uses the execute co
always see timing information (cpu tim

Figure 3: Skeleton obtained by our method with quasi–parallel elimination and subsequent postprocessing

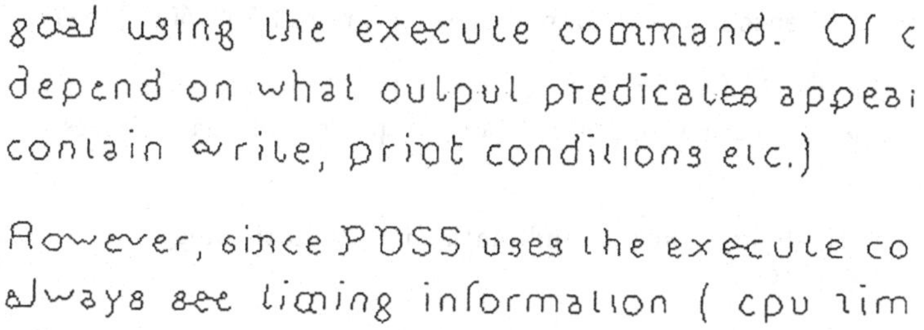

goal using the execute command. Of c
depend on what output predicates appea
contain write, print conditions etc.)

However, since PDSS uses the execute co
always see timing information (cpu tim

Figure 4: Skeleton obtained by using the KONTRON IPS–SYSTEM (LNENDS ON, MARG = 0)

References

[1] BLUM, H.: A transformation for extracting new descriptors of shape. In: Wathen-Dunn W ed.: Models for the Perception of Speech and Visual Form. Cambridge: M.I.T. Press 1967

[2] CALABI, L. and HARTNETT, W. E.: Shape recognition, prairie fires, convex deficiencies and skeletons. American Mathematical Monthly 75,335-342 (1968)

[3] ECKHARDT, U. ed.: Mathematical methods in picture processing. Proceedings of a Minisymposium held at ECMI '90 in Lahti. Hamburger Beiträge zur Angewandten Mathematik, in preparation.

[4] ECKHARDT, U. and MADERLECHNER, G.: Thinning algorithms for document processing systems. IAPR Workshop on Computer Vision – Special Hardware and Industrial Applications, Tokyo 1988, 169-172

[5] ERDÖS, P.: Some remarks on the measurability of certain sets. Bull. AMS 51,728-731 (1945)

[6] HO, S.-B. and DYER, C. R.: Shape smoothing using medial axis properties. IEEE Trans. PAMI-8,512-520 (1986)

[7] HOFER-ALFEIS, J. and MADERLECHNER, G.: Automated conversion of mechanical engineering drawings to CAD models: Too many problems? IAPR Workshop on Computer Vision – Special Hardware and Applications, Tokyo 1988, 206-209

[8] KLEIN, F.: Vollständige Mittelachsenbeschreibung binärer Bildstrukturen mit Euklidischer Metrik und korrekter Topologie. Zürich: Thesis ETH 1987

[9] MINSKY, M. and PAPERT, S.: Perceptrons. An Introduction to Computational Geometry. Cambridge, London: The MIT Press 1969

[10] MOTZKIN, T.: Sur quelques propriétés caractéristiques des ensembles convexes. – Sur quelques propriétés caractéristiques des ensembles bornés non convexes. Atti della Reale Accademia Nazionale dei Lincei, Serie sesta, Rendiconti, Classe di Scienze fisiche, matematiche e naturali 21,562-567, 773-779 (1935)

[11] SANNITI DI BAJA, G.: O(N) Computation of projections and moments from the labeled skeleton. Computer Vision, Graphics, and Image Processing 49,369-378 (1990)

[12] SERRA, J.: Image Analysis and Mathematical Morphology. London etc.: Academic Press 1982

[13] YU, Z.: Regularisierung der Mittelachsentransformation. Informatik-Fachberichte 180,211-218. Berlin etc.: Springer-Verlag 1988

FIELD APPLICATORS FOR MAGNETIC RESONANCE

G.J. Ehnholm and S. Vahasalo

Instrumentarium Corporation/Imaging Division

Teollisuuskatu 27, 00510 HELSINKI, FINLAND

Introduction

The behaviour of the electric and the magnetic field inside the body starts to change as frequencies above 100 MHz are used. The wave nature of the field then has to be taken into account. These frequencies are of interest when doing high field magnetic resonance imaging, hyper-thermia, ESR, and dynamic polarization using the Overhauser effect in the human body.

We have made investigations of field patterns and thermal dissipation in a simple basic system, which consists of a cylindrical "loop-gap" resonator type applicator loaded by spherical bottles containing saline solution. This system is simple enough for reliable calculations to be performed. Although its likeness to the human body is very superficial the results still give insight into the problem of applying electromagnetic signals to the same.

Experimental arrangement

The measurements were made using the arrangement of Fig. 1. A power source, consisting of a frequency synthesizer and a 100W linear VHF amplifier, was matched with a pair of adjustable stubs to the coupling loop. The loop coupled inductively to the loop gap resonator which was made from aluminium tape on a 200 mm diametre acrylic tube. The resonator was 100 mm long and had eight gaps which were formed by overlapping two layers of Al tape with polyethene tape as insulator, the gap width being a few millimetres. The Q-value of an unloaded resonator was typically 100-200. Incident and reflected power was measured with the power meter (P.M.) coupled to the directional coupler (Dir.C.). The resonator was surrounded by a copper sheet screen. The frequency was 280 MHz.

The resonator was loaded by spherical bottles containing 0.45 % saline solution with conductivity s = 0.8 S and a relative dielectric constant = 80. Volumes of 0.5, 1 and 2 l were used and the stubs were adjusted for each bottle for best impedance match. The H-field in the middle of the bottle and the temperature increase of the solution were measured with the field probe and a thermometer, respectively.

In a separate experiment the temperature profile was measured by infrared photography. For this case a separate sphere was prepared by adding a gelling agent to the water to make it semi-rigid. The sphere was cleaved and so the profile inside could be photographed by separating the two halves. The temperature profile of a typical experiment with a 1 l

M. Heiliö (ed.), Proceedings of the Fifth European Conference on Mathematics in Industry, 213–216.
© 1991 *B. G. Teubner Stuttgart and Kluwer Academic Publishers. Printed in the Netherlands.*

sphere is shown in Fig. 2.

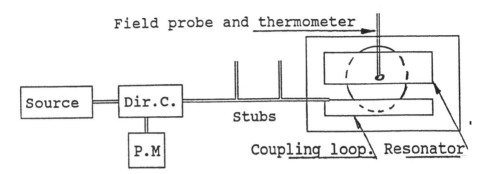

Fig. 1.   Experimental arrangement

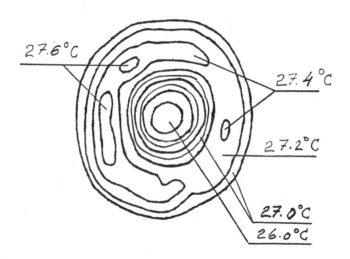

Fig. 2.   Temperature profile of a sphere.

The displayed plane goes through the centre of the sphere perpendicularly to the H-field.
The expected profile has rotational symmetry with temperature increasing along the radius
from the centre to a maximum then again decreasing towards the rim. The measured one
fits this within the experimental accuracy; the deviation from symmetry is probably caused
by a slight angular misalignment. No local hot spots at the gaps in the resonator were
detected, even when placing the sphere to within one centimetre of the resonator wall.

Theoretical calculations

The electromagnetic field was expressed as a series in spherical waves with the centre of the spherical bottle (coincident with the resonator centre) as origin. These are explained in slightly different formalism, both in Schelkunoff and by Morse-Fesbach. The former uses the terms TE-waves and TM-waves (for transverse electric and magnetic, respectively). The latter uses so called vector spherical harmonics to represent the same.

Fields produced by current circling around the Z-axis can be expressed as a series in TE-waves. The waves are characterized by the two indices as TE m,n, where m gives angular dependence for the angle in the x-y plane. For axial symmetry $m=0$. If, moreover (as in our case), the currents are symmetric around the x-y plane then n only takes odd values. The simplest system with these properties is a constant current circular loop around the origin in the x-y plane. Morse-Fesbach (p. 1881) gives the series expansion for the electromagnetic field radiating out, from such a loop.

We have compared the measured field to the computed series expansion, for the field inside a set of 1, 2 or 6 loops symmetrically around the x-y plane. The computations were made by J. Sarvas and co-workers at the Rolf Nevanlinna Institute.

The computed field profiles for a simple loop with 1A current are shown in Figs 3 and 4. The circle in the Figs outlines a 1 litre sphere. The profiles for 2 and 6 loops looked the same within the region bound by about 2 litres, differing only for bigger radii, i.e. closer to the field generating currents. We conclude that the field profile in the spheres is insensitive to the length of the resonator as long as it is less than the radius. This is confirmed by the fact that the computed profile actually closely represents a pure TE01-wave; especially so for loops that form a pair of Helmholz coils or some other similar arrangement.

Comparing Figs 2 and 3 we see that the measured and computed profiles are equal in that both represent an electric field distribution, which has a maximum in the shape of a torus in the x-y plane at a radius of about 4 cm.

In Fig. 5 we show the measured vs. computed values for dissipated power. They coincide within the precision of the measurement. Shown are curves for two different resonators, a simple loop and a six-loop, approximating the arrangement used. For radii below 8 centimetres they are approximately coincident with the TE01 curve (dashed).

Refs.:   Schelkunoff, Electromagnetic Waves, D. van Nostrand, New York 1943.
         Morse-Fesbach, Methods of Theoretical Physics, McGraw-Hill, New York 1953.

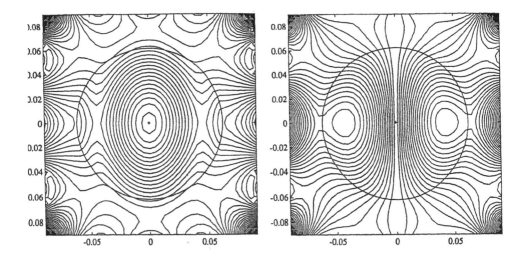

Fig. 3.   Magnetic field equal magnitude   Fig. 4. Electric field equal magnitude
           contour plot. Frequency 280 MHz.   contour plot. Frequency, conductivity
           Conductivity 0.8 Siemens. Single   and source as in Fig. 3.
           loop source. Distances in meters.

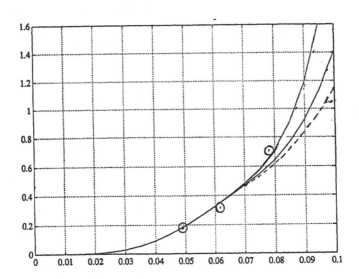

Fig. 5.   Experimental and computed dissipated power in spheres.
           Experiment: 0.5, 1 and 2 litre (points).
           Calculation: Single loop (1), six loops (2) and TE01 wave only (3).

CASE STUDIES FOR A SINGLE
CYLINDER COMBUSTION ENGINE

Gabriele Engl, Peter Rentrop
Mathematisches Institut
Technische Universität München
Arcisstr. 21, Postfach 202420
D 8000 München 2

Abstract

We present results of our work dealing with the coupling of a system of ordinary differential equations and a hyperbolic system of partial differential equations. The equations are part of a mathematical model for a single cylinder combustion engine and describe the gas flow in cylinder, inlet and outlet pipes. Different methods of discrete coupling are discussed. As a main result we show that the various methods lead to quite different solutions. Therefore a careful treatment of the coupling conditions is crucial.

1 Introduction

We consider a four-stroke cycle combustion engine consisting of a single cylinder, inlet and outlet pipes and two valves connecting the cylinder with the pipes as shown in Fig. 1.

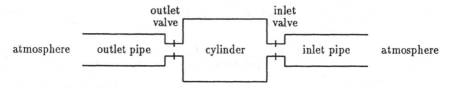

Figure 1: Model of Combustion Engine

An optimal performance of a combustion engine requires the complete replacement of the burnt gas by a fresh fuel-air mixture during the charge cycle. In order to support the gas exchange it is essential to open the inlet valve when there is a high pressure in the inlet pipe. The outlet valve should be opened in case of a low pressure in the outlet pipe. Obviously the pipe diameter and the valve size affect the gas flow in the pipe.

The original problem can be classified as an "Inverse Problem": How should specified technical parameters be chosen in order to achieve an optimal charge cycle.

In the following we treat the "simulation problem" for the simplified model: atmosphere - cylinder (o.d.e. system) - inlet pipe (gas equations) - atmosphere.

M. Heiliö (ed.), Proceedings of the Fifth European Conference on Mathematics in Industry, 217–222.
© 1991 B. G. Teubner Stuttgart and Kluwer Academic Publishers. Printed in the Netherlands.

2   Cylinder and Pipe Equations and their Numerical Treatment

We are not interested in the spatial dependency of the gas flow in the cylinder but only in the balance of inflow and outflow through the valves. The assumption of an ideal gas which does not change its thermodynamical properties during the charge cycle yields a realistic cylinder model which is appropriate for our purpose, see Dubbel [1].

The cylinder volume V_Z is a known function of the crank angle φ, where $\varphi = \omega t$, ω being a given angular frequency and t denoting the time. Mass balance and energy balance yield a system of two ordinary differential equations in time t for mass m_Z and temperature T_Z of the gas in the cylinder:

$$\frac{dm_Z}{dt} = \frac{dm_E}{dt} + \frac{dm_A}{dt} + \frac{dm_B}{dt} \tag{1}$$

$$\frac{dT_Z}{dt} = \frac{1}{m_Z c_v}\left(-p_Z \frac{dV_Z}{dt} + \frac{dm_E}{dt}h_E + \frac{dm_A}{dt}h_A + \frac{dQ_B}{dt} - \frac{dQ_W}{dt} - c_v T_Z \frac{dm_Z}{dt}\right) \tag{2}$$

The pressure p_Z is given by the Ideal Gas Law: $p_Z = \dfrac{R\, m_Z\, T_Z}{V_Z}$.

We use the notations

m_E/m_A	mass flowing through inlet/outlet valve
m_B	mass of fuel
h_E/h_A	enthalpy which is added or removed through inlet/outlet valve
Q_B	combustion energy
Q_W	heat loss caused by heat exchange with the cylinder surface

C_v, c_p, $\kappa = \frac{c_p}{c_v}$ and R are thermodynamic constants.

The right hand side of the o.d.e. system (1),(2) depends on geometric and experimental data and partly discontinuous empirical functions.

E.g., the mass flow through the valves is described by

$$\frac{dm_E}{dt} = -\rho_{ZE}^K\, u_{ZE}^K\, A_E \tag{3}$$

$$\frac{dm_A}{dt} = \begin{cases} -A_A\sqrt{2\rho_Z p_Z}\cdot\psi(p_{AT},p_Z) & \text{for}\quad p_Z \geq p_{AT} \\ A_A\sqrt{2\rho_{AT}p_{AT}}\cdot\psi(p_Z,p_{AT}) & \text{for}\quad p_Z < p_{AT} \end{cases} \tag{4}$$

where $\rho_Z = \frac{m_Z}{V_Z}$ is the gas density in the cylinder, p_{AT} the atmospheric pressure and ψ a given function with a split domain of definition. ρ_{ZE}^K and u_{ZE}^K denote coupling quantities: the gas density and velocity at the boundary between cylinder and pipe.

A_E and A_A are the cross-sections of inlet and outlet valve openings:

$$A_{E/A}(t) = d_{E/A}\, vh_{E/A}(\omega t)\, \pi \sin\beta \tag{5}$$

where the distances d_E, d_A and the angle β are known, and vh_E, vh_A are functions of t as shown in Fig. 2 (U and L denote the upper and lower limit of piston motion respectively).

The enthalpies h_E and h_A have a split domain of definition and depend on coupling quantities. Empirical functions are used for the combustion energy Q_B and the heat loss Q_W at the cylinder surface, see Woschni [6], [7].

For a more detailled description of the cylinder model see Engl [2], Stark et al. [3].

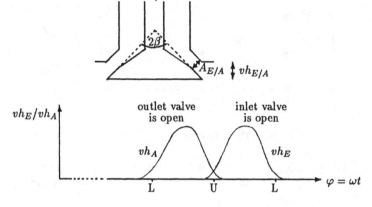

Figure 2: Valve Geometry, Valve Opening and Closing

The gas flow within the inlet pipe is described for the unsteady, one-dimensional, compressible and non-isentropic case which leads to the following hyperbolic system of partial differential equations for density ρ, velocity u and pressure p:

$$\frac{\partial}{\partial t}\begin{pmatrix} \rho \\ u \\ p \end{pmatrix} + \begin{pmatrix} u & \rho & 0 \\ 0 & u & \frac{1}{\rho} \\ 0 & \kappa p & u \end{pmatrix} \frac{\partial}{\partial x}\begin{pmatrix} \rho \\ u \\ p \end{pmatrix} = 0 \qquad (6)$$

(x: space dimension). The boundary values are functions of the cylinder values on one side and atmospheric gas properties on the other.

Since the cylinder model is based on many empirical data and functions, it is not meaningful to look for a solution of high accuracy. A numerical method of high order (using switching functions to handle the discontinuities of the right hand side) does not seem suitable and efficient, see Rentrop [5]. The explicit first order method of Euler was used. The hyperbolic system describing the gas flow in the pipe was solved by two different methods: the Method of Lax-Wendroff and a Method of Lines. Both methods were implemented with mesh adaption techniques taking stability restrictions into account. Besides a shorter calculation time of the second method the difference in the solution of both methods is negligible compared to the effects caused by using different coupling methods.

3 The Coupling Problem

We now consider the coupling of the systems cylinder and inlet pipe on one side, inlet pipe and atmosphere on the other.
The mentioned numerical methods are applied to the cylinder and pipe equations by using uniform discretization in the pipe ($\Delta x = const.$) and variable time steps Δt. Fig. 3 shows the situation for a special time step.
For the following we use the notation $\rho_j^n = \rho(x_j, t_n)$, $p_j^n = p(x_j, t_n)$, etc. .
The atmospheric values are given at any time. We assume that ρ_j^n, p_j^n, u_j^n and T_j^n, $0 \le j \le M$, where ρ, p, u and T denote gas density, pressure, velocity and temperature

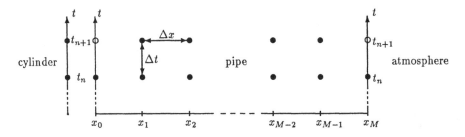

Figure 3: Discretization

respectively, as well as the gas values in the cylinder are known for $t = t_n$. The explicit numerical schemes yield ρ_j^{n+1}, p_j^{n+1}, u_j^{n+1} and T_j^{n+1}, $1 \leq j \leq M - 1$, and the o.d.e. solution for the cylinder values at $t = t_{n+1}$.

The values at meshpoints (x_0, t_{n+1}) and (x_M, t_{n+1}) are not necessarily equal to the cylinder values and atmospheric values respectively, since our model involves discontinuities at the boundaries of the pipe. We were studying three different coupling methods for finding the unknowns at meshpoints (x_0, t_{n+1}) and (x_M, t_{n+1}) which are now presented for the coupling cylinder - inlet pipe.

A Quasisteady Method

Assuming isentropic flow ($p\rho^\kappa = const.$) and conservation of enthalpy ($c_p T + \frac{u^2}{2} = const.$) between cylinder and pipe at any time we obtain the following system of equations for the case of an open inlet valve and $p_1^{n+1} \leq p_Z^{n+1}$:

$$p_0^{n+1} = p_1^{n+1} \ , \ T_0^{n+1} = T_Z^{n+1} (\frac{p_0^{n+1}}{p_Z^{n+1}})^{\frac{\kappa-1}{\kappa}} \ , \ u_0^{n+1} = \sqrt{2c_p(T_Z^{n+1} - T_0^{n+1})} \tag{7}$$

where the subscript "Z" denotes the cylinder values. ρ_0^{n+1} is determined by the Ideal Gas Law.

The other cases (open inlet valve and $p_1^{n+1} > p_Z^{n+1}$; closed inlet valve) yield equations of similar structure which can easily be solved for the unknowns.

A Method of Characteristics

In order to find coupling conditions we now use the characteristic form of the pipe equations

$$\begin{array}{rcl} (p_t + (u + a) p_x) + \rho a (u_t + (u + a) u_x) & = & 0 \\ (p_t + (u - a) p_x) - \rho a (u_t + (u - a) u_x) & = & 0 \\ (p_t + u p_x) - a^2 (\rho_t + u \rho_x) & = & 0 \end{array} \tag{8}$$

where the lower subscripts denote partial derivatives and $a = \sqrt{\kappa \frac{p}{\rho}}$ is the sound speed.

At $x = x_0$, $t = t_n$ all time derivatives are replaced by forward difference quotients, and in each equation the spatial derivatives are approximated by backward (forward) difference quotients if the corresponding characteristic slope at $x = x_0$, $t = t_n$ is positive (negative). For example, for $u_0^n < 0$ the third equation is approximated by

$$\frac{p_0^{n+1} - p_0^n}{\Delta t} + u_0^n \frac{p_1^n - p_0^n}{\Delta x} - (a^2)_0^n \left(\frac{\rho_0^{n+1} - \rho_0^n}{\Delta t} + u_0^n \frac{\rho_1^n - \rho_0^n}{\Delta x} \right) = 0 \tag{9}$$

A positive characteristic slope yields dependence on cylinder values. A linear system of equations has to be solved for the unknowns.

A Combination of the Quasisteady Method and the Method of Characteristics
While the first method uses physical boundary conditions rather than taking the character-istic curves of the hyperbolic system into account, the second method assumes some kind of continuation of the characteristics into the cylinder. Therefore the third method only considers the characteristics with negative slope. That means we only use those equations of the characteristic form for which $\frac{dx}{dt}$ is negative, where $\frac{dx}{dt} = u+a$, $\frac{dx}{dt} = u-a$ or $\frac{dx}{dt} = u$. The system of coupling conditions is completed by equations describing quasisteady gas flow (e.g., conservation of enthalpy is claimed). The cases of inflow and outflow have to be treated separately.

4   Numerical Results

As a main result of our work we present that different coupling methods yield quite different approximate solutions, see Fig. 4.

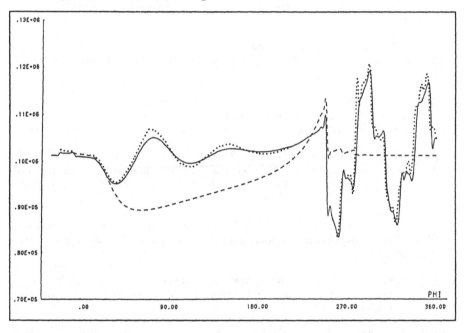

Figure 4: Pipe pressure p as a function of the crank angle φ for fixed $x = \frac{L}{2}$ ($\varphi = \omega t$, L = length of pipe); $\cdots\cdots$ Quasisteady Method, $-----$ Method of Characteristics, $\underline{\qquad}$ Combination Method

As a general interpretation of Figure 4 we see the time history of the pressure at a fixed point in the pipe. As long as the inlet valve is open, the pressure behaves quite smoothly, whereas oscillations occur after the valve is closed at $\varphi = 240^0$ $(t = \frac{\varphi}{\omega})$. This is nicely reproduced when the quasisteady or the combination method coupling conditions are used. The quasisteady flow coupling is based on physical conservation laws, which seems to be a straightforward approach. We did not expect a different solution by just changing the coupling conditions according to the method of characteristics. These effects are independent of the used numerical schemes.

Obviously the coupling problem is of general interest and cannot be solved on the discrete level of numerical schemes as it is usually done in the literature. In Quarteroni, Landriani [4] a similar surprising behavior of the solution of a hyperbolic-parabolic system with different senseful coupling is reported.

A better understanding of the coupling problem for a simple test model is necessary on the analytical and discrete level.

References

[1] Dubbel, *Taschenbuch für den Maschinenbau.* Springer-Verlag, Berlin, Heidelberg, New York, Tokyo, 1983

[2] G.Engl, *Mathematische Modellbildung und numerische Studien beim Verbrennungsmotor.* Diplomarbeit, Mathematisches Institut, Technische Universität München, August 1989

[3] H.-G.Stark, H.Trinkaus and Ch.Jansson, *The Simulation of the Charge Cycle in a Cylinder of a Combustion Engine.* Berichte der Arbeitsgruppe Technomathematik, Bericht Nr.31, Universität Kaiserslautern, September 1988

[4] A.Quarteroni and G.Sacchi Landriani, *Iteration by Subdomains in Numerical Fluid Dynamics.* Proceedings of the Third German-Italian Symposium: "Applications of Mathematics in Industry and Technology", eds.: V.Boffi, H.Neunzert, pp.54-76. Teubner, Stuttgart, 1989

[5] P.Rentrop, *Unterlagen zum Carl-Cranz Kurs V1.18, Rechenverfahren und Programme für die Bewegungsgleichungen mechanischer Systeme.* Oberpfaffenhofen bei München, 29.10.-30.10.1987

[6] G.Woschni and F.Anisitis, *Eine Methode zur Vorausberechnung der Änderung des Brennverlaufs mittelschnellaufender Dieselmotoren bei geänderten Betriebsbedingungen.* MTZ 34, 4, 1973, 106-115

[7] G.Woschni and J.Fieger, *Experimentelle Bestimmung des örtlich gemittelten Wärmeübergangskoeffizienten im Ottomotor.* MTZ 42, 6, 1981, 229-234

A SLOW FLOW PROBLEM ARISING IN CONTINUOUS ELECTRODE SMELTING

A. D. FITT
Faculty of Mathematical Studies
University of Southampton
Southampton SO9 5NH U.K.

ABSTRACT A slow flow model is proposed for the determination of the effective viscosity of a sample of pitch and anthracite electrode material. The case of a tall, thin sample is considered and a simple formula is derived for the viscosity. The boundary layer structure at the edges of the sample is also considered. Finally, numerical calculations are performed using a boundary element method.

1. Continuous Electrode Smelting

The problem considered herein concerns the determination of the 'effective viscosity' of a material used to make continuously consumed electrodes, whose function is to conduct large amounts of electrical energy to the centre of a blast furnace for the production of various kinds of alloy and steel. The material, traditionally known as 'paste', is composed of a mixture of anthracite fines of widely differing sizes, held together by a pitch binder. The mixture is solid at room temperature but begins to flow as the temperature is increased. Although the mixture is clearly a multiphase fluid, and under some circumstances can exhibit segregation, or phase separation (for details of a theoretical study examining this phenomenon see BERGSTROM et al. (1989)) there is much practical interest in determining its effective viscosity by treating it as though it were a single phase mixture. By further assuming that the paste behaves like a Newtonian (highly) viscous fluid, we acknowledge the fact that although this is certainly a large oversimplification, an accurate constitutive law for the mixture would not be easy to propose because of the wide variation in size of the anthracite fines, and sensitive dependence on temperature.

For the purposes of this study we will assume that the temperature is fixed and consider one of the many different tests used, known as the 'velocity test.' Here a sample of the paste is placed on a flat surface, and a moveable plate is placed on the top of the sample. The plate is moved vertically downwards with a prescribed velocity, squashing the sample, and the effective viscosity is inferred from the 'bulge' exhibited by the lower portion of the sample. (There are other tests including those involving prescribed loads which may be treated similarly, but space does not permit their discussion here. [1])

---

[1] Details of other tests may be found in FITT & AITCHISON (1991).

M. Heiliö (ed.), Proceedings of the Fifth European Conference on Mathematics in Industry, 223–228.
© *1991 B. G. Teubner Stuttgart and Kluwer Academic Publishers. Printed in the Netherlands.*

2. A Slow Flow Model

To model the flow of the paste sample, we begin by non-dimensionalizing the Navier-Stokes equations. In order to simplify the analysis presented here, we assume that the sample is a two-dimensional rectangle; normally in practice the sample is cylindrical, and such geometries can be dealt with in a exactly similar fashion to that described here. Taking unit vectors i, j, and assuming that the sample has height h and semi-width L, we scale lengths with h, velocities with a representative velocity U_∞, time with h/U_∞ and pressures and stresses with $\mu U_\infty / h$. This gives the equations

$$Re[q_t + (q.\nabla)q] = -\nabla p + \nabla^2 q - \frac{Re}{Fr} j, \qquad \nabla.q = 0.$$

Here the Reynolds and Froude numbers are defined respectively by $Re = hU_\infty \rho / \mu$, $Fr = U_\infty^2 / gh$. With typical values of $h \sim 1\text{m}$, $U_\infty \sim 1\text{m/hr}$, $\rho \sim 3\text{gm/cm}^3$ and $\mu \sim 10^8$ Pa sec, we find that $Re \sim Fr \sim O(10^{-8})$, leaving the slow flow equations with body force:

$$\nabla p = \nabla^2 q - \alpha j, \qquad \nabla.q = 0.$$

We impose standard no-slip boundary conditions on the bottom surface $y = 0$, stress-free conditions and the usual kinematic constraint on the sample side $x = \xi(y, t)$ say, whilst on the top surface we have $u = 0$, $v = \dot{s}(t)$, corresponding to the imposed velocity of the 'pusher plate'. To attack the full problem posed here a numerical approach is required. However, there are some situations in which analytical results may be obtained.

3. A Tall, thin sample.

Suppose the sample is tall and thin so that $L/h = \epsilon \ll 1$. Making the scalings $x = \epsilon X$, $u = \epsilon U$ and $\xi = \epsilon \eta$, the equations of motion become

$$p_X = U_{XX} + \epsilon^2 U_{yy}, \quad \epsilon^2 p_y = v_{XX} + \epsilon^2 v_{yy} - \epsilon^2 \alpha, \quad U_X + v_y = 0.$$

These may easily be solved by setting $U = U_0 + \epsilon U_1 + \epsilon^2 U_2 + ...$ and using similar expansions for v, p and η. Imposing symmetry conditions $U = v_X = 0$ on $X = 0$, the kinematic boundary condition on the free surface, and considering the y-component of the stress-free boundary condition on the free surface to order ϵ^2, we find that the solution is

$$v = v_0(y, t) + \epsilon v_1(y, t) + \epsilon^2 [\frac{X^2}{2}(\alpha - 3v_{0yy}) + q(y, t)]$$

$$U = -X v_{0y} - \epsilon X v_{1y} + \epsilon^2 [\frac{X^3}{2} v_{0yyy} - X q_y]$$

where η_0 and v_0 satisfy

$$\dot{\eta}_0 + (\eta_0 v_0)_y = 0, \quad 4(\eta_0 v_{0y})_y = \alpha \eta_0.$$

These equations admit two boundary conditions for v_0 and an initial condition for η_0. Clearly the correct ones to impose are $v_0 = 0$ on $y = 0$ and $v_0 = \dot{s}(t)$ on $y = s(t)$; evidently the boundary conditions for u cannot be satisfied and there will be 'inner' regions near to the top and bottom of the sample where a boundary layer analysis will be required to complete the details of the solution. The two equations for η_0 and v_0 are in general not easy to solve, but for small times the viscosity may be estimated from the bulge in the following manner: Assuming that $\eta_0(y,0) = 1$ and $v = -V_T$ on $y = 1$ where V_T is non-dimensional velocity, we find that

$$\eta_0 = 1 + \frac{t}{8}(8V_T - 2\alpha y + \alpha) + O(t^2)$$

$$v_0 = \frac{\alpha y^2}{8} - y\left(V_T + \frac{\alpha}{8}\right) + \frac{t}{192}(\alpha y(2\alpha y^2 - 3\alpha y + \alpha - 24yV_T + 24V_T)) + O(t^2).$$

It has already been noted that this solution is not valid near to $y = 0$. We shall assume however that the maximum 'bulge' is occurs at the boundary layer edge, which corresponds to the point $y = 0$. Re-dimensionalizing shows that if the top plate moves with velocity U_∞ and the maximum semi-width of the sample is BL, then this is related to the effective viscosity by

$$\mu = \frac{h^2 \rho g t}{8((h(B-1) - U_\infty t)}.$$

Unfortunately experimental results are unobtainable, so the best which can be achieved is to compare the theory with a numerical 'experiment'. Figure (1) shows a comparison between the theory and a numerical solution to the full problem which was calculated by a finite element method in which the (p, u, v)-version of the Stokes equations was written in stress-divergence form so that the Galerkin method gave the stress conditions as natural boundary conditions. The topology of the grid was maintained by the time-stepping of exterior and interior nodes. The aspect ratio of the 'experimental' sample was 10:1 and the 'experimental' viscosity was 10^8 Pa Sec. Clearly for small times the theory gives a satisfactory estimate of the viscosity.

Some consideration of the boundary layers which exist near to the top and bottom of the sample is necessary. Considering the bottom of the sample (the top may be treated similarly), it is clear that the additional scalings $y = \epsilon Y$, $v = \epsilon V$ must be made. This leads to the equations

$$P_X = U_{XX} + U_{YY}, \quad p_Y = V_{XX} + V_{YY} - \epsilon\alpha, \quad U_X + V_Y = 0$$

with boundary conditions $U = V = 0$ on $Y = 0$, stress-free conditions on $X = \eta(Y,t)$, and symmetry on $X = 0$. There is also a matching condition which must be satisfied as $Y \to \infty$. For small times the matching condition is known from the solution given above, the free boundary remains vertical to lowest order, and a stream function Ψ may be introduced which satisfies the biharmonic equation. By setting $\Psi = -KYX + \Phi$, where $K = -(V_T + \alpha/8)$, in order to retrieve a problem where $\Phi \to 0$ as $Y \to \infty$, we are finally required to find a function $\Phi(X,Y)$ which is biharmonic on the semi-infinite strip $\{Y \geq 0, X \in [0,1]\}$, vanishes as $Y \to \infty$, and satisfies $\Phi = \Phi_{XX} = 0$ at $X = 0$, $\Phi = 0$, $\Phi_Y = KX$ at $Y = 0$ and $\Phi_{YY} - \Phi_{XX} = \Phi_{XXX} + 3\Phi_{YYX} = 0$ at $X = 1$. This may be accomplished fairly easily by writing Φ as a Papkovitch-Fadle eigenfunction expansion (for fuller details see, for example SPENCE (1978)) in the form

$$\Phi(X,Y) = \sum_n c_n \phi_n(X) e^{-\lambda_n Y}$$

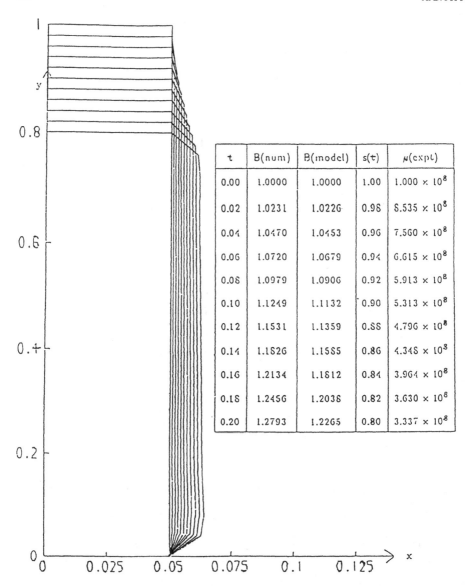

t	B(num)	B(model)	s(t)	μ(expt)
0.00	1.0000	1.0000	1.00	1.000×10^8
0.02	1.0231	1.0226	0.98	8.535×10^8
0.04	1.0470	1.0453	0.96	7.560×10^8
0.06	1.0720	1.0679	0.94	6.615×10^8
0.08	1.0979	1.0906	0.92	5.913×10^8
0.10	1.1249	1.1132	0.90	5.313×10^8
0.12	1.1531	1.1359	0.88	4.796×10^8
0.14	1.1826	1.1585	0.86	4.348×10^8
0.16	1.2134	1.1812	0.84	3.964×10^8
0.18	1.2456	1.2038	0.82	3.630×10^8
0.20	1.2793	1.2265	0.80	3.337×10^8

Figure 1 : Sample shape, numerical calculations and
comparison with theory

where

$$\phi_n(X) = \left(\frac{1}{\lambda_n \cos^2 \lambda_n}\right)(X \cos \lambda_n \cos \lambda_n X + \sin \lambda_n \sin \lambda_n X)$$

(the scaling factor has been included for convenience) and the Papkovitch-Fadle eigenvalues are members of the doubly-infinite family complex solutions to

$$\lambda + \sin \lambda \cos \lambda = 0$$

with real part greater than zero. The relevant eigenvalues may easily be computed using the Newton-Raphson method, for example, and to complete the details of the boundary layer structure it remains only to show that the eigenfunction coefficients c_n may be determined to satisfy the boundary conditions

$$0 = \sum_n c_n \phi_n(X), \qquad KX = \sum_n -\lambda_n c_n \phi_n(X).$$

Ostensibly the task of finding complex numbers c_n to satisfy the above conditions is easy; matters are complicated however by the fact that the eigenfunctions are not mutually orthogonal. In some circumstances this defect may be remedied by constructing so-called biorthogonal functions. To be specific, suppose we define

$$\boldsymbol{f} = (f_1(X), f_2(X), f_3(X), f_4(X)) = (\Phi_{XY}, \Phi_{XX}, Q, P)\,|_{y=0}$$

where $P = \nabla^2 \Phi$ and Q is the harmonic conjugate of P. If we now use the eigenfunction expansion to define functions $\phi_{nk}(k = 1, 4)$ via the expression

$$f_k(X) = \sum_n c_n \phi_{nk}(X),$$

Then it is possible to find 'biorthogonal' functions $\beta_{m1}(X)$ and $\beta_{m3}(X)$ so that

$$\int_0^1 \beta_{m1} \phi_{n1} + \beta_{m3} \phi_{n3} dX = \delta_{mn},$$

so that the coefficients c_n may be calculated directly by quadrature. The same procedure is possible for the functions ϕ_{n2} and ϕ_{n4}, so that when the problem is such that either the f_1 and f_3 or the pair f_2 and f_4 are prescribed as data, calculation of the coefficients is easy. Unfortunately only these two data prescriptions may be dealt with in this way - for other 'non canonical' problems such biorthogonal functions do not exist. Since in our case the data prescribed was Φ and Φ_y, which amounts to f_1 and f_2, the problem is of non-canonical type.

One obvious method calculating the c_n is by collocation, which gives rise to an infinite set of linear equations. It is then possible to proceed by solving a truncated (say $N \times N$) system of equations to approximate the c_n. During this procedure, great care must be taken to ensure that a diagonally dominant matrix is produced, thereby ensuring that as $N \to \infty$ the solutions to the truncated system converge to the solutions of the infinite system. This may however be accomplished by using the 'optimal weighting functions' introduced by SPENCE (1978), so completing the determination of the boundary layer structure.

4. Conclusions

A model has been presented which allows the effective viscosity of materials whose flow is governed by the slow flow equations to be determined. In the case of a tall, thin sample simple estimates for the viscosity may be found which give acceptable agreement with numerical calculations for the full problem performed using the boundary integral method. The boundary layer structure at the top and bottom of the sample leads to a non-canonical biharmonic problem which is of some interest in itself. For the problem with more general geometries however, numerical methods must be used.

Acknowledgements The numerical calculations were performed by Dr. J. Aitchison from R.M.C.S. Shrivenham. The author is also grateful to Dr. P. Wilmott from Oxford for many valuable suggestions and discussions. The problem was first considered at the 1988 Study Group with Industry, Heriot-Watt University, and originated from Dr. Svenn Halvorsen of Elkem Ltd, Norway.

References

[1] Bergstrom, T., Cowley, S., Fowler, A.C. & Seward, P.E. (1989) 'Segregation of carbon paste in a smelting electrode', IMA J. Appl. Math **42** 83-99.

[2] Fitt, A.D. & Aitchison, J.M. (1991) 'Determining the effective viscosity of a carbon paste used for continuous electrode smelting', In preparation.

[3] Spence, D.A. (1978) 'Mixed boundary value problems for the elastic strip : The eigenfunction expansion', Technical Summary Report 1863, MRC, Univ. of Wisconsin, Madison

Faculty of Mathematics, University of Southampton, Southampton SO9 5NH U.K.

NUMERICAL PROBLEMS
OF A CONVECTION-DIFFUSION EQUATION
WITH ARRHENIUS SOURCE TERM

R.J. Gathmann and F.K. Hebeker

The following mathematical problem appears as a strongly simplified, but plausible model of flows in internal combustion engines. Let us assume, a) the flow is one-dimensional and incompressible; b) one exothermic chemical reaction

$$A + B \rightarrow 2P + e \tag{1}$$

(e = heat release) is taken into account; c) for all mole fractions X_i ($i = A,B$) of the reactants we assume

$$X_i + \frac{T}{T_e} = const. \tag{2}$$

during the chemical reactions. Here denotes T, T_e the temperature or temperature equivalent of the heat release, resp. Assumption a) implies the flow velocity v constant in space, moreover, we assume given v. Assumption c) is something like a constant-enthalpy hypothesis. Consequently, the generally complex differential system simplifies to the scalar convection-diffusion equation including a source term:

$$T_t + vT_x = DT_{xx} + \omega(T) , \quad x \in (0,1), \quad t > 0, \tag{3}$$

with an Arrhenius law

$$\omega(T) = B \cdot \left(X_0 - \frac{T - T_0}{T_e} \right)^2 e^{-\frac{T_A}{T}} \tag{4}$$

($X_0 = 0.5$, $T_0 = 1$ = cold temperature, T_A = activation temperature). This source term is arbitrarily smooth, so that the mathematical analysis of (3)-(4) (supplemented with suitable boundary conditions) affirms (locally) existence, uniqueness, and regularity (e.g. [1]). The major problems concerning the numerical treatment are as follows: a) the reaction rate constant is generally large, say

$$B = 10^4 \ldots 10^6 \tag{5}$$

(the differential equation is stiff); b) steep gradients have to be taken into account (shock-capturing methods are required).

M. Heiliö (ed.), *Proceedings of the Fifth European Conference on Mathematics in Industry*, 229–233.
© 1991 *B. G. Teubner Stuttgart and Kluwer Academic Publishers. Printed in the Netherlands.*

1. Interchange of Diffusion and Reaction

The interchange of diffusion and reaction poses complex and mathematically interesting problems. Assume $v = 0$ and a diffusion-reaction wave $T(t,x) = \eta(x + ct)$. Then we are left with this nonlinear two-point eigenvalue problem for the wanted phase velocity c:

$$- D\eta'' + c\,\eta' = \omega(\eta) \quad \text{in } (0,1), \tag{6}$$

with boundary conditions

$$\eta(0) = T_0 , \quad \eta(1) = T_b \tag{7}$$

($T_b = 4 =$ burning temperature). A lot of mathematical contributions are devoted to problem (6),(7) over the past decade (see [6]) . From [6] we cite this result: if autoignition is suppressed (i.e. $\omega(T_0) = 0$), then there exists a unique pair (c, η) solving (6),(7), and, moreover, high energy asymptotics provide us with an asymptotic analytical formula (in case of an infinite domain)

$$c_\infty = \left\{ 2D \lim_{T_A \to \infty} \int_{T_0}^{T_b} \omega(T)\mathrm{d}T \right\}^{1/2} . \tag{8}$$

Since $\omega(T_0)$ is small (4), the assumption above is approximately met here. Concerning the effect of approximating the infinite by a sequence of finite domains see [6] .

2. Numerical Approach - Crucial Task

Due to the essentially different physical phenomena appearing in equation (3), we choose an operator splitting scheme of the second order to solve the problem numerically (cf. [7] , e.g.). Here is the convection step (Φ_C) carried out by MacCormack's scheme including FCT (flux-corrected transport) [3] which has some shock-capturing feature, whereas the diffusion step (Φ_D) is done by Crank-Nicolson's scheme, and the chemical reaction step (Φ_R) requires an adaptive ODE-solver. See [4] for more details. Altogether, the evolution scheme

$$T^{(n+2)} = \Phi_R^{1/2}\Phi_C\Phi_D\Phi_R\Phi_D\Phi_C\Phi_R^{1/2}T^{(n)} \tag{9}$$

($T^{(n)}$ denoting temperature on time level n) is accurate of the second order, if all the well known stability requirements are met.

The following crucial task appears when dealing numerically with coupled convection-reaction problems: if a function with steep gradients is approximated, then the interchange of (indispensible) numerical diffusion with (generally) high reaction rates may introduce an artificial wave from regions of higher to regions of lower temperature, the wave speed of it may destroy all the numerical information. This fundamental task (by no means restricted to a particular numerical scheme) has been pointed out by [2],[5].

We will describe an elementary approach to damp out this artificial wave: estimate its phase velocity and modify the source term appropriately! Let δT_i denote the difference between exact and computed temperature. Clearly,

$$\delta T_i = (T_i - T_{i-1})\lambda \tag{10}$$

(T_i =approximate temperature at mesh point x_i), where

$$\lambda = \frac{c_a}{c_g} = c_a \frac{\Delta t}{\Delta x} . \tag{11}$$

This artificial velocity c_a is estimated via problem (6),(7), where $D = D_a$ is now the (unknown) numerical diffusion of the scheme. Formula (8) gives an approximate relation between D_a and c_a . Together with a second relation $c_a \sim \dfrac{D_a}{L_a}$ ($L_a \sim \Delta x$ denoting a characteristic length) obtained by dimension analysis this results in the formula

$$c_a = \varepsilon B \Delta x , \tag{12}$$

where the proportionality factor ε may be estimated by a single test run, for instance.

This relatively crude method leads to a practically useful modification of the source term depending on steepness and direction of the gradient of the temperature front; instead of ω use

$$\overline{\omega}(T_i) = \begin{cases} \omega((1-\lambda)T_i + \lambda T_{i-1}) : & \text{if } T_{i+1} - T_{i-1} > M \\ \omega((1-\lambda)T_i + \lambda T_{i+1}) : & \text{if } T_{i+1} - T_{i-1} < -M \\ \omega(T_i) & : \quad elsewhere \end{cases}$$

This modification indicates a practical way of handling the problem and might stimulate further work on this subject.

3. Numerical Results

We consider three cases to test the present approach. The computer code was run with a decomposition of the domain (0,1) into 100 equally sized intervals. For the correction procedure the parameters $\varepsilon = 0.004$, $M = 1$ have been chosen.

TEST 1: A diffusion-reaction wave ($v = 0$) has been computed where the parameters

$$T_A = 100, \quad T_e = 2, \quad B = 10^{28}, \quad D = 5, \quad \Delta t = 7.5 \; 10^{-6}$$

are used. The assumption of high activation energy is met, so that the computed phase velocity, $c = 32.51$, should approach the asymptotic value: in fact, $c_\infty = 31.42$.

TEST 2: We neglect diffusive effects for the second example. An analytic solution

$$T(t,x) = \begin{cases} T_b: & 0.4 + 10t \leq x < 0.6 + 10t \\ T_0: & elsewhere \end{cases}$$

is available, if autoignition is suppressed: set $\omega(T) = 0$ when $T < T_0 + 0.01$. Several numerical test runs have shown that, with the help of our modification (10)-(12), the computed phase velocity is nearly exact. See [4] for detailed results.

TEST 3: Finally, the temperature distribution in a combustion chamber has been computed. See the Fig. below. The differential equation (3)-(4) is supplemented with the boundary conditions

$$T = T_0 = 1 \text{ at } x = 0, \quad T_t + vT_x = 0 \text{ at } x = 1,$$

the latter one allowing smooth outflow. Initially, the medium is cold: $T(0,x) = 1$. The parameters used are

$$T_A = 10, \quad T_e = 6, \quad B = 5. \; 10^4, \quad v = 10, \quad D = 0.25, \quad \Delta t = 0.0005.$$

The temperature distribution is plotted all 50 time steps. It shows up clearly that the equilibrium is approached smoothly.

- Further examples are included in [4].

References

1. Bebernes J., Eberly D., Mathematical problems from combustion theory, New York etc. 1989.

2. Colella P., Majda A., and Roytburd V., Fractional step methods for reacting shock waves. In: Ludford G.S.S. (ed.), Combustion and chemical reactors, Vol. 2, pp. 459-477, Providence 1986.

3. Fletcher C.A.J., Computational techniques for fluid dynamics, Vol. 2, Berlin etc. 1988.

4. Gathmann R.J., Hebeker F.K., Modelling a one-dimensional convection-diffusion problem with an Arrhenius source term, IBM Heidelberg Scientific Center, Techn. Rep. 1989.

5. LeVesque R.J., Yee H.C., A study of numerical methods for hyperbolic conservation laws with stiff source terms, NASA Techn. Mem. 1988.

6. Marion M., Mathematical study of a model with no ignition temperature for laminar plane flames. In: Ludford G.S.S. (ed.), Combustion and chemical reactors, Vol. 2, pp. 239-252, Providence 1986.

7. Yee H.C., A class of high resolution explicit and implicit shock-capturing methods, VKI Lecture Series 1989.

R.J. Gathmann
Institut de Mecanique de Grenoble
Grenoble, France

F.K. Hebeker
IBM Scientific Center
Heidelberg, Germany

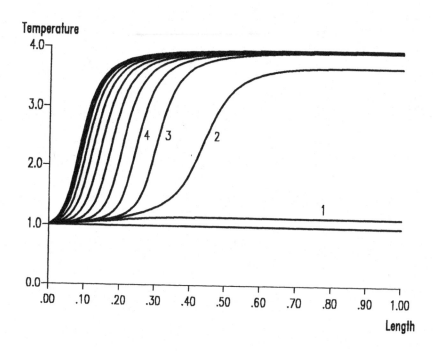

SILICON TECHNOLOGY DEVELOPMENT AND OPTIMIZATION BY INTEGRATED PROCESS, DEVICE AND CIRCUIT SIMULATION SYSTEM SATURN

A. GILG
SIEMENS AG, ZFE SPT 3
Corporate Research and Development
Otto-Hahn-Ring 6, D-8000 München 83, Germany

Abstract. Electronics world market is characterized by rapid innovations in microelectronics. Computer simulation has become an indispensible tool in progress of silicon process technology and chip design. For further improvement of time/cost effectiveness we have developed our SATURN system.
Firstly SATURN is a tool for physical modelling and understanding of physical processes. After modelling and characterization the system is used for prediction. During early stages of development engineers are assisted in identification of optimal process parameters. This reduces the number of time and money consuming fabrication cycles significantly. Even design parameters can be extracted by simulations before fabrication of first test chips. We experienced good agreement of predicted and measured data. So it was possible to parallize technology development and first chip design. This gains further time advantage since a fabrication run typically lasts several weeks or even months. Finally predictive power of simulation has resulted in computer-aided technology optimization. Optimization routines have been made available in SATURN to automate engineering work cycles.
Kernels of these simulators are sophisticated numerical algorithms to solve the underlying large and highly nonlinear systems of PDE numerically [1,2]. Our application problem sizes can be handled by supercomputers e. g. vector processors only, where specially adapted or new algorithms are necessary. Ongoing work includes exploration of simulation tools for further yield enhancement. First results are available for simulation of equipment fabrication effects governed by fluid dynamic models [3].

Simulation Tools:
Process Simulation. We analyze sequences of critical fabrication steps (implantation, oxidation/ diffusion). Modelling of dopant diffusion is described by a system of equations

$$\text{div} (D(C)\text{grad } C \pm C \text{ grad } \psi) = \partial C/\partial t \tag{1}$$

M. Heiliö (ed.), Proceedings of the Fifth European Conference on Mathematics in Industry, 235–238.
© 1991 *B. G. Teubner Stuttgart and Kluwer Academic Publishers. Printed in the Netherlands.*

for each dopant concentration C, where diffusion matrix D depends on temperature and exihibits coupling effects of different dopants. Electrostatic potential ψ is fixed by a Poisson-type equation

$$\varepsilon \operatorname{div} \operatorname{grad} \psi = Q(C), \tag{2}$$

where ε denotes permittivity and Q describes space charge depending on dopant concentrations.

During oxidation this system (1, 2) is solved on a grid with moving boundaries due to volume expansion of silicon oxide. New techniques incorporating smooth transition zones of silicon oxide into a finite element code were developed for our process simulator MIMAS [1,2].

Modelling of more detailed physical effects, e.g. diffusion mechanisms influenced by point defect movement in silicon crystal lattice extend (1) by extra equations and coupling terms giving rise to numerical problems (stiffness). These systems are currently under investigation.

Device Simulation. Necessary input data for electrical characterization of devices include device geometry and doping profiles usually provided by process simulators. Standard device simulators solve Poisson´sequation (2) and electron and hole continuity equations

$$\partial n/\partial t = \operatorname{div} J_n - R, \text{ where } J_n = \mu_n (n \operatorname{grad} \psi - \operatorname{grad} n) \tag{3}$$

$$\partial n/\partial t = \operatorname{div} J_p + R, \text{ where } J_p = \mu_p (n \operatorname{grad} \psi - \operatorname{grad} p), \tag{4}$$

for electron density n and hole density p. These equations are, in general, highly nonlinear. Recombination rate R and the mobilities μ_n and μ_p are themselves sophisticated formulas for complex physical mechanisms. State of the art are computations applying finite difference or finite element methods in two or three spatial dimensions. After discretization usually a nonlinear Gauss-Seidel method, often called Gummel algorithm, and preconditioned conjugate gradient methods are used. Modelling of additional physical effects (e.g. high electrical field modifications called hot electron effects) also leads to supplementary equations. For device simulation sophisticated and reliable software developed at universities is available and integrated in SATURN (MINIMOS [5], GALENE [6]).

Circuit Simulation. Circuit simulators calculate the electric behaviour of circuits consisting of up to several thousands of devices. Single devices are no longer modelled by a system of partial differential equations but by large sets of nonlinear functions. Numerical problems in circuit simulation are to solve a large stiff unstructured implicit system of algebraic differential equations:

$$F(t, y(t), dy(t)/dt) = 0, \tag{5}$$

where y denotes a vector composed of voltages and currents. Problem complexity enforced use of vector processors and implementation of decoupling techniques like multi-rate algorithms in our simulator TITAN [7].

Supercomputing. Industrial applications have always been limited by speed and memory of computers. Thus we have spent quite a lot of effort in improving numerical algorithms for fast computation on our largest (vector) computer. For example the simulation kernel of a circuit simulator computes LU factorizations of sparse unstructured matrices. This usually prevents vectorization gain. But we developed a specialized algorithm (MVAR [4]) reducing computing times by a typical factor of 50 on a VP200 vector processor. All our simulators include specialized algorithms for use on vector computers.

Integrated Simulation System SATURN. Verification of process results is restricted by measuring techniques. Especially doping profiles are not accesssible in two or even three spatial dimensions and in resolution of less than one micron. There is only an indirect way to verify these profiles by investigation of electrical characteristics of the resulting structures and devices.

Thus stand-alone simulation tools required coupling to assist these engineering techniques. Kernels of our SATURN system are simulators from different simulation levels: process, device and circuit. They receive input data from user and data base and store back their results into this database for further processing and retrieval. SATURN user interface (system monitor) is a supervisor system with a UNIX©-shell like structure. It resides on a workstation while access to a high performance computer is fully integrated. SATURN system monitor also reduces computational efforts by a technique based on automatic book-keeping. Intermediate results of previous simulation runs are retrieved automatically if they are reusable for actual computations.

SATURN is used in verification and modelling but it is especially useful for prediction of process parameter modifications and even iterative optimization of technologies. For optimization support and parameter identification a first version of an optimization routine tool box is integrated into the supervisor shell. BFGS-type methods are available for interactive (linear and nonlinear) constrained optimization of objective functions in different simulation levels.

We succesfully predicted 16M DRAM technology and design parameters based on a 4M predecessor technology by use of SATURN: Comparison with fabricated test structures verified our simulation predictions within acceptable tolerances. Thus we reduced the number of expensive and time consuming fabrication cycles and gained several months of development time.

Conclusions. Physical modelling has become impossible without support by simulation tools. Lack of direct verification by measurement enforced coupling of simulators tools in a system. An integrated system even allows predictive optimization for industrial applications with reduced turn around times. Mathematical tasks in this simulation environment are summarized by development and improvement of

- parameter identification methods   for calibration and verification
- new numerical algorithms including
  analysis of new model equations   for modelling
- robust and reliable methods   for prediction and
- fast algorithms   for optimization tasks.

Future Trends. Application of process, device and circuit simulation tools to semiconductor technology development has been successful. Continuing improvement and enhancement of physical models is necessary especially for process modelling. With ongoing miniaturization 3d effects will dominate process results and development of 3d process simulation algorithms becomes essential. Predictive power of SATURN has resulted in first applications of computer-aided technology optimization. Emphasis is now to increase yield by equipment modelling and simulation [3], thus forcing progress in numerical techniques for these specialized fluid dynamic models.

Acknowledgments. The author appreciates numerous excellent contributions to SATURN by his colleagues at SIEMENS Silicon Process Technology Laboratory.

References.
[1]   Paffrath M., Steger K.: Numerical Solution of Diffusion Equations on Time-Variant Domains. In: Proc. NUPAD-III, Honolulu, June 3-4, 1990
[2]   Rank E.: A new finite element approach to the local oxidation of silicon . In: Miller, J.J.H. (ed.): Proc. NASECODE VI. pp.40-51. Dublin: Boole Press 1989
[3]   Ulacia J.I., Howell F.S., Körner H., Werner Ch.: Flow and reaction simulation of a Tungsten CVD reactor. Appl. Surface Sci. *38*, 370-385 (1989)
[4]   Steger K.: Vectorization of the LU-decomposition for circuit simulation. In: Sequin, C. (ed.): VLSI ´87, pp. 363-372. Elsevier, N. Holland 1988
[5]   Thurner M., Selberherr S.: Three-Dimensional Effects Due to the Field Oxide in MOS Devices Analyzed with MINIMOS 5. IEEE Trans. CAD-9 No. 8, 856-867 (1990)
[6]   Engl W.L., Kircher R., Bach K.H., Götzlich J.: Simulation of 3D-Devices with GALENE II. VLSI Process/Device Modelling Workshop, Tokyo (1987)
[7]   Schultz R., Wever U.: Combined Strategies for exploiting Time domain and Iteration Latency in TITAN. In: Proc. ESSCIRC ´90, Grenoble, 1990.

NOVEL ALGORITHMS FOR THE COMPUTATION OF COMPLEX CHEMICAL EQUILIBRIA AND APPLICATIONS

H. Greiner

Introduction: The modelling and simulation of many processes in nature and technology requires the efficient and reliable computation of chemical equilibria. Examples include chemical process design, combustion, metals production, nuclear accidents, chemical vapour deposition, light sources, geochemical phenomena and environmental chemistry. For an overview of the subject we recommend ref.1. Mathematically equilibrium in a closed chemical reaction system is defined by the minimum of the system's total free energy subject to mass balance constraints. Thus it is obtained as the solution of a linearly constrained minimization problem which for simple systems consisting only of a few species and phases can be readily calculated by standard algorithms. For complex systems (see ref.5 for an example), however, which may involve hundreds of species and (possible) phases these algorithms are often of no avail and special methods are called for. The following computational problems arise:

Problem A: a close estimate of the equilibrium has to be furnished in an efficient and reliable way. In particular the stable phase assemblage out of a multitude of possible phase combinations has to be determined.

Problem B: Once this estimate is available the exact equilibrium mole numbers should obtained by a fastly converging method (ideally Newton's method).

In this contribution we present a sketch of two effective algorithms for problems A and B.

Mathematical formulation: consider a chemical reaction system with m chemical elements with mole numbers $\mathbf{b} = (b_1, \ldots, b_m)$. The elements form the components of p phases (gaseous or condensed). Thermodynamically each phase is described by

M. Heiliö (ed.), Proceedings of the Fifth European Conference on Mathematics in Industry, 239–242.
© 1991 B. G. Teubner Stuttgart and Kluwer Academic Publishers. Printed in the Netherlands.

its free energy G^j (j=1,...,p) which depends on the s(j) mole numbers $\mathbf{n}^j = (n_1^j, \ldots, n_{s(j)}^j)$ of its component species possessing chemical potentials μ_i^j. Let N^j denote the total mole number and $\mathbf{x}^j = (x_1^j, \ldots, x_{s(j)}^j)$ the mole fractions of the j-th phase. For pure phases $G^j = \mu_i^j n_i^j$ is a linear function, whereas for multicomponent phases the pure components, the ideal and excess energy of mixing contribute to G^j (energies are expressed in units of RT):

$$G^j(\mathbf{n}^j) = \sum_{i=1}^{s(j)} \mu_i^j n_i^j + \sum_{i=1}^{s(j)} n_i^j \log(x_i^j) + G_{excess}^j(\mathbf{n}^j) \quad (j=1,...,p).$$

The G^j are homogeneous functions and can therefore be written in terms of molar free energies $\overline{G}^j$ as $G^j(\mathbf{n}^j) = N^j \overline{G}^j(\mathbf{x}^j)$. Finally $\mathbf{R}^j$ stands for the (m,s(j)) matrix which for $\mathbf{n}^j$ defines the mole numbers of the elemental atoms $\mathbf{R}^j \mathbf{n}^j$ contained in phase j and $\mathbf{R}$ abbreviates $(\mathbf{R}^1, \ldots, \mathbf{R}^p)$.

With these notations our minimization problem amounts to finding mole numbers $\mathbf{n}^j$ (j=1,...,p) such that

$$G = \sum_{j=1}^{p} G^j(\mathbf{n}^j) = MIN \ \ s.t. \ \ \sum_{j=1}^{p} \mathbf{R}^j \mathbf{n}^j = \mathbf{b}, \ \mathbf{n}^j \geq 0 \ \ j=1,...,p \qquad (1)$$

or equivalently total mole numbers N^j and mole fractions $\mathbf{x}^j$ such that

$$\sum_{j=1}^{p} N^j \overline{G}^j(\mathbf{x}^j) = MIN \ \ s.t. \ \ \sum_{j=1}^{p} N^j \mathbf{R}^j \mathbf{x}^j = \mathbf{b}, \ \ N^j \geq 0 \ \ j=1,...,p. \qquad (2)$$

Problem A: Problem (2) can be viewed as a linear program with infinitely many columns $(\overline{G}^j(\mathbf{x}^j), \mathbf{R}^j \mathbf{x}^j)^t$ and can thus be solved by an analogue of the revised simplex algorithm (generalized linear programming), which involves the repeated determination of mole fractions $\mathbf{x}^j$ minimizing

$$MIN \; (\overline{G}^j(x^j) + u^t R^j x^j) \quad j=1,...,p \tag{3}$$

for given multipliers u for each phase j (pricing operation). Hence the problem is decomposed into the tractable subproblems (3). Although the algorithm requires the $\overline{G}^j$ to be convex (i.e. stable phases), it can be extended to systems with nonconvex $\overline{G}^j$ (i.e. phases with miscibility gaps). In this case the absolute minimum of problem (3) has to be determined. Computational experience has demonstrated that generalized linear programming is a very efficient means for obtaining a close estimate of the equilibrium in multiphase systems. More details are provided in ref.2,3.

Problem B: To calculate the exact mole numbers the system of nonlinear equations (we can suppose that the pure phases have been eliminated, see ref.6) corresponding to problem (1)

$$\nabla G^j(n^j) + R^{j\,t} \, u = 0 \quad j=1,...,p \;\; and \;\; \sum_{i=1}^{p} R^j n^j = b \tag{4}$$

has to be solved for the n^j and Lagrange multipliers u. For systems with many species the solution of the corresponding linearized equations

$$\begin{bmatrix} \nabla^2 G & R^t \\ R & 0 \end{bmatrix} \begin{bmatrix} \Delta n \\ \Delta u \end{bmatrix} = \begin{bmatrix} -\nabla G - R^t u \\ b - Rn \end{bmatrix} \tag{5}$$

can in practice only be achieved by so-called range-space methods, which require the blockdiagonal matrix $\nabla^2 G$ to be invertible. But unfortunately we have $\nabla^2 G^j(n^j) n^j = 0$ due to homogeneity! To circumvent this difficulty we introduce the total mole numbers N^j as additional independent variables and define modified free energies

$$\hat{G}^j(n^j) = G^j(n^j) + \left(\sum_{i=1}^{s(j)} n_i^j \right) \log \left(\sum_{i=1}^{s(j)} n_i^j \right). \tag{6}$$

The linearization of the augmented equations

$$\nabla \hat{G}^j(\mathbf{n}^j) + \mathbf{R}^{j\,t}\,\mathbf{u} - (1+\log(N^j),\,\ldots,\,1+\log(N^j))^t = 0 \quad j=1,\ldots,p \qquad (7)$$

$$\sum_{j=1}^{p} \mathbf{R}^j \mathbf{n}^j = \mathbf{b}\,, \qquad \sum_{i=1}^{s(j)} n_i^j - N^j = 0 \quad for\ j=1,\ldots,p\,.$$

is now amenable to the range-space method: As we have shown in ref.4 the matrix blocks $\nabla^2 \hat{G}^j$ are necessarily positive-definite for stable phases. Furthermore the blocks corresponding to ideal phases become diagonal, which is a particular advantage for gaseous phases with many species. Thus the matrix $\nabla^2 \hat{G}$ can be readily inverted and an efficient implementation of Newton's method is accomplished. In comparison the so-called intermediate algorithms (ref.1) which approximate the Hessian $\nabla^2 \hat{G}^j$ in the linearized equations (7) by its ideal part show only linear convergence for highly nonideal systems. For more information on these matters see ref.4.

References

1) Smith, W.R. and Missen, R.W. (1982) Chemical Equilibrium Analysis, John Wiley

2) Greiner, H. (1988) "Computing complex chemical equilibria by generalized linear programming", Mathematical and Computer Modelling 10, 529-550

3) Greiner, H. (1988) "The Chemical Equilibrium Problem for a Multiphase System Formulated as a Convex Program", CALPHAD 12, 155-170

4) Greiner, H. (1991) "An efficient implementation of Newton's method for nonideal chemical equilibria", Computers&Chemical Engineering 15, 115-123

5) Greiner, H. and Schnedler, E. (1990) "Modelling High Temperature Transport Reactions", High Temperature Science 27, 199-208

6) Greiner, H. "An Algorithm for Linearly Constrained Programs with a Partly Linear Objective Function", Mathematical and Computer Modelling, to be published
Author's address: H.Greiner, Philips Forschungslabor GmbH, Weisshausstrasse, D-51 Aachen

Numerical Simulation of a Countercurrent Heat Exchanger
by Piecewise Linearization

G. Hendorfer, L. Peer, W. Stöger, Hj. Wacker *

1. Introduction

In [2] a model was presented for the numerical simulation of a countercurrent heat exchanger:

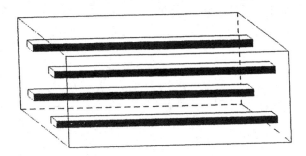

Though the model provided an approximation of a satisfying accuracy two draw-backs were observed. Firstly, pure substances could not be dealt with because the enthalpy $H(T)$ is not defined in the case T_b (bubble point) $= T_d$ (dew point): H is discontinuous. For mixtures we have continuity. Secondly the numerical effort was by no means neglectable: 5' and more (COMPAREX 7/78) depending on the complexitiy of the problem. For comparison: to optimize a flowsheet of an oxygen plant [1] consisting of 37 units and 55 streams needs approximately the same CPU-time. The following proposal could dispose of both drawbacks.

2. The Model

2.1. Assumptions and Basic Model

We refer to the same assumptions as met in [2]:

A1: No exchange of material between the tubes

A2: The pressure of each stream decreases linearly with l resp. $(L - l)$

---

* all: Math. Deptm. of Math., Univ. Linz

M. Heiliö (ed.), Proceedings of the Fifth European Conference on Mathematics in Industry, 243–247.

A3: Because of the high conductivity of the material of the heat exchanger (e.g. metal) we may assume the wall temperature to be constant over the whole crosscut at each l

A4: We neglect heat transfer by radiation and heat transfer by conduction in the mass stream direction

A5: No chemical reactions within the tubes

A6: Steady state conditions

However, we allow for phase transitions.

We get the following two point boundary value problem for the temperature of tube i in dependence on l.

$$\frac{dT_i(l)}{dl} = \frac{K_i \cdot U_i \cdot (T_w(l) - T_i(l)) - \frac{\partial H(T_i(l), P_i(l), (f_{ij}))}{\partial P_i} \cdot \frac{dP_i(l)}{dl}}{\frac{\partial H(T_i(l), P_i(l), (f_{ij}))}{\partial T}} \qquad (1 \le i \le N)$$

$$(1)$$

with

$$T_w(l) = \frac{\sum\limits_{i=1}^{N} K_i \cdot U_i \cdot T_i(l) \cdot d_i + K_w \cdot U_w \cdot T_A}{\sum\limits_{i=1}^{N} K_i \cdot U_i \cdot d_i + K_w \cdot U_w} \qquad (2)$$

shortly

$$\begin{cases} \frac{dT(l)}{dl} = f(l, T(l)) & T(l) = (T_i(l), \cdots T_N(l)) \\ T_i^f = \begin{cases} T_i(0) & d_i > 0 \\ T_i(L) & d_i < 0 \end{cases} & 1 \le i \le N \end{cases} \qquad (3)$$

2.2. A Simplified Model: Piecewise Linearization of $\overline{T}$

We made the additional assumption:

A7: The temperature T_i of stream i between two discretization points can be approximated by a piecewise linear function.

This is justified at least for a sufficiently fine discretization.

T_{ij} :   temperature of stream i at l_j

T_{wj} :   temperature of wall at l_j

We have by A7: $T_i(l) = T_{ij} + \frac{l-l_j}{\Delta l_j}(T_{ij+1} - T_{ij})$ $l \epsilon [l_j, l_{j+}$

$T_w(l) = T_{wj} + \frac{l-l_j}{\Delta l_j}(T_{w,j+1} - T_{w,j})$ $l \epsilon [l_j, l_{j+}$

We get the following nonlinear equational system for the $M(2N+1)$ variables H_{ij}, T_{ij}, T_{wj}.

heat transfer:

$$0 = h_{ij} \equiv H_{i,j+1} - H_{i,j} - \tilde{K}_i \int\limits_{l_j}^{l_{j+1}} (T_w - T_i)(l)dl \qquad \begin{array}{l} 1 \le i \le N \\ 1 \le j \le M \end{array} \qquad (4.1.)$$

Remark: T_w, T_i are piecewise linear; however phase transitions are to be considered.

enthalpy-temperature relationships:

$$0 = e_{ij} \equiv \begin{cases} T_{ij} - T(H_{i,j}; P_{ij}; f_{i1}) & \text{pure component streams} \quad 1 \le i \le N \\ H_{i,j+1} - H(T_{ij}; P_{ij}; (f_{ij})) & \text{else} \quad 1 \le i \le N, \ 1 \le j \le M \end{cases}$$
$$(4.2.)$$

By this formulation - using the inverse function - we have already disposed of the problem for pure substances arising from a discontinuous enthalpy at points where a phase transition takes place .

overall energy balance:

$$0 = o_j \equiv \sum_{i=1}^{N} |\tilde{K}_i|(T_{wj} - T_{ij}) + \bar{K}_w \cdot (T_{wj} - T_A) \qquad 1 \le j \le M \quad (4.3.)$$

boundary conditions:

$$0 = b_j \equiv \begin{cases} T_{i,j} - T_i^f & d_i = +1 \\ T_{i,M} - T_i^f & d_i = -1 \end{cases} \qquad 1 \le i \le N \qquad (4.4.)$$

3. Numerical Solution

3.1. Solution of the nonlinear system (4)

We use Newton's method with a damping strategy and exploit the sparse structure of the Jacobian. For semi globalization we use imbedding with respect to the K-values.

The Jacobian has a structure which is similiar to that resulting from Multiple Shooting technique:

$$J = \begin{pmatrix} A_1 & B_1 & & & & \\ & A_2 & B_2 & & & \\ & & A_3 & B_3 & & 0 \\ & & & \ddots & \ddots & \\ & 0 & & & \ddots & \\ & & & & A_{M-1} & B_{M-1} \\ & C & & & & A_M \end{pmatrix}$$

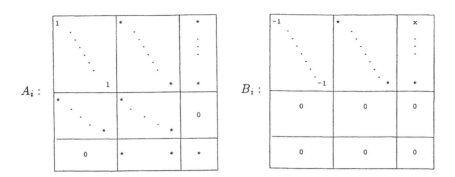

C and A_M are of a similiar (even sparser) structure.

After linearization the resulting system is solved by a block L-R decomposition exploiting the substructure indicated above. We have the recursion:

$$L_i \cdot A_i = -L_{i-1}B_{i-1}$$

The L-R decomposition of A_i is done explicitely and can be performed quite efficiently. Some care has to be taken w.r.t. stability: e.g. assuming 30 discretization points $(L \sim 1)$ we get for the largest eigenvalue of the R_M block $\lambda(B_{i-1} \cdot A_i^{-1})^{30}$ whereas the smallest eigenvalues lay between 0 and 1.

3.2. Example 1: Heat exchange between two H_2O-streams

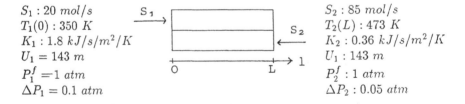

$S_1 : 20 \; mol/s$ $S_2 : 85 \; mol/s$
$T_1(0) : 350 \; K$ $T_2(L) : 473 \; K$
$K_1 : 1.8 \; kJ/s/m^2/K$ $K_2 : 0.36 \; kJ/s/m^2/K$
$U_1 = 143 \; m$ $U_1 : 143 \; m$
$P_1^f = 1 \; atm$ $P_2^f : 1 \; atm$
$\Delta P_1 = 0.1 \; atm$ $\Delta P_2 : 0.05 \; atm$

The solution was got within 5".

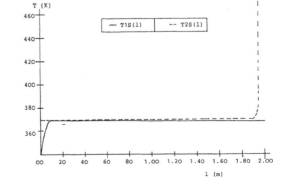

Example 2 Heat Exchange between 7 $Ar - N_2 - O_2$ **shows:**

7 streams (three $\rightarrow$, four $\leftarrow$) exchange heat without changing phase. Using a discretization of 20 points a solution was got - CPU-time: 11" (COMPAREX 7/78). A maximum relative error of $1.5 \cdot 10^{-5}$ w.r.t. temperature was observed between the old method [2], [3] and the one presented here.

Summary: the new numerical model can be solved satisfactorily both with respect to accuracy and CPU time. In addition there are no troubles with pure substances.

NOMENCLATURE

N	number of feed streams
M	number of discretization points
T_i	temperature, K
K	heat transfer coefficient, $kJ/s/m^2/K$
U	circumference of the tube, m
H	enthalpy, kJ
L	length of the heat exchanger, m
P	pressure, bar
T_A	temperature outside, K

Indices

subscripts:	i	number of stream
	j	discretization point l_j
	w	wall
superscripts:	f	feed

REFERENCES

[1] F. Kokert, L. Peer Hj. Wacker: Mathematical Simulation and Optimization of Chemical Plants, to appear in ECMI 5, proc. of ECMI conference, Glasgow, 1988, (ed. J. Manley et al.), Teubner/Kluwer, pp.75-90 (1990) (1990)

[2] F. Kokert, L. Peer, Hj. Wacker: A numerical model for calculation of a multistream heat exchanger with phase transitions, to appear in proc. ECMI IV, Strobl (eds. Hj. Wacker, W. Zulehner), Kluwer/Teubner, pp.1-10 (1990)

[3] M. Leitner: Numerical Simulation of the Steady State of Counter Current Heat Exchangers, Dipl.-Thesis, Univ.Linz, Math.Dept. (1988), pp.1-42 (in German)

THE MATHEMATICAL MODELLING OF TWO-PHASE FIRE-SPRINKLER INTERACTION

N.A. Hoffmann and E.R. Galea

Abstract

This paper extends the field modelling approach to the simulation of enclosure fires to include fire-sprinkler interaction. The Eulerian-Eulerian approach is used to simulate the two phases. The hot fire gases represent one phase while the liquid water particles the second. The mathematical model takes into account the three modes of interaction between the phases; drag, heat- and mass-transfer. The resulting finite-difference equations are solved for using the computer package PHOENICS, which employs the two-phase algorithm IPSA. Comparisons between experimental and predicted results are presented. These indicate a 'close' agreement near to the sprinkler source, which deteriorates in the far field. Current research is directed towards improving the efficiency of both the algorithms and their implementation.

Introduction

During 1989, fire claims cost the Association of British Insurers in excess of £600 million and the lives of about 1000 people. The installation of sprinkler systems within occupied enclosures has always been seen as an effective means of reducing these fire losses. Full-scale physical experimentation with its excessive demands on human and material resources can be extremely expensive or even impossible. Mathematical modelling provides a means to overcome these difficulties.

Mathematical Modelling

Mathematical field modelling of fires and smoke spread in enclosures has been underway for a number of years [1-3]. At its heart lies the numerical solution of the Navier-Stokes equations. The considerably more complicated problem of fire-sprinkler

M. Heiliö (ed.), Proceedings of the Fifth European Conference on Mathematics in Industry, 249–253.

interaction also lends itself to this form of analysis. There are now two physical phases, the gas phase involving the general fluid circulation of the hot combustion products and the liquid phase, representing the evaporating water droplets. The numerical procedure is adjusted to take into account these interacting phases. This set of equations including the interphase processes of drag, heat and mass transfer are solved using the procedures SIMPLEST and IPSA as implemented within the computer program PHOENICS. Due to space restrictions the reader is referred to previous studies by the authors [2-3].

The Modelling of a Sprinkler System

In order to evaluate the validity of the model, experimental data of a fire-sprinkler scenario was obtained [4]. The compartment represents an office with an open doorway situated at the far end of the room. This flow domain was fitted with a 12x11x17 regular cartesian grid. The fire was represented by a ramped volumetric heat source (max. of 40kW) located in a corner opposite the doorway. A sprinkler was located towards the centre of the office, 0.1m below the ceiling. The water was released at a rate of 0.596x10-3m3/s at an angle of 70° from the sprinkler head line of symmetry. It was assumed that droplets with a uniform average diameter of 1mm were released with an initial temperature of 10°C. The room conditions prior to sprinkler activation were obtained by modelling only the fire. This single-phase simulation which lasted for 175 seconds was used as the initial values for the two-phase fire-sprinkler study. During this second stage, which lasted for 25 seconds, the fire was assumed to be still in progress with a heat release rate of 30kW. The simulations were performed using 1 second time-steps.

Results

During the experiment gas temperatures were monitored at two main locations.

These were within the centre of the room and towards the open doorway, just below the ceiling. Comparisons between experimental and numerical gas temperatures are presented in figure 1. These indicate that the correct trends in temperature variation have been captured. Experimental uncertainties concerning the fire load and initial water temperature contribute towards the observed discrepancies.

The following diagrams show a vertical plane along the length of the room passing through the sprinkler and the centre-line of the doorway. Figure 2 depicts the gas temperatures before and 25 seconds after sprinkler activation. These clearly show how the stable temperature stratifications are disturbed by the water spray. The injected water removes heat from the gas, reducing its temperature. In addition the downward motion of the spray creates a type of water curtain which confines the hot gases to the fire end of the room. This can more clearly be seen in figure 3. Finally, the spread of the water droplets in terms of their volume fractions are shown in figure 4 after 25 seconds.

A further observation to emerge from these simulations is that this approach is very expensive in terms of computing time. The simulations were performed on a Norsk Data ND-5900 machine, roughly the equivalent to a VAX 11/780. The calculations required 27 hours of CPU time to simulate the 175 seconds prior to sprinkler activation. The next 25 seconds consumed 63 hours. These calculations were performed on a relatively coarse grid. Clearly a means must be found of reducing this enormous computing effort. Parallel computing techniques offer a way of achieving this. At Thames Polytechnic we have modified the fluid flow package HARWELL-FLOW3D to make efficient use of the multi-processor architecture offered by the INMOS Transputer. Early work [5] reveals that on test problems involving up to 17,640 computational cells a 13 fold speed-up is achieved utilising 15 transputers.

Translating this performance to the above fire-sprinkler simulation it is expected that the 90 hour simulation could be performed in about 7 hours.

Conclusions

The results presented above indicate that time-dependent fire-sprinkler scenarios can be simulated using the outlined volume fraction approach. Further studies are currently in progress to investigate the effect of flow rate and droplet size. Furthermore, detailed grid-refinement studies need to be undertaken. It is intended that these calculations will be pursued on parallel architecture machines.

Acknowledgement

The authors would like to thank Professor N.C. Markatos for many useful discussions in the early stages of this project, CHAM for allowing the use of PHOENICS, as well as the SERC and Ove Arup and Partners for funding this project.

References

[1] E.R. Galea, J. of Fire Prot. Eng., Vol.1(1), 1989, pp.11-22

[2] N.A. Hoffmann, E.R. Galea and N.C. Markatos, Appl. Math. Mod., Vol.13, 1989, pp.298-306

[3] N.A. Hoffmann, PhD. Thesis, Thames Polytechnic, London, 1990

[4] L.Y. Cooper and D.W. Stroup, NBSIR 87-3633, Sept. 1987

[5] M. Cross and S. Johnson, to appear in Appl. Math. Modelling

Author's Address:

Centre for Numerical Modelling and Process Analysis

Thames Polytechnic

Wellington Street

London SE18 6PF

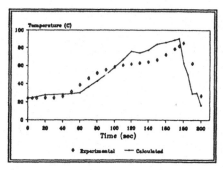 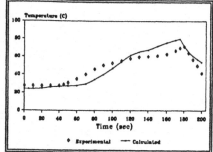

a) towards the centre of the room, 0.05m below the ceiling

b) near the door, 0.24m below the ceiling

Figure 1 Predicted and measured gas temperatures

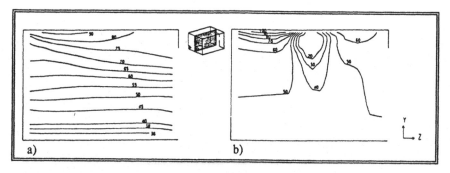

Figure 2 Temperature Contours on the vertical central plane
a) fire only (175 sec.)   b) fire and sprinkler (200ssec.)

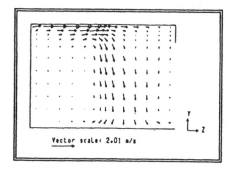 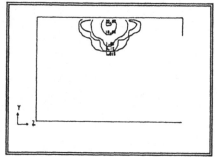

Figure 3 Velocity vectors of the gas phase on the vertical central plane after 25 seconds

Figure 4 Water volume fractions on the vertical central plane after 25 seconds

Modelling the hardenability of steel by using the generalized linear hypothesis

K. Huhtala
University of Jyväskylä
Department of Mathematics
PL 35
SF-40351 Jyväskylä, Finland

1. Introduction

To evaluate the hardenability of steel the so called *Jominy-test* can be used. A steel-bar of fixed dimensions is heated up to about 850 °C and then quenched from one end by water spray. Hardness is then measured by increasing distances from the quenched end. For M steel bars with n_r, $r = 1, \ldots, M$, measurements each, we have $N = \sum_{r=1}^{M} n_r$ observations $(\mathbf{y}, \mathbf{t}) = \{(y_{ri}, t_{ri})\}$, where y_{ri} is the hardness of the rth steel bar at the ith Jominy-distance t_{ri}.

The observations $(\mathbf{y}, \mathbf{t})$ may be fitted by some descending curve $\mu = \mu(t; \boldsymbol{\eta})$ the shape parameters $\boldsymbol{\eta} \in \mathbf{R}^{p \times 1}$ being functions of the alloying elements $\mathbf{X}$ used. In this paper we assume that $\boldsymbol{\eta} = \tilde{\mathbf{X}}\boldsymbol{\beta}$ with $\boldsymbol{\beta} \in \mathbf{R}^S$ to be estimated and

$$\tilde{\mathbf{X}} = \begin{bmatrix} \mathbf{X}_1 & 0, \ldots, 0 & \ldots & 0, \ldots, 0 \\ 0, \ldots, 0 & \mathbf{X}_2 & \ldots & 0, \ldots, 0 \\ \vdots & & & \\ 0, \ldots, 0 & \ldots & \ldots & \mathbf{X}_p \end{bmatrix} \in \mathbf{R}^{p \times S},$$

where $\mathbf{X}_i$, $i = 1, \ldots, p$, are the alloying elements for η_i. We also assume that the measurements y_{ri} has independently some *exponential family distribution* with density $f(y, \boldsymbol{\beta}) = \exp((y(\theta(\boldsymbol{\beta}) - b(\theta(\boldsymbol{\beta})))/a(\phi) + c(y, \phi))$, where ϕ is some known dispersion parameter.

Algorithm to compute the *maximum likelihood estimates* for $\boldsymbol{\beta}$ is developed in Huhtala (1989) and is now shortly presented.

2. Estimation and diagnostics

Let the measurements be numbered consequtively by $1, \ldots, N$. The logarithm of the likelihood function for the whole sample is $\ell(\boldsymbol{\beta}) = \log f(\mathbf{y}, \boldsymbol{\beta}) = \sum_{r=1}^{N}((y_r \theta_r(\boldsymbol{\beta}) - b(\theta_r(\boldsymbol{\beta})))/a(\phi_r) + c(y_r, \phi_r))$ and the updating equations of the *Fisher's scoring method* are $-E(\partial^2 \ell(\boldsymbol{\beta}^{(k)})/(\partial \boldsymbol{\beta}^T \partial \boldsymbol{\beta}))\, (\boldsymbol{\beta}^{(k+1)} - \boldsymbol{\beta}^{(k)}) = \partial \ell(\boldsymbol{\beta}^{(k)})/\partial \boldsymbol{\beta}$, or analogously

$$\tilde{\mathbf{X}}^T \mathbf{D}^{(k)} V^{-1} (\mathbf{D}^{(k)})^T \tilde{\mathbf{X}} \boldsymbol{\beta}^{(k+1)} = \tilde{\mathbf{X}}^T \mathbf{D}^{(k)} V^{-1} (\mathbf{D}^{(k)})^T \left[\boldsymbol{\eta}^{(k)} + ((\mathbf{D}^{(k)})^+)^T (\mathbf{y} - \boldsymbol{\mu}^{(k)}) \right],$$

where $\tilde{\mathbf{X}} = (\tilde{\mathbf{X}}_1^T, \ldots, \tilde{\mathbf{X}}_N^T)^T \in \mathbf{R}^{Np \times S}$, $\mathbf{D} = \text{blockdiag}(\partial \mu_r / \partial \boldsymbol{\eta}_r; r = 1, \ldots, N) \in$

255

M. Heiliö (ed.), *Proceedings of the Fifth European Conference on Mathematics in Industry*, 255–258.
© 1991 B. G. Teubner Stuttgart and Kluwer Academic Publishers. Printed in the Netherlands.

$\mathbf{R}^{Np \times N}$, $\mathbf{D}^+$ is the Moore-Penrose inverse of $\mathbf{D}$ (Stewart, 1973) and V is the diagonal variance matrix of measurements.

These equations are the normal equations of the linear regression

$$\mathbf{Z} = \tilde{\mathbf{X}}\boldsymbol{\beta}^{(k+1)} + \zeta, \tag{1}$$

with weight matrix

$$\mathbf{D}^{(k)}V^{-1}(\mathbf{D}^{(k)})^T \tag{2}$$

and $\mathbf{Z} = \boldsymbol{\eta}^{(k)} + ((\mathbf{D}^{(k)})^+)^T(\mathbf{y} - \boldsymbol{\mu}^{(k)})$. This means, that the updated estimate $\boldsymbol{\beta}^{(k+1)}$ is found by solving the linear regression problem (1) & (2) and thus the maximum likelihood estimate $\hat{\boldsymbol{\beta}}$ is found by iterating (1) & (2) until the convergence (see also McCullagh & Nelder, 1989).

It can be shown (Dacunha-Castelle & Duflo, 1986) that under quite reasonable assumptions the maximum likelihood estimator $\hat{\boldsymbol{\beta}}$ is *asymptotically normally distributed*: $\sqrt{N}(\hat{\boldsymbol{\beta}} - \boldsymbol{\beta}) \xrightarrow{D} \mathrm{N}_S(0, I(\boldsymbol{\beta})^{-1})$ as $N \to \infty$, where $I(\boldsymbol{\beta})$ is the Fisher information matrix with estimate $N^{-1}E(\partial^2\ell(\mathbf{y}, \hat{\boldsymbol{\beta}})/(\partial\boldsymbol{\beta}^T\partial\boldsymbol{\beta}))$. For small samples this limit may be unusable. However, by sampling the original sample with replacement we can produce K bootstrap-samples of size N each and compute the estimates $\hat{\boldsymbol{\beta}}^k$, $k = 1, \ldots, K$. The $(1 - \alpha)$-level *bootstrap confidence interval* for β_i is then $[\hat{\beta}_i^{[\alpha/2]}, \hat{\beta}_i^{[1-\alpha/2]}]$, where $\hat{\beta}_i^{[a]}$ is the $[aK]^{\text{th}}$ smallest bootstrap estimate of β_i. Further, $\sqrt{N}(\hat{\mu}(t_{ri}, \mathbf{X}_r) - \mu_0(t_{ri}, \mathbf{X}_r)) \xrightarrow{D} \mathrm{N}(0, J^T I(\boldsymbol{\beta})^{-1}J)$ as $N \to \infty$, where $J = \partial\mu(t_{ri}, X_r)/\partial\boldsymbol{\beta}$ and the bootstrap confidence interval for $\mu(t_{ri}, \mathbf{X}_r)$ could be formed as above.

3. Jominy-curve

A set of Jominy-test results for low alloyed special steels was offered by Ovako Steel oy-ab, Imatra, Finland. From this set a sample of $M = 48$ Jominy-curves with total amount of $N = 639$ observations was collected. Details are obtainable from the author.

As a model for the Jominy-curve we used $\mu(t) = (\eta_1 \exp(\eta_3 - t^{\eta_4}) + \eta_2)/(1 + \exp(\eta_3 - t^{\eta_4}))$, which gave adequate fit, when fitted to the observed Jominy-curves.

Assuming normally distributed errors a series of nested models with 13 – 20 β-parameters was fitted. The model with $S = 19$ β-parameters gave good fit: likelihood ratio test statistic, now equal to residual sum of squares, $D = 2143.4$ with $df = 639 - 19 = 620$ degrees of freedom. Adding covariates gave no significant reduction in D. The residual variance was estimated by $\hat{\sigma}^2 = D/df \approx 3.46$ and the residual standard deviation by $\hat{\sigma} \approx 1.86$, which is quite good.

Scaled residuals $(y_r - \hat{\mu}_r)/\hat{\sigma}$ are plotted against predictions $\hat{\mu}$ and Jominy-distance $\mathbf{t}$ in figures 1(a) and (b). Some pattern is visible in figure 1(a), but residuals are otherwise so small that no action is inevitable.

A small bootstrap-experiment was carried on with 50 re-samples. From the estimates $\hat{\boldsymbol{\beta}}^k$, $k = 1, \ldots, 50$, we computed the 95% confidence intervals, which are given in table 1 together with the final estimates $\hat{\boldsymbol{\beta}}$ and the 95% asymptotic confidence intervals. The difference between the two confidence intervals varies from

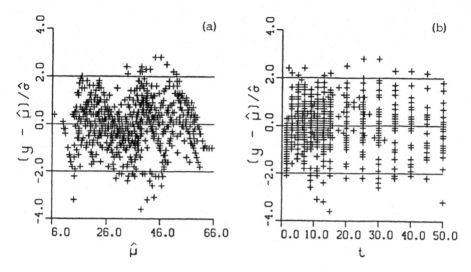

Fig. 1: Standardized residuals plotted against (a) the predictions $\hat{\mu}$ and (b) the Jominy-distance **t**.

Table 1: Final estimates and their 95% asymptotic and bootstrap confidence intervals

	$\tilde{X}$	$\hat{\beta}$	asymptotic		bootstrap	
η_1	1	49.68	[46.2,	53.1]	[45.7,	52.9]
	C	29.60	[21.7,	37.5]	[21.6,	35.0]
η_2	1	−12.66	[−14.7,	−10.6]	[−14.5,	−10.5]
	C	69.12	[65.3,	72.9]	[64.0,	71.6]
	Mn	8.63	[6.7,	10.5]	[6.0,	11.9]
	Ni	13.57	[12.9,	14.3]	[12.4,	14.3]
	Mo	11.66	[9.1,	14.2]	[6.2,	14.4]
	Ti	−1011.0	[−1276.,	−747.]	[−1500.,	−656.]
	Cu	17.37	[12.6,	22.2]	[10.7,	21.5]
η_3	1	1.35	[1.10,	1.60]	[1.17,	1.55]
	C	6.41	[5.51,	7.31]	[5.74,	7.27]
	Cr	−.25	[−.41,	−.09]	[−.54,	−.09]
η_4	1	1.071	[1.03,	1.11]	[1.04,	1.12]
	C	−.096	[−.16,	−.03]	[−.18,	−.04]
	Mn	−.249	[−.28,	−.22]	[−.29,	−.21]
	Cr	−.311	[−.33,	−.29]	[−.35,	−.30]
	Mo	−.406	[−.44,	−.37]	[−.48,	−.37]
	Ti	−17.13	[−14.8,	−19.5]	[−20.9,	−14.6]
	B	42.18	[29.7,	54.7]	[31.7,	53.2]

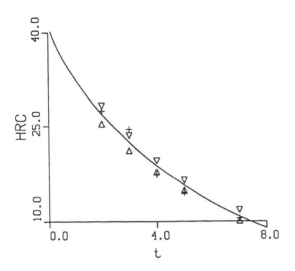

Fig. 2: One observed Jominy-curve (+), its prediction (solid line) and point-wise 95 % bootstrap confidence intervals (lower bound △, upper bound ∇).

parameter to parameter; however, this experiment 'shows' that the distribution of $\hat{\beta}$ is already nearly normal though sample size is 'only' 639.

As an example we plotted one observed Jominy-curve with its prediction and pointwise 95 % bootstrap confidence intervals in figure 2.

4. Summary

We have shortly presented a procedure to compute the maximum likelihood estimates for the parameters of the Jominy-curve and some asymptotic results. Method is applied to modelling the hardenability of low alloyed special steels and a small bootstrap-experiment shows how well the estimates behave. We feel that this method is quite usable for various problems of this kind. Method, once programmed, is easy to apply and convergence is quite stable.

References

Dacunha-Castelle D. & Duflo M. (1986), *Probability and statistics, vol. II*, New York: Springer-Verlag.

Huhtala K. (1989), Estimation of response curves: Representation of shape parameters by using generalized linear hypothesis, *Reports from the Department of Statistics, University of Jyväskylä*.

McCullagh P.& Nelder J.A. (1989), *Generalized Linear Models, second edition*, London: Chapman and Hall.

Stewart G.W. (1973), *Introduction to Matrix Computations*, New York: Academic Press.

A NUMERICAL INVESTIGATION OF AN ELASTIC STRIP IN FREE ROLLING WITHOUT SLIP

M. J. JAFFAR
Department of Mathematical Sciences
Leicester Polytechnic
P O Box 143
Leicester LE1 9BH
United Kingdom

ABSTRACT. The problems of an elastic strip which either passes freely between identical rigid rollers (unbonded) or is attached to the surface of one roller (bonded) are investigated numerically. It is assumed that the contacting surfaces are adhered over the entire contact region. The unknown contact stresses and the given displacements are presented in terms of Chebyshev expansions and hence the governing equations are reduced to a system of linear equations.

The influence of the strip thickness and Poisson's ratio ν on contact stresses, contact width, penetration depth (the deformation at the contact centre), and creep ratio is considered. The present method showed close agreement with the existing asymptotic solutions when the strip becomes either a half plane or very thin.

1. Introduction

Many processes involve the passage of a strip or sheet of material of low elastic modulus through the nip between two cylinders. This paper is concerned with a numerical solution of the stresses and the deformations produced in an elastic strip of finite thickness in free rolling (no resultant tangential force transimitted across the contact) with no slip (stick) over the entire contact region, i.e. the strain difference is constant within the contact. In principle slip between the surfaces can be prevented entirely by a sufficiently high coefficient of friction μ ($\mu \rightarrow \infty$) at the interface. Two different geometries are investigated where the strip either resting (unbonded) or is attached (bonded) to a rigid foundation and indented by a rigid cylinder of radius R as shown in Fig. 1.

Bentall and Johnson [1] examined the free rolling contact problems of bonded and unbonded strips by using a numerical method where the surface stresses were approximated by overlapping triangle elements. Nowell and Hills [2-3] considered different contact problems for a cylinder indenting respectively an unbonded and bonded strips using the numerical method developed in Ref. 1. Jaffar and Savage [4] have investigated numerically the frictionless line contact problems involving bonded and unbonded strips. Both the unknown pressure and the given displacement (over the contact region) were expanded in terms of Chebyshev polynomials. The numerical

259

M. Heiliö (ed.), Proceedings of the Fifth European Conference on Mathematics in Industry, 259–265.
© 1991 *B. G. Teubner Stuttgart and Kluwer Academic Publishers. Printed in the Netherlands.*

results showed an excellent agreement with the asymptotic solutions reported by Meijers [5]. The purpose of the present work is to extend the technique further to analyse the stick case.

2. Formulation

The problems outlined above and illustrated in Fig. 1 should satisfy the following boundary conditions:-

$$\sigma_y (x, 0) = \sigma_{xy} (x, 0) = 0, \qquad\qquad\qquad\qquad (|x| > a)$$

$$u (x, t) = v (x, t) = 0,$$
$$\qquad\qquad\qquad\qquad\qquad\qquad\qquad \text{(bonded strip)}$$
$$\sigma_{xy} (x, t) = v (x, t) = 0, \qquad\qquad\qquad\qquad \text{(unbonded strip)}$$

where a is the contact half width, u and v are the tangential and normal displacements respectively. The boundary conditions on u and v inside the contact region ($|x| \leq a$) are given in Ref. 1 as

$$u (x, 0) = C + \zeta x,$$

$$v (x, 0) = v^* + \frac{ex}{R} - \frac{x^2}{2R}$$

where C, e, v^* and ζ are the cumulative slip, the contact eccentricity, the penetration depth and the creep ratio respectively. In addition, the shear stress q(x) within the stick zone is limited by the coefficient of friction, i.e.

$$|q(x)| \leq \mu |p(x)|,$$

where p(x) is the normal stress. It is to be noted that since the cylinder is rigid then its creep velocity $\delta f_1 = 0$ and $\partial u_1 / \partial x = 0$. Hence, $\zeta = - \delta f_2 / V$ using the analysis given by Johnson ([6], p.244).

The tangential and normal surface displacements due to the tangential and normal forces were formulated in Ref. 1. Assume that p(x), q(x), u(x) and v(x) can be represented by

$$2\theta p(x) = \frac{1}{\sqrt{a^2 - x^2}} \sum_{n=0}^{N} a_n T_n (X), \quad \text{and} \quad 2\theta q(x) = \frac{1}{\sqrt{a^2 - x^2}} \sum_{n=0}^{N} b_n T_n (X), \quad (1)$$

$$v(X) = \sum_{n=0}^{N} c_n T_n (X), \quad\quad\quad \text{and} \quad u(X) = \sum_{n=0}^{N} d_n T_n (X), \quad (2)$$

where $\theta = (1 - v^2) / E$ and $X = x / a$, then the governing equations can be reduced to a set of algebraic equations (see Appendix).

Since the surface stresses fall to zero at the contact ends, therefore, (2) gives

$$\sum_{\substack{n=0 \\ (n \text{ even})}}^{N} a_n = 0, \quad \sum_{\substack{n=1 \\ (n \text{ odd})}}^{N} (-1)^n a_n = 0, \quad \sum_{\substack{n=0 \\ (n \text{ even})}}^{N} b_n = 0, \quad \text{and} \quad \sum_{\substack{n=1 \\ (n \text{ odd})}}^{N} (-1)^n b_n = 0 \tag{3}$$

Let $\quad b_i = a_{i+N+1}$ and $d_i = c_{i+N+1}$ $\hspace{3cm}$ $(i = 0, 1 \ldots N)$

then system (A1) and Eq. (3) provide $2N + 6$ linear equations for the unknowns $(a_0, a_1, \ldots, a_{2N+1}, C, e, v^*, \zeta)$ which were solved using the F04ATF routine from the NAG LIBRARY. The values of these unknowns will determine the full solution.

3. Results and Discussion

3.1. BONDED STRIP

When the strip becomes a half plane ($\gamma = 0$), the numerical results obtained in the present work agreed well with the exact solutions derived by Bufler [7]. In the case of a thin bonded strip ($a \gg t$) with $v < 0.5$, the following asymptotic solutions were established in Ref.1

$$p(X)/p_0 = \frac{(1-v)^2}{1-2v} \frac{a}{a_H} \gamma(1-X^2), \quad \text{and} \quad q(X)/p_0 = 0, \tag{4}$$

$$a/a_H = \frac{3\pi}{8\gamma} (1-2v)(1-v)^{-2}, \tag{5}$$

$$Rv^* / a = 0.5, \tag{6}$$

where p_0 is the maximum Hertzian pressure and a_H is the Hertzian contact half width.

For values of $\gamma \geq 1$ the stress distributions scaled with p_0 are plotted in Figs. 2 and 3 respectively for $v = 0.3$ and 0.5. It is clear that the shear stresses induced for a compressible strip with $v = 0.3$ are of opposite sign to those obtained for $v = 0.5$. Thus, the present results are consistent with those reported in Ref. 3. Figure (2) displays also that the results obtained using Eq. (4) are in close agreement with the numerical results when $\gamma \geq 5$.

The ratio a/a_H is calculated for different values of Poisson's ratio and compared in Fig. 4 with those obtained from Eq. (5). The figure shows that contact width decreases as γ and/or v are increased. Moreover, it tends to zero more rapidly as the strip becomes incompressible ($v = 0.5$). Thus, it will be extremely difficult in practice to obtain such large values of γ. It should be noted that Eq. (5) must be used with care, especially for comparison purposes, taking into account the value of v and the range of γ. In fact, good agreement over a wide range of γ is achieved when $v = 0.4$.

The variations of creep ratio and the penetration depth with strip thickness are illustrated in Figs. 5 and 6 respectively in the full range $0 \leq v \leq 0.5$. In both cases there is a rapid change in the behaviour when the strip becomes thin and the value of v is increased from 0.495 to 0.5. Fig. 5 shows that rolling creep at a large values of γ changes sign at Poisson's ratio between 0.2 and 0.3 which is similar to that reported in Ref. 1. This situation can be explained using the Hertz theory which predicts negative strains within the contact at low values of v. On the other hand, strips of finite thickness and high values of v yield positive strains to conserve volume (Johnson [8]). It is clear from Fig. (6) that the asymptotic value given in Eq. (6) showed satisfactory agreement with the present results at high values of γ and in the range of $0 < v \leq 0.45$.

3.2. UNBONDED STRIP

Numerically, this problem is similar to the bonded strip. Some results, therefore, will be presented in this section. Figs. 7 and 8 show the stress distributions for $v = 0.3$ and 0.5 respectively. It is interesting to note that the shear stresses are of same sign which is quite different from those calculated for a bonded strip.

4. Conclusions

The numerical method for frictionless contact problems reported in Ref. 4 has been modified and used for solving the stick problem. The method exhibited all features of the problem which were observed by other workers. Also, it is not sensitive to Poisson's ratio for values which are close to 0.5. The numerical results presented in the present work showed good agreement with the existing asymptotics solutions.

REFERENCES

[1]   BENTALL, R. H. and JOHNSON, K. L. (1968) "An elastic strip in plane rolling contact", Int. J. Mech. Sci., 10, 637 - 663.
[2]   NOWELL, D. and HILLS, D. A. (1988) "Contact problems incorporating elastic layers", Int. J. Solids Structures, 24 (1), 105 - 115.
[3]   NOWELL, D. and HILLS, D. A. (1988) "Tractive rolling of tyred cylinders", Int. J. Mech. Sci., 30 (12), 945 - 957.
[4]   JAFFAR, M. J. and SAVAGE, M. D. (1988) "On the numerical solution of line contact problems involving bonded and unbonded strips", J. Strain Analysis, 23 (2), 67 - 77.
[5]   MEIJERS, P. (1968) "The contact problem of a rigid cylinder on an elastic layer", Appl. Sci. Res., 18, 353 - 383.
[6]   JOHNSON, K. L. (1985) Contact Mechanics, 1st Edn., Cambridge University Press, UK.
[7]   BUFLER, H. (1959) "Zur theorie der rollendon reibung", Ing. Arch., 27, 137-152.
[8]   JOHNSON, K. L. (1990) Private Communication.
[9]   GLADWELL, G. M. L. (1980) Contact Problems in the Classical Theory of Elasticity, Alphen ann den Rijn: Sijthoff and Noordhoff, The Netherlands.

APPENDIX. In this Appendix the governing equations given in Ref. 1 will be reduced to a system of linear equations. Writing

$$L_{ii}\,\omega = -\,F\,(\omega) + G_{ii}\,(\omega) \quad \text{and} \quad L_{ij}\,(\omega) = 1 + G_{ij}\,(\omega) \quad \text{where}$$

$$F\,(\omega) = e^{-2\omega} - 1, \quad G_{ij}\,(\omega) = L_{ii}\,(\omega) + F\,(\omega), \qquad\qquad (i = 1, 2)$$

$$G_{ij}\,(\omega) = L_{ij}\,(\omega) - 1, \qquad\qquad (j = 1, 2 \ \& \ i \neq j)$$

$$L_{ii}\,(\omega) = \frac{2\kappa \sinh 2\omega + 4(-1)^{-i}\omega}{2\kappa \cosh 2\omega + 4\omega^2 + \kappa^2 + 1}, \quad \text{and} \quad L_{ij}\,(\omega) = \frac{2\kappa\,(\cosh 2\omega - 1) - 4\omega^2/(1 - 2\nu)}{2\kappa \cosh 2\omega + 4\omega^2 + \kappa^2 + 1}$$

for the bonded strip with $\kappa = 3 - 4\nu$, and

$$L_{ii}\,(\omega) = \frac{\cosh 2\omega + (-1)^{-i}}{\sinh 2\omega + 2\omega} \quad \text{and} \quad L_{ij}\,(\omega) = \frac{\sinh 2\omega - 2\omega/(1-2\nu)}{\sinh 2\omega + 2\omega}$$

Substituting the expansions (1) and (2) into the governing equations, using the orthogonality relation and the Gauss-Chebyshev quadrature formula for $T_n(x)$ with formulae (6.11.2) and (7.3.12) given by Gladwell [9], yields

$$\frac{a_m^*}{2m} + \sum_{n=0}^{N} a_n B_{nm}^{ii} - \alpha \left[\frac{b_0 R_m}{\pi} + \sum_{n=0}^{N} b_n D_{nm}^{ij} \right] = \frac{c_m}{2}$$

$$\alpha \left[\frac{a_0 R_m}{\pi} + \sum_{n=0}^{N} a_n D_{nm}^{ij} \right] + \frac{b_m^*}{2m} + \sum_{n=0}^{N} b_n B_{nm}^{ii} = \frac{d_m}{2}$$

$$(m = 0, 1, ...N) \qquad (A1)$$

where

$$B_{nm}^{ii} = \frac{1}{N} \sum_{i=1}^{N} A_n^{ii}\,(X_i)\,T_m(X_i), \qquad R_m = \int_{-1}^{1} \frac{\sin^{-1}(X)\,T_m\,(X)}{\sqrt{1 - X^2}}\,dX,$$

$$D_{nm}^{ij} = \frac{1}{\pi} \int_{-1}^{1} U_{n-1}\,(X)\,T_m\,(X)\,dX + \frac{1}{N} \sum_{i=1}^{N} A_n^{ij}\,(X_i)\,T_m\,(X_i),$$

$$A_n^{ii}\,(X) = \frac{1}{N} \sum_{i=1}^{N} H_{ii}\,(X, S_i)\,T_n\,(S_i), \quad A_n^{ij}\,(X) = \frac{1}{N} \sum_{i=1}^{N} H_{ij}(X, S_i)\,T_n\,(S_i),$$

$$H_{ii}\,(X, S) = \frac{1}{2} \ln \left| \frac{4 + (X - S)^2 \gamma^2}{\gamma^2} \right| + \int_0^{\infty} \omega^{-1} G_{ii}\,(\omega)\,\cos[(X - S)\gamma\omega]\,d\omega$$

$$H_{ij}\,(X, S) = \int_0^{\infty} \omega^{-1}\,G_{ij}\,(\omega)\,\sin[(X - S)\gamma\omega]\,d\omega$$

$\alpha = (1 - 2\nu)/2\,(1 - \nu)$ and $S = s/a$. The asterisks in Eqs. (A1) denote that the first terms in the Chebyshev expansions of the first and the second kinds are $\ln 2$ and $\sin^{-1}(x)$ respectively.

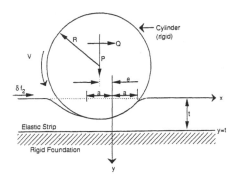

Figure 1 Equivalent Geometry of Contact

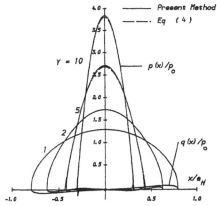

Fig. 2 Stress distributions for a bonded strip when γ = 0.3.

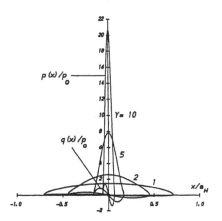

Fig. 3 Stress distributions for a bonded strip when γ = 0.5.

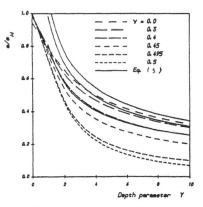

Fig. 4 Variation of a/a_H with Y.

(BONDED STRIP)

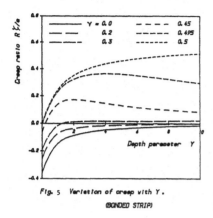

Fig. 5   Variation of creep with Y.
(BONDED STRIP)

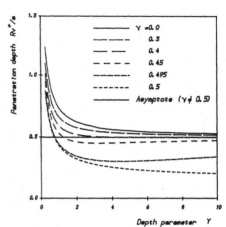

Fig. 6   Variation of penetration depth with Y
for different values of γ.
(BONDED STRIP)

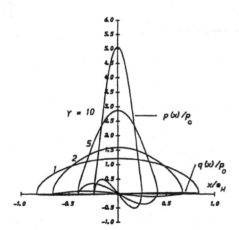

Fig. 7   Stress distributions for an unbonded
strip when γ = 0.3.

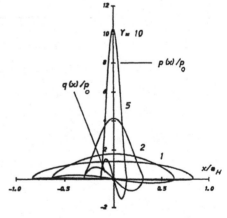

Fig. 8   Stress distributions for an unbonded
strip when γ = 0.5.

FORMATION OF OSCILLATION MARKS ON CONTINUOUSLY CAST STEEL

J.R.KING, A.A.LACEY, C.P.PLEASE, P.WILMOTT

ABSTRACT. A common method of casting steel is to solidify it continuously in a water-cooled mould. As the steel is drawn from the bottom of the mould the mould is oscillated vertically and the steel and mould are kept apart by a lubricating layer of "slag" which is supplied at the top. The finished solid steel has regularly spaced marks in the form of indentations with period given by the speed of the steel divided by the frequency of oscillation. The final shape of the steel surface is expected to be given by the interaction of a thin solid or mushy steel layer with the slag as this flows slowly in the lubricating channel. This preliminary paper describes a possible model for the flow of the slag and the change in location of the steel surface which is one sort of free boundary. The model is a coupled hyperbolic and elliptic system involving two further free boundaries. Completion of the model requires further work to fix some conditions, including those at the free boundaries. To proceed further more information will be required on the thermal and mechanical properties of steel near its melting point.

Physical Problem and Mathematical Model

In the process of continuous casting liquid steel is supplied to, and frozen by, a water cooled copper mould. To help the process, e.g. to prevent sticking, a lubricating powder, slag, which melts and goes between the steel and the mould, is also supplied, and the mould is oscillated vertically (fig. 1).

As the steel is drawn downwards it freezes and contracts. Unfortunately the surface of the steel has marks, in the form of small indentations, with a spacing given by the speed of the steel and period of oscillation of the mould. It is aimed to understand the formation of these marks, in particular to find which are the key factors influencing the size of the marks, with the eventual intention of being able to reduce their size.

M. Heiliö (ed.), Proceedings of the Fifth European Conference on Mathematics in Industry, 267–270.
© 1991 B. G. Teubner Stuttgart and Kluwer Academic Publishers. Printed in the Netherlands.

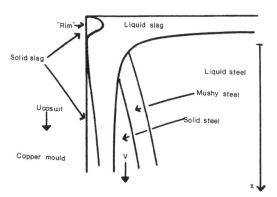

Figure 1. Side of mould.

As the solid layer moves downwards it thickens and hardens, so that it becomes more resistant to bending and the region most important, with respect to the formation of the marks, will be near the top of the steel in the vicinity of the mould (fig. 2).

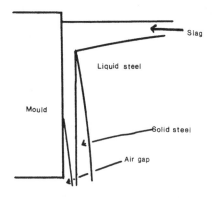

Figure 2. Region of first solidification of steel.

A number of possible mechanisms have previously been considered for mark formation including: (i) thermo-mechanical effects in a relatively large meniscus region; (ii) thermo-mechanical effects of the solid slag rim; (iii) tearing and healing of the solid steel. Disadvantages of these appear to be: (i) unrealistically sized surface tension; (ii) a higher slag viscosity (associated with higher melting point and hence a larger rim) would give larger marks, in conflict with observation; (iii) no sign of such tearing is observed in the microstructure of the fully solidified steel.

We therefore propose to consider a new model, one which may give increased mark size with increased viscosity. The experimental observations that are available indicate that the slag occupies a narrow channel between the steel and the mould so that classical lubrication theory models the flow of liquid slag down the mould. This then gives viscosity dominated pressure fluctuations in the slag. For these to cause variation in the steel surface it is required that the (at least partially) solidified steel layer near the slag is not fully rigid and may be relatively easily bent. Although it is possible to consider a fairly complicated model for this steel layer it may, under a reasonably wide class of assumptions on how the mechanical properties of the steel vary near melting temperature, be reduced to taking it to resist shear stress but not pressure difference, except near some transition point where the layer becomes effectively rigid.

The present paper gives only the preliminary work as the model is, as yet, far from complete. A number of boundary conditions are still to be imposed, one of which is to describe cavitation in lubrication theory and has been the subject of discussion for several years. Also more information is still required on the physical properties of steel near its melting point and the main heat transfer mechanisms are not clear.

The mechanism for mark formation that we consider here is to take, above some point $x = s$, the solid/mushy steel to act as a "floppy" skin which can sustain shear stress but not pressure differences. Below this point, $x > s$, the steel is considered to be totally rigid. Between the steel and the mould there is a thin lubricating layer of slag where the mass flow is given by

$$Q = \frac{h}{2}(U \cos\omega t + V) - \frac{h^3}{12\mu} \frac{\partial p}{\partial x} , \qquad (1)$$

where h denotes the width of the slag channel, $U\cos\omega t$ is the velocity of the mould, V is the velocity of the steel (both measured downwards), p is the pressure difference from "hydrostatic" in the slag, and μ is its viscosity.

For conservation of mass we require

$$\frac{\partial Q}{\partial x} + \frac{\partial h}{\partial t} = 0. \qquad (2)$$

For the floppy part the pressure p is just given by that of the much less viscous liquid steel, i.e. pressure is ferrostatic:

$$p = \Delta\rho g x, \quad x < s. \qquad (3)$$

Combining (1), (2), (3) h satisfies the first order hyperbolic equation

$$\frac{\partial h}{\partial t} + \frac{\partial}{\partial x} \left[\frac{h}{2}(U\cos\omega t + V) - \frac{h^3}{12\mu} \Delta\rho g \right] = 0. \qquad (4)$$

Meanwhile in the rigid section we simply have that the convective derivative of h is zero:

$$\frac{\partial h}{\partial t} + V\frac{\partial h}{\partial x} = 0. \qquad (5)$$

It should be noted that in equation (5) we have ignored terms due to contraction of the solid steel, tapering of the mould, and freezing or remelting of the slag. If these effects are included the right hand side becomes some $R(x,t) \neq 0$.

At the top of the steel it is unclear as to what boundary condition should be imposed. If the characteristics of (4) are moving upwards we impose no condition otherwise we may required the "natural" condition

$$h^2 = \frac{2\mu}{\Delta\rho g} (U\cos\omega t + V), \quad x = 0 \tag{6}$$

to hold. Lower down there is a free boundary at some x = L due to cavitation:

$$p = -\rho g L, \quad x = L. \tag{7}$$

It is not certain as to what the other condition at this point should be.

At x = s where the steel layer changes to become fully rigid we only require continuity of pressure:

$$[p] = 0, \quad x = s; \tag{8}$$

and conservation of mass:

$$[Q] = [h]\frac{ds}{dt} , \quad x = s. \tag{9}$$

It should again be noted that these conditions may be insufficient if the characteristics of (4) are directed away from x = s.

To complete the model we have still to (a) resolve the difficulties regarding the boundary conditions, in particular an alternative to (6) at x = 0 may be given by thermal properties and there needs to be another condition at x = l., (b) predict the location of the transition point x = s(t) which again depends on a coupled thermal problem together with the mechanical properties of hot steel, (c) include the extra terms on the right hand side of (5) as these should fix the typical slag channel width (these terms involve the change of phase of the slag and the thermal contraction of the steel so again thermal effects and mechanical properties of steel have to be considered).

J.R. KING
Department of Theoretical Mechanics
University of Nottingham
University Park
Nottingham NG7 2RD

A.A. LACEY
Department of Mathematics
Heriot-Watt University
Riccarton
Edinburgh EH14 4AS

C.P. PLEASE
Faculty of Mathematical Studies
University of Southampton
Southampton SO9 5NH

P. WILMOTT
Mathematical Institute
Oxford University
24-29 St Giles
Oxford OX1 3LB

TECHNIQUES FOR FINITE ELEMENT ANALYSIS OF MAGNETOSTATIC AND EDDY CURRENT PROBLEMS

Jorma Luomi

Abstract: The paper discusses techniques used in the numerical analysis of magnetic field problems. Magnetic vector potential and reduced vector potential formulations of the field are reviewed. The use of the vector potentials in the finite element and boundary element solution of the magnetostatic field is illustrated by examples. Attention is also given to the coupled magnetic field and electric circuit analysis of time-dependent problems.

Introduction

Numerical analysis of electromagnetic fields has drawn increasing attention in recent years. In many cases, the solution of the field is based on the magnetic vector potential A defined by

$$B = \nabla \times A \tag{1}$$

where B is the magnetic flux density. The Coulomb gauge $\nabla \cdot A = 0$ is commonly used to ensure the uniqueness of the vector potential. Depending on the problem it may be reasonable to divide the vector potential into two parts

$$A = A_s + A_m; \qquad A_s(r) = \frac{1}{4\pi\nu_0} \int_{V_J} \frac{J(r')}{|r - r'|} \, dV' \tag{2}$$

where J is the current density, ν_0 is the reluctivity of free space and the region of integration V_J is supposed to include all source currents [1]. The part A_s is produced by source currents in free space and can be calculated analytically or semianalytically, and a numerical solution is needed only for the reduced vector potential A_m which is due to the induced magnetization.

Finite element and boundary element solutions of the magnetostatic field

The partial differential equation of the vector potential is obtained by substituting the magnetic field strength $H = \nu B$ and Eq. (1) in the equation $\nabla \times H = J$, giving

$$\nabla \times (\nu \nabla \times A) = J; \qquad \nabla \cdot A = 0 \qquad \text{in } V \tag{3}$$

The boundary S of the problem region V is divided into two parts S_1 and S_2 in accordance with the boundary conditions

$$A \times n = \bar{A}_1 \qquad\qquad \text{on } S_1 \tag{4}$$

271

M. Heiliö (ed.), *Proceedings of the Fifth European Conference on Mathematics in Industry*, 271–276.
© 1991 *B. G. Teubner Stuttgart and Kluwer Academic Publishers. Printed in the Netherlands.*

$$(v\nabla \times A) \times n = \bar{C}; \quad A \cdot n = \bar{A}_2 \quad \text{on } S_2 \tag{5}$$

where n is the outward unit normal vector of the boundary. The boundary conditions are usually homogeneous and the known functions $\bar{A}_1$, $\bar{C}$ and $\bar{A}_2$ vanish. The problem given by Eqs. (3) – (5) has a unique solution [2]. In addition, interface conditions are needed if discontinuities of the reluctivity v are present. The boundary value problem in terms of the reduced vector potential is [1]

$$\nabla \times (v\nabla \times A_m) = \nabla \times [(v_0 - v)\nabla \times A_s]; \quad \nabla \cdot A_m = 0 \qquad \text{in } V \tag{6}$$

$$A_m \times n = \bar{A}_1 - A_s \times n \qquad \text{on } S_1 \tag{7}$$

$$(v\nabla \times A_m) \times n = \bar{C} - (v\nabla \times A_s) \times n; \quad A_m \cdot n = \bar{A}_2 - A_s \cdot n \qquad \text{on } S_2 \tag{8}$$

The finite element method has become the standard technique for the numerical solution of magnetic field problems. The problems (3) – (5) or (6) – (8) are discretized by using the Galerkin method and isoparametric finite elements.

An alternative approach is the boundary element method [1,2,3]. The problem is formulated as a boundary integral equation

$$\frac{\Omega}{4\pi} A(r) = \int_V G(r|r') J(r') \, dV' + \oint_S G(r|r')[v\nabla' \times A(r')] \times n' \, dS'$$

$$- \oint_S \left\{ [n' \times v\nabla' G(r|r')] \times A(r') + [n' \cdot v\nabla' G(r|r')] A(r') \right\} dS' \tag{9}$$

of the vector potential. The reluctivity is assumed to be homogeneous, Ω is the solid angle included by S at point r, and

$$G(r|r') = \frac{1}{4\pi v |r - r'|} \tag{10}$$

is the free space Green's function in the three-dimensional space. The Galerkin method and isoparametric boundary elements are used for the discretization, resulting in a system of equations for the nodal values of A and $(v\nabla \times A) \times n$. These are partly known from the boundary conditions (4) and (5), and the unknown values can be solved from the system of equations. Correspondingly, the boundary integral equation of the reduced vector potential is [1]

$$\frac{\Omega}{4\pi} A_m(r) = \int_V \frac{v_0 - v}{v} G(r|r') J(r') \, dV' + \oint_S G(r|r')[v\nabla' \times A_m(r')] \times n' \, dS'$$

$$- \oint_S \left\{ [n' \times v\nabla' G(r|r')] \times A_m(r') + [n' \cdot v\nabla' G(r|r')] A_m(r') \right\} dS' \tag{11}$$

The boundary element method offers several advantages when compared with the finite element method: only the boundaries are discretized, the number of unknowns is reduced, exterior regions are easy to handle, and the solutions are analytical functions inside the region. However, the finite element method is better suited to solving the field problem in inhomogeneous and nonlinear media. In some problems it may be of advantage to combine both methods in a hybrid method [1,4].

The hybrid solution and the choice of the vector potential are illustrated by an axisymmetric example concerning the magnetically shielded coil system of a superconducting homopolar motor. The geometry, field plot and hybrid discretization of a 50 kW motor are shown in Fig. 1. The finite element discretization is practically restricted to the saturable iron parts, and the boundary element method is used for the solution of the exterior field and nonmagnetic inner part. The differences between the computed and measured values of the magnetic flux were less than one percent. A comparison of the formulations is shown in Fig. 2. In this case, choosing the reduced vector potential

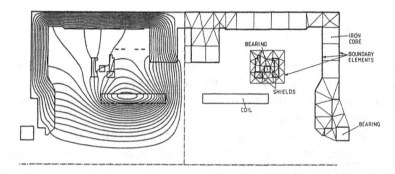

Figure 1. Axial cross-section, field plot and discretization of a 50 kW superconducting homopolar motor. The reduced vector potential and the hybrid method with quadratic finite and boundary elements have been used for the solution.

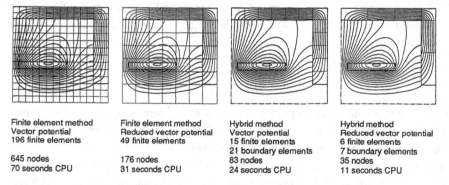

Finite element method	Finite element method	Hybrid method	Hybrid method
Vector potential	Reduced vector potential	Vector potential	Reduced vector potential
196 finite elements	49 finite elements	15 finite elements	6 finite elements
		21 boundary elements	7 boundary elements
645 nodes	176 nodes	83 nodes	35 nodes
70 seconds CPU	31 seconds CPU	24 seconds CPU	11 seconds CPU

Figure 2. Finite element and hybrid discretizations of a superconducting homopolar motor, giving approximately equal accuracy.

results in considerable savings in computing time and memory requirements. This is due to the fact that the reduced vector potential A_m behaves in this case better and is considerably easier to solve numerically than the vector potential A_s produced by the coils. The most efficient solution is obtained by using the combination of the hybrid method and reduced vector potential.

Solution of coupled magnetic field and circuit equations

Eq. (3) is valid even in time-dependent problems when the displacement current can be neglected. The current density is given by

$$J = - \sigma \frac{\partial A}{\partial t} - \sigma \nabla V \tag{12}$$

in solid conductors exposed to eddy currents, where σ is the conductivity and V is an electric scalar potential. The partial differential equation is then

$$\nabla \times (v \nabla \times A) + \sigma \frac{\partial A}{\partial t} + \sigma \nabla V = 0 \tag{13}$$

Solution of time-dependent problems is rather straightforward, at least in two-dimensional cases, if the exciting current densities are known and the solid conductors exposed to eddy currents are short-circuited. Otherwise, the magnetic field has to be constrained by the electric circuit equations of the windings and conductors.

In two-dimensional cases the gradient of the scalar potential is constant on the cross-section S_c of a conductor, and the partial differential equation of the single-component vector potential can be written in the form

$$- \nabla \cdot (v \nabla A) + \sigma \frac{\partial A}{\partial t} + \sigma \frac{\Delta V}{l_c} = 0 \tag{14}$$

where l_c is the length of the conductor and ΔV is the potential difference induced between the ends of the conductor. Eq. (14) can also be written in terms of the total current i of the conductor:

$$- \nabla \cdot (v \nabla A) + \sigma \frac{\partial A}{\partial t} - \frac{1}{S_c} \int_{S_c} \sigma \frac{\partial A}{\partial t} dS - \frac{i}{S_c} = 0 \tag{15}$$

The finite element discretization of Eq. (14) leads to a system of equations of the form $\mathbf{Sa} + \mathbf{T\dot{a}} = \mathbf{Eu}$ where $\dot{\mathbf{a}}$ is the time derivative of the nodal value vector $\mathbf{a}$ and the vector $\mathbf{u}$ contains the voltages between the ends of the conductors. Similarly, Eq. (15) leads to a system of equations $\mathbf{Sa} + \mathbf{K\dot{a}} = \mathbf{Fi}$ where the vector $\mathbf{i}$ contains the currents of the conductors. Contributions of the same types are obtained also for multiturn coils with no skin effect. The field can be solved from these equations if the induced voltages or currents are known, respectively. Otherwise, the loop equations or nodal voltage equations of the electric circuits have to be solved simultaneously with the field equations [5,6].

A method for the combination of the circuit equations with the field equations was presented by Arkkio for the magnetic field analysis of cage induction motors [5,7]. Loop equations were chosen for the stator windings, and nodal voltage equations were chosen for the cage winding of the rotor. Even the equation of motion can be combined with the field and circuit equations. The discretization of a 15 kW cage induction motor is shown in Fig. 3. The three-dimensional end-region fields can be approximately modelled by end-winding impedances in the circuit equations. The connections of the rotor winding are shown in Fig. 4. The resulting system of ordinary differential equations is solved by the Crank-Nicholson method, and the Newton-Raphson method can be used for the solution of nonlinearities at each time step. An example of the computed and measured currents of an inverter-fed cage induction motor is shown in Fig. 5.

Following on Arkkio's approach, a general method for the solution of problems involving windings, solid conductors with eddy currents, voltage and current sources, external impedances with arbitrary connections, and rotating parts was developed based on a systematic application of the nodal voltage equations to the electric circuits [6]. Considerable savings in computing time are achieved if the field can be assumed to be time-harmonic. The field can in that case be solved in terms of complex variables, and no time-stepping is needed. The saturation of iron can be approximately taken into account by using an effective reluctivity or permeability [8].

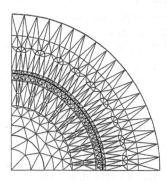

Figure 3. Finite element discretization of a 15 kW cage induction motor (904 quadratic finite elements).

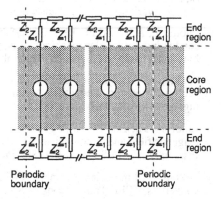

Figure 4. Electrical connections of the rotor cage winding.

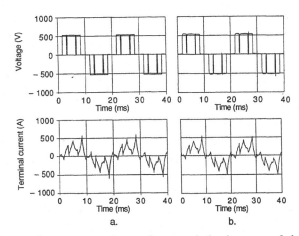

Figure 5. Voltage and current waveforms of an cage induction motor fed from a voltage-source inverter: a) computation, b) measurement.

References

1 Luomi, J., Magnetic field analysis of superconducting homopolar machines. Acta Polytechnica Scandinavica, Electrical Engineering Series No 53, Helsinki 1984.

2 Li, L., Luomi, J., Three-dimensional magnetostatic field analysis using vector variables. In: Brebbia, C.A. (ed.), Electromagnetic applications. Topics on Boundary Element Research Vol. 6. Springer-Verlag, Berlin & Heidelberg 1989, p. 25–46.

3 Li, L., Luomi, J., On three-dimensional boundary element methods for magnetostatics in vector variables. IEEE Transactions on Magnetics Vol. 24 (1988) No. 1, p. 19–22.

4 Luomi, J., A hybrid FE-BE method for axisymmetric magnetostatic fields of superconducting homopolar machines. International Conference on Electrical Machines, Lausanne, Switzerland, 18–21 September 1984, Proceedings Part 3, p. 1110–1113.

5 Arkkio, A., Analysis of induction motors based on the numerical solution of the magnetic field and circuit equations. Acta Polytechnica Scandinavica, Electrical Engineering Series No 59, Helsinki 1987.

6 Lindfors, H., Luomi, J., A general method for the numerical solution of coupled magnetic field and circuit equations. International Conference on Electrical Machines, Pisa, Italy, 12–14 September 1988, Proceedings Vol. I, p. 141–146.

7 Arkkio, A., Finite element analysis of cage induction motors fed by static frequency converters. IEEE Transactions on Magnetics Vol. 26 (1990) No. 2, p. 551–554.

8 Luomi, J., Niemenmaa, A., Arkkio, A., On the use of effective reluctivities in magnetic field analysis of induction motors fed from a sinusoidal voltage source. International Conference on Electrical Machines, München, 8–9–10 September 1986, Proceedings Part 3, p. 706–709.

Dr Jorma Luomi

Helsinki University of Technology, Laboratory of Electromechanics

Otakaari 5, SF-02150 Espoo, Finland

FINITE ELEMENT THERMAL ANALYSIS OF CLOSELY
SPACED UNDERGROUND POWER CABLES

K. G. Mahmoud[1] and S. Abdel-Sattar[2]
Faculty of Engineering, Assiut University, Egypt

ABSTRACT

Steady-state temperature distribution at the core, insulation, and the surrounding soil of a closely spaced underground cables is studied. A finite element method is applied to a variational functional equivalent to the conduction heat equation. The continuum is represented by a finite number of triangular elements and the isoparametric concept is used in formulating the linear approximating model of the temperature. The temperature distribution at the core, insulation, and around the vicinity is calculated and the results are presented in graphs. The results are in a good agreement with the previous work.

INTRODUCTION

Heat transfer analysts often calculate temperature distributions as a first step in the selection of materials or for the design of an engineering problem that may experience abnormal temperature levels during its service life. Analysts must be able to model realistically environmental boundary conditions, and represent complicated geometry. Underground power cables is an example of these problems where the maximum permitted insulating temperature is often a limitation to its current rating. The full utilization of the cable material is only possible if the temperature can be calculated for different loading conditions. Several methods have been introduced for solving the problem of thermal heat flow in buried cable system. Usually the interest is in calculating the temperature rise of an underground cable system buried directly in the soil. Such temperature rise depends on the load current, electrical resistance of the cable system and the thermal resistances of the surrounding medium [1-3].

The determination of the distribution of the temperature-rise at the vicinity of underground power cables and in the surroundinding medium is very important and it is difficult problem specially for unbounded region. Various numerical methods, such as finite difference, thermal simulation technique and finite elements techniques have been employed in the past to resolve the problem but at different cases [1-9]. Kennely [4] has been derived an approximate expression to calculate the surface tem-

---

[1]ÖAD fellow, Institute of Mathematic, Johannes Kepler University, Linz, Austria
[2]Alexander von Humboldt fellow, TU Hamburg-Harburg, Postfach 901052, D-2100 Hamburg 90

M. Heiliö (ed.), Proceedings of the Fifth European Conference on Mathematics in Industry, 277–282.
© 1991 B. G. Teubner Stuttgart and Kluwer Academic Publishers. Printed in the Netherlands.

perature rise for a three-phase single core flat arranged cable system. His approach has been developed based on the assumptions that the spacing between conductors and the depth of burial are large compared with the diameter. Such assumptions are not generally fullfilled by many actual underground cable systems. Another attempt, to improve Kennely's approach, has been descirbed by Binns [5]. He used digital computer program based on iteration technique to calculate the temperature rise of the system. But while results were good, some difficulties were encounted in obtaining convergance. Abdel-Sattar with others [6] used the thermal simulation technique for calcualting the temperature distribution around the vicinity of underground cables.

Recently, Lyall [9] has studied the transient temperature-rise of a three core cable. He developed a transient heat flow computer model of a cable system for the heating time of less than two hours. The fundamental concepts of both finite difference and finite element methods have been used in such analysis [9]. On the other hand, Tarasiewicz et al. [7-8] have calculated the equivalent thermal resistance and the temperature rise of different cable systems.

Non of the previous work can predict the temperature distribution within the core itself. In this work, a finite element method together with variational formulation will be used to study steady-state temperature distribution for underground power cables. A general and more realistic mathematical model will be developed to analyize the thermal field produced by loading the underground power cable system. A more realistic assumptions will be used in building such model. Different arrangements of underground power cables will be studied. The cases analyized in this study are triangular, flat, and vertical arrangements.

VARIATIONAL PRINCIPLES

Some of the early development in the finite element heat transfer analysis appear in references [11-15]. Consider steady-state heat transfer in a two-dimensional anisotropic solid Ω bounded by a surface Γ.

$$\frac{\partial}{\partial x}\left(k_x \frac{\partial T}{\partial x}\right) + \frac{\partial}{\partial y}\left(k_y \frac{\partial T}{\partial y}\right) + \dot{q} = 0 \tag{1}$$

The symmetry character of the domain around the vertical axis has been taken into consideration during mesh generation, Fig. 1. In Fig. 2, if we assume that H and h are large enough that the temperatures of the cables does not affect the temperature $T(x, y)$ of the soil at this region, then we will have the following boundary conditions.

$$T(x, 0) = T_s \quad on \quad \Gamma_1, \quad T(h, y) = T_o(y) \quad on \quad \Gamma_2, \tag{2}$$

$$T(x, H) = T_c \quad on \quad \Gamma_3, \quad \frac{\partial T}{\partial x} = 0 \quad on \quad \Gamma_4$$

where

k thermal conductivity($k = k_x = k_y$) are assumed to be constant within an element and equals to core, insulation or soil thermal conductivity, depending material of the element, see Fig. 1., $Wm^{-1}C^{-1}$.

$\dot{q}$ represents the electric power transformed into heat, so it will be equal to the power loss per unit volume of the cable, Wm^{-3}.

It can be shown [16-17] that the temperature, T(x,y) that satisfies equations (1)-(2) also minimizes the functional

$$I(T(x,y)) = \frac{1}{2}\int_{\Omega^e}\left[k_x\left(\frac{\partial T}{\partial x}\right)^2 + k_y\left(\frac{\partial T}{\partial y}\right)^2 - 2\dot{q}T\right]dxdy \qquad (3)$$

The problem now is to find a function, T(x,y), that will give a minimum value of $I(T(x,y))$. To make the functional $I(T)$ extremum with respect to $T(x,y)$, requires $\delta I(T) = 0$. Suppose that the solution domain Ω is divided into M triangular elments of r nodes each. By the usual procedure we may express the behavior of the unknown temperature T(x,y) within each element as

$$T^e(x,y) = \sum_{i=1}^{r} N_i T_i = \{N\}^t\{T\}^e \qquad (4)$$

where T_i is the nodal temperature at node i. The discretized form of the functional for one element is obtained by substituting (4) into (3). Then the minimum condition $\delta I(T) = 0$ for one element becomes

$$\frac{\partial I(T^e)}{\partial T_i} = 0, \quad i = 1,2,..r \qquad (5)$$

Combining all of the equations for all of the nodes of the element gives the following set of element equations:

$$[K]^e\{T\}^e = \{R\}^e \qquad (6)$$

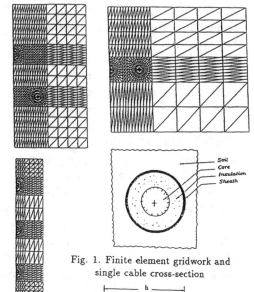

Fig. 1. Finite element gridwork and single cable cross-section

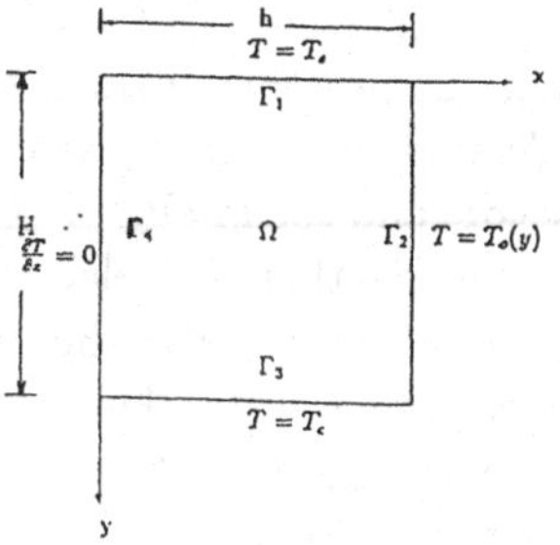

Fig. 2. Solution domain and boundary conditions

where

$$k_{ij} = \int_{\Omega^e}\left(k_x\frac{\partial N_i}{\partial x}\cdot\frac{\partial N_j}{\partial x} + k_y\frac{\partial N_i}{y}\cdot\frac{\partial N_j}{\partial y}\right)d\Omega^e, \qquad (7)$$

$$R_i = \int_{\Omega^e}\dot{q}N_i d\Omega^e \qquad (8)$$

Assembly of these elements equations to obtain the system equations follows the standard procedure. Then, the following set of equatins can be obtained

$$[K]\{T\} = \{R\} \qquad (9)$$

To save computer time and storage, the symmetric and banded character of the coefficient matric, [K], is utilized in solving Eq.(9).

RESULTS AND DISCUSSION

The input data used by Tarasiewicz et al. [7] is employed in this study. The computer program is written in a general way, so any boundary conditions can be applied. Figure 3 shows the temperature distribution around the cables system in triangular arrangement. The cables are placed at depth L=1.0 m, distance between centers S= 150 mm, and single core copper conductor area=630mm^2. The power loss (heat input) rate per unit length in each cable equals to 32.4 W/m, the value of thermal conductivity of the copper, insulation and soil were assumed constant and equal to 400, .20, .34 W/m °C re-

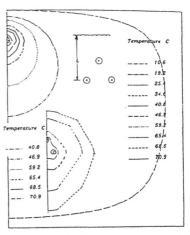

Fig. 3. Isotherms for triangular arrangement.

spectively. The relsults compared well with the results obtained by Tarasiwicz et al. [7]. The same parameters which were used in triangular arrangement are used for vertical arrangement, and the results are presented in the form of isothermal lines in Fig. 4. It is obvious from this figure, as it would be anticipated, that the outer cable temperatures are lower than the middle one.

The same data and the same power loss are used for flat arrangement, the maximum temperature of the middle cable is higher than that of the outer one, see Fig. 5. As it is clear from these figures, the maximum temperature is the temperature of the core and the minimum temperature is very close to the boundary temperatures. This proves that the selected values of h and H justifies the assumption that they are large enough that the temperature at the boundaries are not affected by the cables temperature. This is applicable also to Fig. 4&5. The temperature pattern in the area close to the cores is shown in the lowest left corners of Fig. 3,4 & 5.

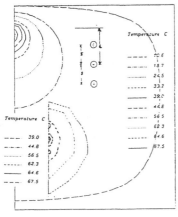

Fig. 4. Isotherms vertical arrangement.

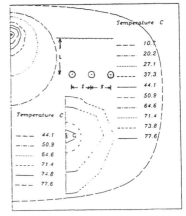

Fig. 5. Isotherms for flat arrangement.

The effect of the distance between core centers, S, on the maximum temperature for the three arrangements discussed previously is presented in Fig. 6. The results are plotted in a normalized dimensions. As it would be anticipated, the maximum temperature decreases as the spacing, S , increases in the three cases. We can see that the maximum temperature in case of triangular arrangement is more dependent on S than that in case of vertical arrangement. The maximum temperature in the vertical arrangement is more depenent on S than the flat one. The effect of the depth L on the maximum temperature can be studied. The lack of real data, specially the temperature distribution in the soil, make it difficult to make such study.

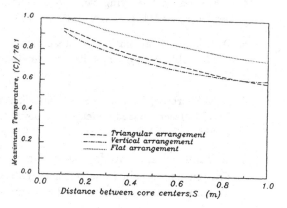

Fig. 6. Maximum temperature VS. Spacing, S.

CONCLUSION

A finite element thermal analysis model has been developed for closely spaced underground cables. In this model, the finite element method is applied to an variational functional equivalent to a two dimensional heat conduciton equation. Comparing to the reported work, the model is more general and more realistic, because realistic assumptions are used. Non of the previous work can predict the temperature distribution in the core. This model can predict the temperature distribution within the core, the insulation and the surrounding siol. The resutls presented in this work are beneficial to high voltage cable designers in choosing the optimal cable arrangement and cable spacing.

ACKNOWLEDGEMENT

This work has been done on the computational facilities of the working group for *Numerical Analysis and Optimization, Mathematic Institute, Johannes Kepler University Linz, Austria.* The stay of K. Mahmoud is supported by ÖAD and S. Abdel-Sattar is supported by Alexander Von Humboldt.

REFERENCES

1. J.H.Neher,"The temperature rise of buried cables and pipes," AIEE Trans., **73**, pp 406-412,(1954).

2. J.H.Neher and M. McGrawth,"The calculation of the temperature rise and load capability of cable system," AIEE Trans., 76, pp752, (1957).

3. A.Lombardi,C. Taralli, and C. Tencer, "Heat transfer in forced cooled cables," IEEE Trans. on power Delivery, 5, pp 8-13, (1990).

4. A.E.Kennely, "Carrying capacity of electric cables submerged, buried or sus-bended in air," Elect. World, **22**, p. 183, 1893.

5. D.F.Binns, "Calculation of steady-state temperature rise of water cooled buried cables using a new iterative method," Proc. IEEE,**116**, pp. 101-106, (1969).

6. M.El-Sherbiny, M. Abdel-Salam and S. Abdel-Sattar, "Digital study of heat flow for closely spaced underground power transmission cables using simulation technique," IEEE paper no. A78 553-0, pp. 1-9, (1978).

7. E. Tarasiewicz, E. Kuffel and S. Grzybowski, "Calculation of temperature distributions within cable tench backfill and the Surrounding Soil, " IEEE Trans. on Power Apparatus and Systems, **104**, pp 1973-1978, (1985).

8. E. Tarasiewicz et al., "Generalized coefficient of external thermal resistance for ampacity evaluation of underground multiple cable systems ," IEEE Trans. on Power Delivery, **2**, pp. 15-20, (1987).

9. J. Lyall, "Two dimensional modelling of three core cable transient temperature rise," IEEE Trans. on Power Delivery, **5**, pp 21-25, (1990).

10. V. Privezentsev et al., "Fundamental of Cable Engineering," Mir Publishers, Moscow, (1973).

11. E.L.Wilson and R.E.Nickell, "Application of the Finite Element Method to Heat Conduction Analysis," Nucl. Eng. Des.,4, (1966).

12. A.F.Emery and W.W.Carson, "An Evaluation of the Use of the Finite Element Methods in The Computation of Temperature," J. Heat Transfer,**93**, pp 136-153, (1971).

13. R.H.Gallagher and R.H. Mallett, "Efficient Solution Process for Finite Element Analysis of Transient Heat Conduction," J. Heat Transfer,**93** , pp 257-263, (1971).

14. J. Heuser, "Finite Element Method for Thermal Analysis," Goddard Space Center, Greenbelt, Md., NASA TN D-7274, (1973).

15. E.L.Wilson, K.J.Bathe, and F.E.Peterson, "Finite Element Analysis of linear and Nolinear Heat Transfer," Nucl. Eng. Des.,**29**, pp 110-114, (1974).

16. S.G. Mikhlin, "Variational Methods in Mathematical Physics," Macmillan Company, New York, (1964).

17. R.S.Schechter, "The Variational Method in Engineering," McGraw- Hill Book Company, New York, (1967).

Heat propagation by hot gas flow through a pebble bed

Jürg T. Marti

Abstract

Thermal energy storage requires the computation of the time-dependent temperature in a cylindrical container filled with solid balls which are heated up to high temperatures by an axial flow of hot gas. The moving heat front has been computed by the use of a time dependent model based on a finite element method including the heat conduction between the solid balls and the gas stream. The numerical problem becomes extremely stiff and requires special effort to reduce the computing time.

The storage of solar thermal energy in a pebble bed heated up (or cooled down) by hot air leads to the following system of partial differential equations [1]

$$\partial_t z = \frac{2}{r}\partial_r z + \partial_r^2 z \,, \;\; z(r,x,0) \;=\; 0, \;\; r \in [0,1]\,, \;\; x \in [0,\infty), \tag{1}$$
$$\partial_r z(0,x,t) \;=\; 0, \;\; x,t \in [0,\infty),$$
$$\partial_r z(1,x,t) \;=\; \alpha[u(x,t) - z(1,x,t)]\,, \;\; x,t \in [0,\infty)$$

for $z\colon [0,1] \times [0,\infty)^2 \to \mathbb{R}$ and

$$\partial_x u = -\gamma \partial_r z(1,\cdot,\cdot)\,, \;\; u(x,0) \;=\; 0\,, \, x \in (0,\infty), \tag{2}$$
$$u(0,t) \;=\; 1\,, t \in [0,\infty),$$

for $u\colon [0,\infty)^2 \to \mathbb{R}$. The first equation gives the time-dependent temperature distribution z in the spherical pebbles of diameter 2 and the second gives the temperature distribution u of the one-dimensional air stream, where $\partial_x u$ is the decrease of u in the air stream direction x due to convection. Estimates show that in this situation radiation and heat currents from pebble to pebble may be neglected. Since the velocity of the gas stream in the problems considered is several magnitudes higher than the velocity of the wave front creeping through the pebble bed, eqn. (2) is stationary and a diffusion term may be neglected. A fictive time scale for t allows the pebbles to be unit balls in $\mathbb{R}^3$.

In order to solve the initial and boundary value problem (1) and (2) we introduce a basis $b_0,\ldots,b_n$ of linear combinations of cubic B-splines such that the b's form a

M. Heiliö (ed.), Proceedings of the Fifth European Conference on Mathematics in Industry, 283–286.

partition of unity and all except b_0 satisfy the homogeneous boundary conditions corresponding to (1). If S_n is the vector space of cubic splines with knots $0, h, 2h, \ldots, 1$, $h := 1/(n-1)$ and $a = (a_1, \ldots, a_{n+2})^T$ is the vector of the usual translated cubic B-splines a_i with support $[0 \vee (i-4)h, ih \wedge 1]$ on $[0,1]$ (which, again, form a partition of unity of $[0,1]$ it follows that $b_0 = \beta a_{n+2}$ and that $b = (b_1, \ldots, b_n)^T$ is given by $b = Fa$, where $F = diag(F_1, F_2)$ is the 2×2 block diagonal matrix in $\mathbb{R}^{n \times (n+2)}$ defined by $F_1 := [e_2 e_1 e_2 e_3 \ldots e_{n-2}]$ ($e_i = i$th standard basis vector of $\mathbb{R}^{n-2}$),

$$F_2 := \begin{bmatrix} \beta & \frac{3}{2} - \frac{\beta}{2} & 0 \\ 1 - \beta & -\frac{1}{2} + \frac{\beta}{2} & 1 - \beta \end{bmatrix}$$

and $\beta := 6\alpha h/(3 + \alpha h)$. The weak boundary value problem corresponding to (1) is now

$$\partial_t(z, w) = -\langle z, w \rangle + \alpha[u(.,.) - z(1,.,.)]w(1) , \quad w \in H^1(0,1) ,$$

where

$$(v, w) := \int_0^1 r^2 v(r) w(r) dr , \quad v, w \in L_2(0,1) ,$$

$$\langle v, w \rangle := \int_0^1 r^2 v'(r) w'(r) dr , \quad v, w \in H^1(0,1)$$

and $H^1(0,1)$ denotes the Sobolev space of (weakly) differentiable real functions on $(0,1)$. The choice $z = ub_0 + y^T b$, $y : [0,\infty)^2 \to \mathbb{R}^n$ and $w = b_1, \ldots, b_n$ gives the system of Galerkin equations

$$\partial_t(y^T b, b) = -\langle y^T b, b \rangle + \alpha[u - y^T b(1)]b(1) - u[\langle b_0, b \rangle + \alpha b_0(1)b(1)] .$$

The equivalent matrix formulation, using

$$B := F(a, a^T)F^T , \quad C := F[\langle a, a^T \rangle + \alpha a(1)a(1)^T]F^T \in \mathbb{R}^{n \times n} ,$$

is

$$\partial_t By = -Cy + \beta u F[(2h)^{-1} a(1) - \langle a_{n+2}, a \rangle] .$$

Again using $D := B^{-1}C$ and $d := \beta B^{-1} F[(2h)^{-1} a(1) - \langle a_{n+2}, a \rangle]$ it follows that

$$\partial_t y = -Dy + ud , \quad y(x,0) = 0 , \quad x \in [0,\infty) . \tag{3}$$

Moreover, (2) and the above assumption on z yield

$$\partial_x u = (2h)^{-1}\beta\gamma(y_{n-1} - u) , \quad u(x,0) = 0 , \quad x > 0 , \quad u(0,t) = 1 , \quad t \geq 0 . \tag{4}$$

Since the system of partial differential equations (3) and (4) is extremely stiff, since $u(x,0)$ is discontinuous in x and since in practical cases t becomes very large, the solution of (3) and (4) needs either an astronomic amount of computing time or an idea to manage the inherent stiffness. Since (3) and (4) are of the hyperbolic type we can use the ideas for the solution of systems of conservation laws (see e.g. [2]) for hyperbolic equations. We thus make use of the following transformation of the solution $(y, u)^T : [0, \infty)^2 \to \mathbb{R}^{n+1}$ into a new pair of functions p and $q : [0, \infty)^2 \to \mathbb{R}^{n+1}$ given by

$$p(x, t) = \int_0^t \begin{bmatrix} y \\ u \end{bmatrix} (x, t') dt' \qquad (5)$$

$$q(x, t) = \int_0^x \begin{bmatrix} y \\ u \end{bmatrix} (x', t) dx' . \qquad (6)$$

Integration of (3) and (4) with the corresponding initial and boundary value conditions then yields

$$\partial_t p = \partial_x q = Ap + (1 + c^T q) e_{n+1} , \qquad (7)$$

where $\{e_1, \ldots, e_{n+1}\}$ is the standard basis of $\mathbb{R}^{n+1}$, $c := (2h)^{-1} \beta \gamma (e_{n-1} - e_{n+1})$ and

$$A := \begin{bmatrix} -D & d \\ 0 & 0 \end{bmatrix} \in \mathbb{R}^{(n+1) \times (n+1)} .$$

The initial and boundary conditions for (7) are now obtained from (5) and (6) and the corresponding conditions in (3) and (4): $p(x, 0) = q(x, 0) = 0$ $x > 0$, $q(0, t) = 0$ and

$$p(0, t) = \begin{bmatrix} D^{-1}(td - w) \\ t \end{bmatrix}, t > 0,$$

where w is the solution of the system of ordinary differential equations

$$w' = -Dw + d , \quad w(0) = 0 .$$

An adequate method for the solution of (7) is the trapezoidal rule along the characteristics of (7) in the $x - t$-plane:

$$p_{j+1,k+1} = p_{j+1,k} + \tfrac{1}{2} \Delta t [A(p_{j+1,k} + p_{j+1,k+1})$$
$$+ (2 + c^T q_{j+1,k} + c^T q_{j+1,k+1}) e_{n+1}]$$
$$q_{j+1,k+1} = q_{j,k+1} + \tfrac{1}{2} \Delta x [A(p_{j,k+1} + p_{j+1,k+1})$$
$$+ (2 + c^T q_{j,k+1} + c^T q_{j+1,k+1}) e_{n+1}] .$$

The results of a computation with $\Delta x = \Delta t = 0.0625$, $n = 4$ and constants $\alpha = 0.5$ and $\gamma = 20$ corresponding to the realistic case of granite pebbles of diameter 0.025 heated (cooled) by air is shown in Fig.1. In this example the velocity of the air is 2.2 m/s.

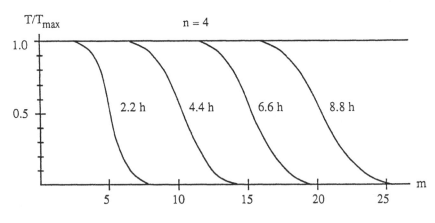

Fig. 1

References

[1] W. Durisch, E. Frick, P. Kesselring. Heat storage in solar power plants using solid beds. ETH-EIR Report TM 13-86-08 (1986).

[2] R.J. LeVeque. A large time step generalization of Godunov's method for systems of conservation laws. SIAM J. Numer. Anal. 22, 1051-1073 (1985).

Jürg T. Marti
Institute for Applied Mathematics
ETH Zürich
CH-8092 Zürich, Switzerland

THERMAL EFFECTS DURING CEMENTATION OF AN ENDOMEDULLARY INFIBULUM

S. Mazzullo, G. Paganetto, T. Simonazzi
HIMONT ITALIA S.r.l.
Centro Ricerche "G.Natta"
44100 FERRARA (Italy)

ABSTRACT. A 1-dimensional mathematical model of a femoral prosthesis implant fixation has been developed. Surgical fixing procedures call for resectioning the femoral head of the patient and preparing an adequate femoral medullar cavity into which the stem of the infibulum is cemented. The actual complex geometry has been idealized as a composite cylinder of infinite length having a multishell bone/cement/stem structure. Attention is focused on the cement, which is a polymer that polymerizes "in situ", inside the medullar channel and generates heat.
The "cementation" phenomenon is described by the heat diffusion Fourier equation coupled with the polymerization kinetics. The numerical solution of the model has been accomplished by a finite difference explicit scheme. An important aspect of this work was to carry out experiments in our laboratory, to provide data for the model.
The two most relevant results are the following:
1. The conversion of monomer into polymer is never 100% under the imposed initial and boundary conditions.
2. Bone/cement interface temperature is a function of the interface heat transfer coefficient.
Depending on such a parameter, ranging from 10. to 1000.($Wm^{-2}°K^{-1}$), the temperature may reach the maximum values of 80°C and 40°C, respectively. Therefore, the model predicts relatively high temperature values at the bone/cement interface. Such values confirm the criticality of this type of surgical implant.

1. INTRODUCTION

Cementation technique of hip prostheses by polymethylmethacrylate (PMMA) and its copolymers started less than 30 years ago |6|. Total hip prostheses generally consist of two pieces: the femoral component and the aceta

M. Heiliö (ed.), Proceedings of the Fifth European Conference on Mathematics in Industry, 287–291.
© 1991 B. G. Teubner Stuttgart and Kluwer Academic Publishers. Printed in the Netherlands.

bular component (Fig.1a). We shall deal with the cementation thermal
effects on the femoral component: the cement is, in fact, a polymer which
polymerizes "in situ", inside the medullar channel, and generates heat.
It is pretty natural to idealize the complex geometry of a femoral implant
as a bone/cement/stem composite cylinder (Fig.1b). Assuming that the stem
length/medullar channel ratio is generally greater than 5:1, the composite
cylinder can be considered of infinite length. Thus the study shall be
limited to the heat flow in the radial direction only.

2. MATHEMATICAL MODEL

The cementation phenomenon is described by the heat diffusion Fourier
equation coupled with the polymerization kinetics. If $T(r,t)$ is tempera
ture and $X(r,t)$ the dimensionless conversion of polymerization, we
obtain in radial coordinates:

$$\varrho \, Cp \frac{\partial T}{\partial t} = K \left(\frac{\partial^2 T}{\partial r^2} + \frac{1}{r} \, \frac{\partial T}{\partial r} \right) + \varrho \, \lambda \frac{dX}{dt} \tag{1}$$

$$\frac{dX}{dt} = k \, (T) \, f \, (T,X) \tag{2}$$

where $\varrho = 1.19 \times 10^3$ kgm^{-3} is density, $Cp = 1.6 \times 10^3$ J Kg^{-1}°K^{-1} is
specific heat, $K = 0.17$ W m^{-2}°K^{-1} is thermal conductivity and $\lambda = 193.$
J Kg^{-1} is enthalpy of polymerization of cement. For simplicity all these
quantities are assumed independent of temperature.
Although polymerization of methyl-methacrylate (MMA) is qualitatively
well known, not all kinetic parameters are always available for the
more common cements is use [3], [4]. Therefore we have decided to carry
out isothermal polymerization experiments (Fig.2) in order to identify
all the actual kinetic parameters of a commercial cement. Experimental
data suggest an Arrhenius behaviour for the kinetic constant:

$$k(T) = k_o \, exp - E/RT \tag{3}$$

where $k_o = 2.64 \times 10^8$ sec^{-1} and $E = 62.966 \times 10^3$ J mol^{-1} and the following functional form for the conversion of polymerization:

$$f(T,X) = \begin{cases} \dfrac{\beta}{\eta(T)} \, X^{1-\frac{1}{\beta}} (\eta(T)-X)^{1+\frac{1}{\beta}} & ; \quad X < \eta(T) \\ \\ 0 & ; \quad X \geq \eta(T). \end{cases} \tag{4}$$

where $\beta = 9.2$ is an empirical parameter.
The function $\eta(T)$ is the equilibrium conversion of monomer to polymer.
It behaves as follows [1]:

$$\eta(T) = \begin{array}{ll} \dfrac{T}{Tg} & ; \qquad 0 \le T \le Tg \\[2mm] 1 & ; \qquad\quad T > Tg \end{array} \tag{5}$$

where Tg = 378 °K is the glass transition temperature of cement.
To complete the model, appropriate initial and boundary conditions are
required. The temperature of the cement, before its introduction into
the femoral cavity, is either known or can be determined experimentally.
The same applies also to the initial conversion. Therefore, as initial
conditions we have:

$$T(r,o) = T_o \tag{6}$$

$$X(o) = X_o \tag{7}$$

At the stem/cement boundary, r = a, we assume perfect contact between
the cement and a perfect heat conductor (the metal infibulum) |2|:

$$\pi a^2 \varrho_m \, Cp_m \, \frac{\partial T}{\partial t} = 2 \pi a \, K \, \frac{\partial T}{\partial r} \tag{8}$$

where ϱ_m = 7.85 x 10^3 Kg m^{-3}, and Cp_m = 0.46 x 10^3 J kg^{-1}°K^{-1}.
At the cement/bone boundary, r = b, we assume linear heat transfer
between the cement and the sorrounding medium (the bone) at body temperature T_B = 37°C:

$$K \, \frac{\partial T}{\partial r} = - \, U(T - T_B) \tag{9}$$

where U(W m^{-2}°K^{-1}) is the surface heat transfer coefficient. The actual
value of U strongly depends on the preparation of the femoral cavity
by the surgeon |3|.

3. NUMERICAL SOLUTION

On physical grounds, the model formulated in section 2 seems well posed.
From a mathematical view point the model belongs to the class of Stefan
problems with a relaxation dynamics for the phase variable X,|5|. More
formal sufficient conditions of existence and uniqueness of the solution
can be found in |5|. The numerical solution of the model has been obta-
ined in three steps as follows: i) transformation of the problem to
dimensionless form, ii) change of the coordinate system from hollow
cylindrical "r" to rectangular "x", by setting r = exp x. iii) numerical
solution of the cartesian problem by explicit finite difference schemes.
Use has been made of forward difference operators to approximate time
derivatives, and central difference operators to approximate space deri-

vatives. Stability restrictions on time step can be easily obtained.

4. RESULTS

Numerical simulations shall be limited to a sensitivity analysis with respect to variations of the heat transfer coefficient, at the cement/bone interface. The U parameter takes values in the range 10 ÷ 1000 $(W \ m^{-2} {}^{o}K^{-1})$. The metallic infibulum has a 16 mm diameter and the cement a 6 mm thickness. Temperature at the cement/bone interface, as a function of time, is shown in Fig.3. The temperature profile always reaches a maximum, T max, whose intensity increases when U decreases. The maximum conversion, X max, shows a similar trend. It should be mentioned that the temperature behaviour is basically in agreement with the results of Huiskes' model |3| which is numerically solved by finite elements. An important and so far unique feature emphasized by the numerical simulations, is that in no case a 100% conversion is achieved. Therefore, presence of residual unreacted monomer is always noticed in the cement. Fig. 4 provides the radial conversion profile at fixed times, in case the U parameter assumes the value of 1000. The maximum radial conversion is 82.1%, while maximum cement/bone interface temperature is 41.7°C.

5. REFERENCES

|1| Burnet, G.M., Duncan G.L. (1962):'High conversion polymerization of vinyl systems. I Methyl-methacrylate'.Die Makromoleculare Chem. 51, 154-170.

|2| Carslaw H.S.; Jaeger J.C. (1959): Conduction of heat in solids. Oxford Univ. Press, Oxford.

|3| Huiskes R.; (1979): Some foundamental aspects of human joint replacement. Acta Orthop. Scand.; Supp. 185 - Copenhagen.

|4| Jefferis C.D.; Lee A.J.C.; Ling R.S.M. (1975):'Thermal aspects of self curing PMMA '.J. Bone Joint Surg. 578, 511-518.

|5| Verdi C.; Visintin A. (1987):'Numerical analysis of the multidimensional Stefan problem with supercooling and superheating'.Boll. U.M.I. 1-B 795-814.

|6| Pipino F. (1987): Il punto sulla cementazione degli impianti protesici. OIC Medical Press, Firenze.

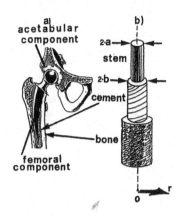

FIG. 1

Hip joint replacement, a) and femoral geometric model, b)

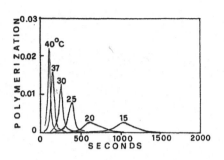

FIG. 2

Exper. isotherms of polymerization of a cement. Polymerization (experimental rate of polymerization) is expressed in 1/sec.

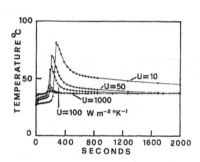

FIG. 3

Cement/bone interface temperature

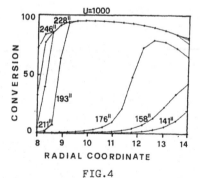

FIG.4

Radial conversion at fixed times.

A LINEAR MIXED-INTEGER PROGRAMMING MODEL FOR THE
OPTIMIZATION LOGICAL OF THE MODULES BURN-IN TEST

Dr. Anita Merli
Ing.Giancarlo Poli
Advanced Design Dept.
IBM Italy Manufacturing
VM 365 367
via Lecco 61
20059 Vimercate - ITALIA

PROBLEM DESCRIPTION

Logical modules are very small but very important logical components of
cards, which are the more technologically advanced parts of a computer.
The production process is very complex, requires advanced technology,
and can be roughly described in three steps:
* kitting of the components from the warehouse to be sent to the
assembly dept.
* assembly of the components
* test, several tests are performed on logical modules, and will pay
attention to one of them, "Burn-in" test.

BURN-IN TEST

Burn-in phase is a stress test: its target is to ensure to the yield of
electronic components an high standard of reliability. In fact have
been established, by many tests and analysis on materials, that modules
behavior had a well defined trend:
* "some" of the sample of the module under study break in the very
early hours of life; this phenomena is called "infantile death"
* another group of the sample lot, the so called "freak population"
reveals some failure in the time lag included between some hundreds
and some thousands of hours of run
* at last the "main population", which ends its life after more
10.000 hours of running.

Of course, being known the trend of life of every component, it is
possible to simulate its running life in test phase for a time lag
large enough, and to get rid of that part of them which 'd break by
"infantile death". So burn-in test is performed in order to make more
reliable the produced modules.

Burn-in test consists of:
* thermal cycling
* power on-off
* over voltage and voltage variation.

In Vimercate IBM plant the steps of the process are the following:

M. Heiliö (ed.), Proceedings of the Fifth European Conference on Mathematics in Industry, 293–296.

* the modules are put on fire-proof trays, called Burn-In-Boards
(BIB in the following), bibs contains all the electric circuits for
sending to modules all the required signals; the sequence of signals
is proper to each module, and each bib can be used for a set of
different modules
* bibs are put inside the oven: oven is partitioned in some areas
(usually 3 or 6), and in each are there are 20 slots (electric
contacts) where the bibs are inserted; in all the areas the same
thermal cycle is performed, but is possible to address to each area
different electric cycles; the electric cycles are managed by control
software on a PC
* the oven is shut and the thermal cycle begins: it is about 8 hours
lasting and each thermal cycle is proper of a set of modules (usually
different from the set by bib, of course).

THE LOGISTIC PROBLEM

The assembly machines require a not small set-up time to pass from a
module to another one, and also the testers to be used after the
burn-in phase require set-up, so that lots have a huge dimension. As
the bib are available in a very limited number (for they are very
expensive and are to be scratched when a module is no more produced)
and for the things told above, burn-in department would prefer small
lots. A lot cannot be divided: from an administrative point of view the
division of a lot cannot be managed; so a lot may not leave the
department until it is wholly processed. So burn-in department is a
bottle neck in the module production process: the time of crossing is
very high because resources are limited, and technical constrains are
so many that the human operator cannot, besides respect them, to
maximize the loading of each oven.

The application provides a help to human operator by proposing a
maximized loading with respect of the following condition:
* only one cure family for each oven load
* only one lot for each oven area
* bibs/modules compatibility
* bib availability
* the objective function is weighted by a priority criteria defined
according to the needs of the user.

THE APPLICATION ARCHITECTURE

The software systems which manage the logical modules logistic data are
on an AS400, while MPSX is installed on a 3090. Therefore the
application must also allow the communication between different
environment in order to integrate them: the user friendly environment
for the user, the problem data base environment and the optimization
environment. The user front-end allows a whole visibility of data, the
possibility to modify some of them, it contains a model generator, that
is a routine which writes a file using MPS format using the model and
current data, it allows to see the output of optimization and to update

data properly. The front-end have been written in APL2 under VM/CMS
with APE panels. The communication between the different environments
is kept by an expert system which works as an administrative robot
(ARENA, developed inside IBM plant of Vimercate): such administrative
robot opens and closes the different sessions as necessary, and
transfers files from a session to another. The optimizer is used when
an oven is available, and maximizes its utilization.

THE MODEL

- Variables

$X(i)$ number of modules of lot i to be processed

$Z(ijk)$ decision to process lot i belonging to cure family j in oven
           area k   0-1 variable

$F(j)$ active cure family
           variable 0-1 with sum 1 (Special Ordered Set)

$BIB(il)$ number of bibs of kind l to be used for lot i
           integer variable

- Objective Function

$$(0) \text{ MAX } \sum_i \sum_k P(i).Z(ijk) + \sum_i Q(i).X(i)$$

i lot pointer, k cure family pointer,
$P(i)$ and $Q(i)$ are weight vectors chosen by the user, and reflect the
lot priority
pointer j, which means belonging to a cure family, depends univocally
by i
- Constrains

$$(1) \quad X(i) - A.\sum_k Z(ijk) >= 0 \quad \text{for each i}$$

A is the minimum within the lot remaining quantity
                         the capacity of suitable and available bibs
                         the capacity (in modules) of each oven area

$$(2) \quad \sum_j F(j) = 1$$

one and only one cure family is to be chosen

$$(3) \quad \sum_i \sum_k Z(ijk) <= ZONE$$

ZONE is the number of areas of the oven to be filled
The number of used areas may not exceed the number of available areas

(4)   Sum X(i) <= disp(il) . capbib(il)   for each l
         i

i runs on the lots which may use the same bib, and the inequality is
written only for the values of l which have at least one bib suitable for the queued lots the link impose that the number of processed modules don't exceed the correct bib availability multiplied
by their capacity

(5)   0 <= capbib(il).BIB(il) - X(i) <= capbib(il) - 1   for each i

the number of the empty places on a bib must be
strictly smaller than the capacity of a bib

(6)   Sum BIB(il) <= 20.ZONE
         l

the number of bibs altogether used must not exceed the number
of available slots, which is 20 in each oven area

(7)   H.Sum Sum Z(ijk) < = F(j) < = Sum Sum Z(ijk)   for each j
           i   k   i   k

H is a number small enough
this double inequality links the choice of the lot with
the choice of the cure family

(8)   Sum Z(ijk) <= 1   for each k

no more than one lot can be put in each area

(9)   20.Sum Z(ijk) - BIB(il) <= 19   for each i
            k

the capacity of an area must be wholly used before beginning to fill
the next one.

PERFORMANCES

This model gets quite good fillings of the ovens. In less than 1 mi-
nute of CPU time of a 3090 a good integer solution is found for 150-160
integer variables by MPSX/370 V2. The optimality is not always proved,
in order to contain run time, but the solution found is good enough
for production purposes.

THE SPECTRUM OF AN INHOMOGENEOUS MEMBRANE

J. Molenaar

1. Introduction

In this report we analyze the spectrum of a vibrating membrane with inhomogeneous density and search for a geometry with overtone frequencies which fit harmonically to a fundamental frequency. Then, the membrane will have a clear pitch, which makes it attractive for application in a certain class of musical instruments, namely the kettle drums. The present investigations are part of a study after the relationship between the performance of these instruments and their (many) geometrical parameters. Kettle drums consist of a circular membrane stretched over a kettle. The vibration modes of the system as a whole are the result of the interactions between the vibration modes of membrane, kettle and enclosed air. For an introduction into the physics of kettle drums we refer to [1]. In most kettle drums membranes of homogeneous density are used. The determination of the modes of such membranes (without kettle and *in vacuo*) may be considered as one of the standard problems in mathematical physics, which has been solved already long ago. The corresponding spectrum follows from the zeros of Bessel functions and is clearly inharmonic. In a certain Indian type of kettle drums, the so-called tabla, membranes of non-uniform density are applied. It is generally assumed that the presence of the inhomogeneities improves the harmonicity of the instrument. To focus on this latter idea, we study here a model in which the kettle and the air load are ignored.

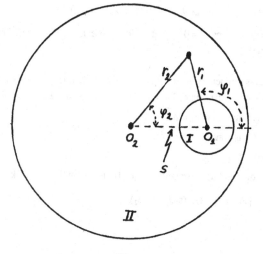

Fig. 1.

M. Heiliö (ed.), Proceedings of the Fifth European Conference on Mathematics in Industry, 297–302.
© 1991 B. G. Teubner Stuttgart and Kluwer Academic Publishers. Printed in the Netherlands.

2. The inhomogeneous membrane

We consider a circular membrane with a circular inhomogeneity, the centre of which does not coincide with that of the membrane. Then, no circular symmetry is present, which has far reaching consequences. The geometry is drawn in Fig. 1, wherein also some symbols to be used in the following are indicated.

The circular membrane has radius R, centre O_2 and homogeneous density ρ except for the circular inhomogeneity, which has homogeneous density $\delta\rho$ with multiplication factor $\delta > 0$. It has radius αR, with $0 < \alpha < 1$, and centre O_1. The vector $s = O_2 - O_1$ connecting the centres has length βR with $0 \leq \beta \leq (1-\alpha)$. The transversal displacement u is assumed to satisfy the linear wave equation

$$\gamma^2 u_{tt} = \Delta u \tag{2.1}$$

with $\gamma^2 = \gamma_1^2 = \delta\rho / T$ in region I and $\gamma^2 = \gamma_2^2 = \rho / T$ in region II (see Fig. 1) with T the constant and uniform tension. We separate time and space variables and search for basis solutions of the form $u(t,r,\phi) = f(t)\, g(r,\phi)$. The spatial part g satisfies

$$g_{rr} + \frac{1}{r}\, g_r + \frac{1}{r^2}\, g_{\phi\phi} + \gamma^2 \lambda^2\, g = 0 . \tag{2.2}$$

The admissible values for the separation constant $\lambda \in I\!R^+$ constitute the spectrum of the membrane. These values are determined by the boundary conditions, which are given underneath. In terms of the polar coordinates (r_1, ϕ_1) as indicated in Fig. 1 we find the following expressions for g:

$$
\begin{aligned}
g(r_1, \phi_1) &= \sum_L A_L\, J_L(\gamma_1 \lambda r_1, \phi_1) && \text{in I} \\
&= \sum_L (B_L\, J_L(\gamma_2 \lambda r_1, \phi_1) + C_L\, Y_L(\gamma_2 \lambda r_1, \phi_1)) && \text{in II}
\end{aligned}
\tag{2.3}
$$

where we use the shorthand notations $L = (n,k)$ with $n \in I\!N$ and $k \in \{0,1\}$ and

$$J_L(r,\phi) = J_n(r)\, cs_L(\phi) \quad , \quad Y_L(r,\phi) = Y_n(r)\, cs_L(\phi)$$

with $\quad cs_{0,0} = (2\pi)^{-\frac{1}{2}}$, $cs_{0,1} = 0 \quad$ and $\quad$ for $\quad n \geq 1, \quad cs_L(\phi) = \pi^{-\frac{1}{2}} \cos n\phi$, $k = 0$, and $cs_L(\phi) = \pi^{-\frac{1}{2}} \sin n\phi$, $k = 1$.

J_n and Y_n are Bessel functions of the first and second kind, resp. In region I Y_n is not admissible because it is singular at the origin. The other conditions on g are

i) g is continuous at $r_1 = \alpha R$

ii) $\partial g / \partial r_1$ is continuous at $r_1 = \alpha R$

iii) $g = 0$ at $r_2 = R$.

The application of i) and ii) is straightforward. To deal with iii) we invoke the expansion rules

$$J_L(r_1, \phi_1) = \sum_{L'} M_{LL'}(s)\, J_{L'}(r_2, \phi_2) \tag{2.4a}$$

$$Y_L(r_1,\phi_1) = \sum_{L'} M_{LL'}(s)\, Y_{L'}(r_2,\phi_2)\,. \tag{2.4b}$$

Note that the matrix M only depends on the shift of origins $s = O_2 - O_1$ as denoted in Fig. 1. Expansion (2.4b) holds under restrictions, which are always satisfied in the present applications. Expansion (2.4a) holds without geometrical restrictions. For a derivation of (2.4a,b) see [2]. In [2] it is also derived that a non-trivial solution for the coefficients A, B and C in (2.3) exists if the following conditions hold:

$$\det \mathbf{G}^0(\lambda) = 0 \quad , \quad \det \mathbf{G}^1(\lambda) = 0 \tag{2.5}$$

with $\mathbf{G}^0$ and $\mathbf{G}^1$ matrices of infinite dimensions. $\mathbf{G}^0$ corresponds with the equations with $k = 0$ and $\mathbf{G}^1$ with the equations with $k = 1$. These matrices are built up out of 3×3 blocks, which we label as $\mathbf{G}^0_{n,n'}$ and $\mathbf{G}^1_{n,n'}$. The diagonal blocks are given by

$$\mathbf{G}^0_{n,n} = \begin{bmatrix} -J_n(\gamma_1 \alpha R \lambda) & J_n(\gamma_2 \alpha R \lambda) & Y_n(\gamma_2 \alpha R \lambda) \\ -\gamma_1 J_n'(\gamma_1 \alpha R \lambda) & \gamma_2 J_n'(\gamma_2 \alpha R \lambda) & \gamma_2 Y_n'(\gamma_2 \alpha R \lambda) \\ 0 & J_n(\gamma_2 \lambda R)\, M^T_{LL}(\gamma_2 \lambda s) & Y_n(\gamma_2 \lambda R)\, M^T_{LL}(\gamma_2 \lambda s) \end{bmatrix} \tag{2.6a}$$

with $L = (n, 0)$. $\mathbf{G}^1_{n,n}$ has a similar form with $L = (n, 1)$. The off-diagonal blocks are given by

$$\mathbf{G}^0_{n,n'} = \begin{bmatrix} 0 & 0 & 0 \\ 0 & 0 & 0 \\ 0 & J_n(\gamma_2 \lambda R)\, M^T_{LL'}(\gamma_2 \lambda s) & Y_n(\gamma_2 \lambda R)\, M^T_{LL'}(\gamma_2 \lambda s) \end{bmatrix} \tag{2.6b}$$

with $L = (n, 0)$, $L' = (n',0)$. $\mathbf{G}^1_{n,n'}$ has a similar form with $L = (n, 1)$ and $L' = (n',1)$. From (2.5) we obtain a discrete spectrum λ_m, $m = 1,2,3,\dots$. In practice one has to cut off the $\mathbf{G}^0$ and $\mathbf{G}^1$ matrices in order to be able to calculate the determinants. Reliable values for this cut-off value must be determined numerically and will, of course, depend on the accuracy required.

3. Numerical results

We have implemented conditions (2.5) numerically and calculated the spectra of several geometries. In all examples we fixed the radius, tension and density of the membrane at the values $R = 1$, $T = 1$ and $\rho = 1$, respectively. The parameters α, β and δ (see Fig. 1) are varied successively. In all cases a cut-off val ue of $n = 5$ for the matrices in (2.6a,b) is used. We present results for four cases. In Figures 2, 3, 4, 5 the corresponding spectra are given. The (1,1) frequency is in general the fundamental one of a kettle drum, because the (0,1) mode, which has a lower frequency, damps out much faster. A musical instrument conveys a clear sense of pitch if it has several harmonic overtones, the frequencies of which are related to the fundamental in whole-number ratios. From the latter figures one can find whether the presence of an inhomogeneity makes the membrane more or less "harmonic".

- $0 \le \alpha \le 1$, $\beta = 0$, $\delta = 2$.

See Fig. 2. This geometry has circular symmetry. For $\alpha = 0$ and $\alpha = 1$ the membrane is homogeneous. In these limiting cases the spectrum is given by the zeros of the functions $J_n(\lambda)$ and $J_n(\sqrt{2}\,\lambda)$ respectively. The spectrum is given in Fig. 2 as a function of α. For $n = 0$ the eigenvalues are non-degenerate; for $n \geq 1$ they are two-fold degenerate. It is seen, that for a small, centrally positioned inhomogeneity ($\alpha < \frac{1}{2}$), the eigenvalues with $n = 0$ are most influenced. For $\frac{1}{2} < \alpha \leq 1$ all eigenvalues behave in nearly the same way. Going from $\alpha = 0$ to $\alpha = \frac{1}{2}$ the total mass of the membrane is increased by a factor of 5/4.

– $\alpha = 0.5$, $0 \leq \beta \leq 0.5$, $\delta = 2$.

See Fig. 3. In this case the inhomogeneity is shifted away from the central position. For $\beta > 0$ the circular symmetry is disturbed, so that the two-fold degeneracy of the eigenvalues with $n \geq 1$ as present in Fig. 2 is absent here. The $\beta = 0$ values in Fig. 3 correspond to the $\alpha = 0.5$ values in Fig. 2.

– $0 \leq \alpha \leq 0.5$, $\beta = 0.5$, $\delta = 2$.

See Fig. 4. Here, the inhomogeneity is positioned decentrally and its radius is varied. For $\alpha = 0$ the membrane is homogeneous and the degeneracies are present. Increasing α leads to a small decrease of the eigenvalues because the average density is enhanced. The (1,1) and (1,2) eigenvalues are mostly affected, especially in the $\alpha > 0.25$ region.

– $\alpha = 0.5$, $\beta = 0.5$, $1 \leq \delta \leq 4$.

See Fig. 5. In this case the influence of the density of an asymmetrically placed inhomogeneity of fixed size is studied. For $\delta = 1$ we have a homogeneous membrane. Going from $\delta = 1$ to $\delta = 4$ the average density of the membrane is enhanced by a factor of 7/4. As expected, most of the eigenvalues decrease for increasing δ. This happens in quite a regular way except for the (4.1) and the (2,2) frequencies which show an increasing behaviour for $\delta < 1.5$.

Our general conclusion is negative. In the usual kettledrum with homogeneous membrane the (1,1), (2,1), (3,1), and (4,1) frequencies form a harmonic series. For a membrane in vacuo we have not found a geometry in which these frequencies show a tendency to approach a harmonic series due to the introduction of an inhomogeneity. It is not clear whether the results for a membrane in vacuo may also be drawn with respect to the system consisting of membrane, kettle and air. If this is so we conclude that the inhomogeneities in the membranes of indian tablas play a role only in the fine tuning of these instruments and are not essential for the harmonicity of their spectra.

References

1.   Rossing, T.D., The Physics of Kettledrums, Scientific American, November 1982, pp. 172-178.

2.   Molenaar, J., The Spectrum of an inhomogeneous Membrane, Report IWDE 90-06, july 1990 (available on request).

Author's address: J. Molenaar, Instituut voor Wiskundige Dienstverlening (IWDE), Technische Universiteit Eindhoven (TUE), Postbus 513, 5600 MB Eindhoven, The Netherlands.

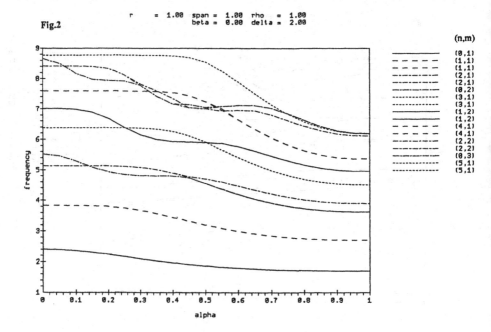

Fig.2

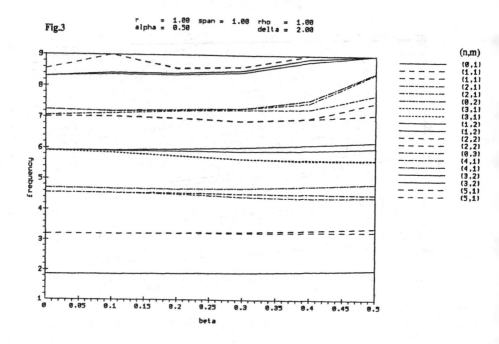

Fig.3

Fig.4

r   = 1.00 span = 1.00 rho = 1.00
beta = 0.50 delta = 2.00

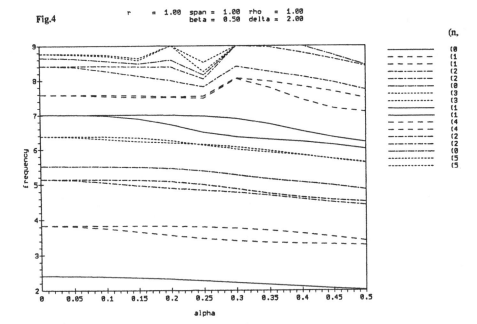

Fig.5

r   = 1.00 span = 1.00 rho = 1.00
alpha = 0.50 beta = 0.50

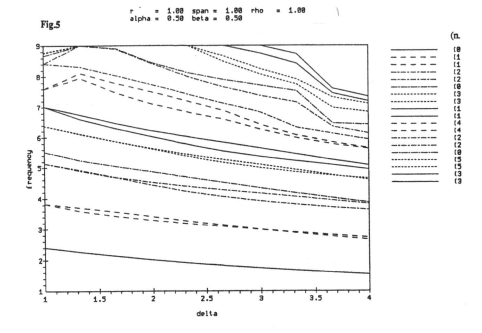

NONSMOOTH OPTIMIZATION
WITH INDUSTRIAL APPLICATIONS

MARKO M. MÄKELÄ and TIMO MÄNNIKKÖ
University of Jyväskylä
Department of Mathematics
P.O. Box 35
SF-40351 Jyväskylä
Finland

1. Introduction

The mathematical model of industrial processes can often be formulated as an optimal control problem. The classical mathematical tools for such problems are based on strong smoothness assumptions. However, in real world applications we often get into the situation when our problem is nonsmooth, i.e. the functions are not necessarily differentiable.

The standard approach to avoid the difficulties of nonsmoothness is to regularize the problem by using some penalization technique (see diagram 1). In this way we obtain a smooth problem $[\mathbf{P}_\varepsilon(\mathbf{x})]$ which tends to $[\mathbf{P}(\mathbf{x})]$ whenever $\varepsilon \to 0$. Discretization of the problem by using, for example, finite element (FEM) method leads to a smooth optimization problem $[\mathcal{P}_{\varepsilon,h}(x)]$. It tends to the problem $[\mathbf{P}_\varepsilon(\mathbf{x})]$ whenever $h \to 0$, and can be solved by standard optimization methods.

However, this regularization method has all drawbacks of the penalty technique: for large ε the problem $[\mathbf{P}_\varepsilon(\mathbf{x})]$ is stable but approximates $[\mathbf{P}(\mathbf{x})]$ badly, while for small ε we obtain a good approximation but unstable problem.

To avoid the errors caused by the regularization one can always try to apply smooth optimization methods directly to the nonsmooth problem $[\mathcal{P}_h(x)]$. However, the lack of nonsmooth, implementable optimality conditions gives no theoretical guarantees to this approach. Moreover, the gradient and function values are obtained as solutions of nonlinear systems, which can be solved only iteratively with a certain prescribed accuracy. Hence there always exists some errors in calculations and this often leads to failure of smooth methods.

In our direct approach we apply the methods of nonsmooth optimization in order to avoid the errors caused by the regularization and inaccurate function and gradient evaluations. We shall consider an optimal shape design problem and the process of continuous casting of steel, which are typical large-scale problems arising in industrial applications.

2. Nonsmooth Optimization

2.1. METHODS OF NONSMOOTH OPTIMIZATION

The discretization of the nonsmooth optimal control problem leads to a nonlinear optimization problem

$$\begin{cases} \text{Minimize} & f(x) = J(x, y(x)) \\ \text{subject to} & x \in G, \end{cases} \qquad (\mathcal{P})$$

303

M. Heiliö (ed.), *Proceedings of the Fifth European Conference on Mathematics in Industry*, 303–306.
© 1991 *B. G. Teubner Stuttgart and Kluwer Academic Publishers. Printed in the Netherlands.*

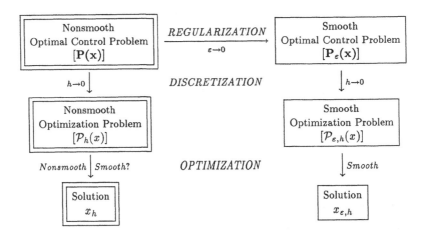

Diagram 1. Solution processes of nonsmooth optimal control problems.

where $y(x)$ is obtained as a solution of some state system. If the state $y : \mathbb{R}^n \to \mathbb{R}^m$ is locally Lipschitz continuous and the function $J : \mathbb{R}^n \times \mathbb{R}^m \to \mathbb{R}$ is continuously differentiable, then the objective function $f : \mathbb{R}^n \to \mathbb{R}$ is locally Lipschitz continuous on the feasible set $G \subset \mathbb{R}^n$. Note that f need not to be differentiable or convex.

The nonsmooth optimization methods for solving $(\mathcal{P})$ can be divided into two main classes: Kiev subgradient methods and bundle methods. The guiding principle behind Kiev subgradient methods (see Shor [12]) is to generalize smooth gradient and quasi-Newton methods by replacing the gradient by an arbitrary subgradient to obtain so called subgradient and space dilatation methods. This simple idea leads, however, to difficulties with a priori choice of the step size in line search operation and the lack of an implementable stopping criterion.

In our nonsmooth approach the problem $(\mathcal{P})$ is solved by using the Proximal Bundle method (see Mäkelä [5]), which is one of the most efficient methods of nonsmooth optimization at the moment. The basic idea is the following. We suppose that at each point $x \in \mathbb{R}^n$ we can evaluate the function value $f(x)$ and one arbitrary subgradient $\xi(x) \in \partial f(x)$ (see Clarke [1]). At each iteration the search direction $d_k \in \mathbb{R}^n$ from the current iteration point x_k is obtained as a solution of the quadratic optimization problem

$$\min_{x_k+d \in G} \hat{f}^k(x_k + d) + \tfrac{u_k}{2}\|d\|^2, \qquad (\mathcal{Q})$$

where the piecewise linear approximation $\hat{f}^k$ is defined by

$$\hat{f}^k(x) := \max\left\{ f(y_j) - \langle \xi(y_j), x - y_j \rangle \mid j \in J_k \right\} \qquad \text{for all} \quad x \in \mathbb{R}^n.$$

Here $u_k > 0$ is a weight parameter, $J_k \subset \{1, \ldots, k\}$ is a nonempty index set, $y_j \in \mathbb{R}^n$ are some auxiliary points and $\xi(y_j)$ are arbitrary subgradients from $\partial f(y_j)$ for $j \in J_k$. The problem $(\mathcal{Q})$ is solved by the quadratic solver QPDF4 derived in Kiwiel [3].

2.2. CALCULATION OF THE SUBGRADIENT

In order to apply nonsmooth optimization methods to problem $(\mathcal{P})$ we have to be able to calculate one subgradient $\xi(x) \in \partial f(x)$ at each point $x \in \mathbb{R}^n$. This can be done by using the following result.

THEOREM. *Let f be defined by $f(x) = J(x, y(x))$, where the mapping y is locally Lipschitz and the function J is continuously differentiable. If $Z_y(x) \in \partial y(x)$, then*

$$\nabla_x J(x, y(x)) + Z_y(x)^{\mathrm{T}} [\nabla_y J(x, y(x))] \in \partial f(x).$$

The computation of the matrix $Z_y(x)$ from the generalized Jacobian $\partial y(x)$ is normally highly complicated and available results concern only special cases (see Outrata [11].) To avoid this disadvantage it is typical to use the so called adjoint state technique (see Lions [4]).

3. Optimal Shape Design

The main goal in optimal shape design is to find the best possible design within a prescribed objective and a given set of limitations. Shape optimization has a very wide field of applications for example in automobile, marine and aerospace industries.

In our example the control variable x represents the shape of the body and the state $y(x)$ represents a thin membrane over the body. The state is defined by a unilateral Dirichlet-Signorini boundary value problem. After discretization $y = y(x)$ for fixed $x \in G$ is obtained as a solution of the problem

$$\begin{cases} \text{Minimize} & g(y) = \frac{1}{2}\langle y, A(x)y \rangle - \langle F(x), y \rangle \\ \text{subject to} & y_i \geq 0 \quad \text{for} \quad i \in I_\Gamma, \end{cases} \tag{S}$$

where $A(x)$ is $m \times m$ stiffness matrix, $F(x)$ is m-dimensional force vector and I_Γ is the set of boundary node indexes. The nonsmoothness caused by the unilateral boundary condition is normally regularized by using the outer penalty method

$$\min_{y \in \mathbf{R}^m} \left\{ g(y) + \frac{1}{2\varepsilon} \sum_{i \in I_\Gamma} \left(\max(0, y_i) \right)^2 \right\} \tag{S_ε^{out}}$$

or the inner (barrier) penalty method

$$\min_{y \in \mathbf{R}^m} \left\{ g(y) + \varepsilon \sum_{i \in I_\Gamma} B(y_i) \right\}, \tag{S_ε^{in}}$$

where $B(y_i) = \frac{1}{y_i}$ or $B(y_i) = \ln(y_i)$. These type of state problems can be solved by projected modification of well-known SOR (Successive Over Relaxation) method (see Haslinger and Neittaanmäki [2], Appendix I).

We have compared our direct nonsmooth approach with regularization techniques of Haslinger and Neittaanmäki [2] (outer penalty method), Neittaanmäki and Stachurski [10] (inner penalty method). The numerical results show the superiority of the nonsmooth optimization approach (see Mäkelä [5], Mäkelä and Neittaanmäki [6]).

4. Continuous Casting of Steel

In simulation of the continuous casting of steel our aim is to control the secondary cooling water sprays. The control variable x represents now a heat transfer coefficient, which has an effect on the temperature distribution $y(x)$ (the state) of the steel strand. This physical situation is modelled by the Stefan multiphase problem, which after discretization leads to the nonlinear state problem

$$TH(y(x)) + AK(y(x)) + G(x, y(x)) = F(x), \qquad (S)$$

where T and A are $m \times m$ block diagonal matrices, $F : \mathbf{R}^n \to \mathbf{R}$ and $G : \mathbf{R}^n \times \mathbf{R}^m \to \mathbf{R}$ are nonlinear functions, and $H : \mathbf{R}^m \to \mathbf{R}^m$ and $K : \mathbf{R}^m \to \mathbf{R}^m$ are the discretized enthalpy function and Kirchhoff's transformation, respectively. In this case the nonsmoothness of the state mapping $x \mapsto y(x)$ is caused by the fact, that due to phase change, H and K are nonsmooth (piecewise linear) mappings. Because the system (S) is highly nonlinear we use a combination of a modified SOR method and Newton-Raphson method.

Due to the numerical test runs performed on a Cray supercomputer we may state that nonsmooth optimization methods can be successfully applied to very large problems. For more details we refer to Männikkö [7], Mäkelä [5], Männikkö and Mäkelä [8], [9].

References

1. Clarke, F. H. (1983) "Optimization and Nonsmooth Analysis", Wiley, New York.

2. Haslinger, J. and Neittaanmäki, P. (1988) "Finite Element Approximation of Optimal Shape Design: Theory and Applications", Wiley, Chichester.

3. Kiwiel, K. C. (1986) *A Method for Solving Certain Quadratic Programming Problems Arising in Nonsmooth Optimization*, IMA Journal of Numerical Analysis **6**, 137–152.

4. Lions, J. L. (1971) "Optimal Control of Systems Governed by Partial Differential Equations", Springer-Verlag, Berlin-Heidelberg-New York.

5. Mäkelä, M. M. (1990) "Nonsmooth Optimization: Theory and Algorithms with Applications to Optimal Control", Dissertation, Report 47, Univ. of Jyväskylä, Dept. of Mathematics.

6. Mäkelä, M. M. and Neittaanmäki, P. (1990) *Nonsmooth Optimization in Optimal Shape Design*, in "Proc. of the International Conference on Operations Research", Vienna, Austria, August 28–31, 1990 (to appear).

7. Männikkö, T. (1991) "Optimal Control of the Continuous Casting Process", Reports on Applied Mathematics and Computing 1/91, Univ. of Jyväskylä, Dept. of Mathematics.

8. Männikkö, T. and Mäkelä, M. M. (1990) *On the Nonsmooth Optimal Control Problem Connected with the Continuous Casting Process*, in "Proc. of the First Conference on Advanced Computational Methods in Heat Transfer", (eds. Wrobel et al.), Portsmouth, UK, July 17–20, 1990, Computational Mechanics Publications, Southampton, pp. 67–78.

9. Männikkö, T. and Mäkelä, M. M. (1990) *Nonsmooth Penalty Techniques in Control of the Continuous Casting Process*, in "Proc. of the Conference on Numerical Methods for Free Boundary Problems", Jyväskylä, July 23–27, 1990 (to appear).

10. Neittaanmäki, P. and Stachurski, A. (1989) *Solving Some Optimal Control Problems Using the Barrier Penalty Function Method*, in "Proc. of 14th IFIP Conference on System Modelling and Optimization", Leipzig, GDR, July 3–7, 1989 (to appear).

11. Outrata, J. V. (1990) *On the Numerical Solution of a Class of Stackelberg Problems*, ZOR – Methods and Models of Operations Research **34**, 255–277.

12. Shor N. Z. (1985) "Minimization Methods for Non-Differentiable Functions", Springer-Verlag, Berlin-Heidelberg.

COMPUTER AIDED DESIGN OF GROUNDING GRIDS: A BOUNDARY ELEMENT APPROACH

F. NAVARRINA, L. MORENO, E. BENDITO, A. ENCINAS,
A. LEDESMA (†) and M. CASTELEIRO
Depto. de Matemática Aplicada III and Depto. de Ingeniería del Terreno (†)
E.T.S. de Ingenieros de Caminos. Universidad Politécnica de Cataluña
C/ Gran Capitán S/N. 08034 Barcelona. Spain

ABSTRACT. In a grounding system design two basic goals must be achieved: a) human safety must be preserved, by limiting potential gradients at the earth surface when a fault condition occurs, and b) integrity of equipment and continuity of service must be granted, by ensuring fault currents dissipation into the earth [Sverak et al., 1981]. In this paper, an approach is presented to compute both, the ground surface potencial distribution due to a fault current and the equivalent resistance of the system, by solving a Dirichlet Exterior Problem on the basis of several integral operators. This leads to integral equations that can be solved using a Boundary Element Method (BEM) formulation. Several classical and widespread intuitive formulations, generically known as Average Potential Methods (APM), can be understood as particular cases of this general approach when suitable assumptions are made in order to reduce computational cost. Application examples are presented, and the most relevant numerical problems that arise are briefly discussed.

1. Introduction

Physical phenomena underlying to fault currents dissipation into the earth can be modelled by means of Maxwell's Electromagnetic Theory, although no suitable solutions are available for practical cases. Constraining the analysis to the obtention of the electrokinetic steady-state response, and after widely accepted simplifications (homogeneus and isotropic ground properties, negligible resistivity of buried conductors), the problem can be described by means of the Laplace equation with mixed boundary conditions [Durand, 1966]. In most practical cases (flat ground surface), symmetry allows to rewrite the problem in terms of a Dirichlet Exterior Problem.

Although this classical mathematical problem has been rigorously studied by means of diverse integral equations [Dautray and Lions, 1988], and its solution can be efficiently obtained in many technical applications by means of standard numerical techniques (FEM and Finite Differences), additional difficulties appear in our case. Due to the type of domain (3D infinite) and the complexity of the

M. Heiliö (ed.), Proceedings of the Fifth European Conference on Mathematics in Industry, 307–314.
© 1991 B. G. Teubner Stuttgart and Kluwer Academic Publishers. Printed in the Netherlands.

boundary (the surface of a grid of horizontally buried conductors supplemented by a number of vertical ground rods), the obtention of sufficiently accurate solutions should imply unaffordable computing requirements. On the other hand, a number of simplified formulations (most of them based on intuitive ideas, such as superposition of punctual current sources) have been proposed [Garrett et al., 1985] and are widely used.

2. Problem Statement

By means of suitable simplifications on Maxwell's Equations, one can write the electrokinetic stationary problem associated to an electrical current derivation to earth as

$$\text{div}\,(\gamma\ \mathbf{grad}\,V) = 0 \quad \text{in} \quad D,$$

$$\frac{\partial V}{\partial n} = 0 \quad \text{in} \quad \Gamma_t, \qquad V = 1 \quad \text{in} \quad \Gamma, \qquad V \longrightarrow 0 \quad \text{if} \quad \boldsymbol{x} \to \infty,$$

where domain D is the earth and γ its conductivity, Γ_t is the ground surface and $\boldsymbol{n}$ its normal exterior unit field, Γ is the electrical generator (grounding grid) surface and V is the potential at an arbitrary point.

In general one may assume that earth conductivity is constant. If one further assumes that ground surface is horizontal, the solution of the previous problem can be considered as a restriction to D of the solution of the Dirichlet Exterior Problem

$$\Delta V = 0 \quad \text{in} \quad \Omega,$$

$$V = 1 \quad \text{in} \quad \Gamma \cup \Gamma', \qquad V \longrightarrow 0 \quad \text{if} \quad \boldsymbol{x} \to \infty,$$

where Γ' is the symmetric of Γ with respect to the ground surface, and Ω is the exterior domain to Γ and Γ'

Applying Green's Identity [Stakgold, 1970] one gets the following expression for the potential V in $\mathbb{R}^3$ in terms of the unknown leakage current density σ in the grid boundary:

$$V(\boldsymbol{x}) = \frac{1}{4\pi\gamma} \int\!\!\int_\Gamma \left(\frac{1}{r} + \frac{1}{r'}\right)\, \sigma \; d\Gamma$$

where

$$r(\boldsymbol{x},\boldsymbol{\xi}) = |\boldsymbol{x} - \boldsymbol{\xi}|, \qquad r'(\boldsymbol{x},\boldsymbol{\xi}) = |\boldsymbol{x} - \boldsymbol{\xi}'|,$$

$$\boldsymbol{x} = (x,y,z) \in R^3, \qquad \boldsymbol{\xi} = (\xi,\eta,\zeta) \in \Gamma, \qquad \boldsymbol{\xi}' = (\xi,\eta,-\zeta) \in \Gamma'.$$

Furthermore, the system equivalent resistance can be written as

$$R_{eq} = \left[\int\!\!\int_\Gamma \sigma \; d\Gamma\right]^{-1}.$$

To evaluate the leakage current density one imposes the boundary condition $V = 1$ in the above potential representation, and the result is the Fredholm integral equation of the first kind on Γ

$$1 = \frac{1}{4\pi\gamma} \int\!\!\int_\Gamma \sigma(\boldsymbol{\xi}) K(\boldsymbol{x},\boldsymbol{\xi})\, d\Gamma \qquad \forall \boldsymbol{x} \in \Gamma$$

where

$$K(x, \xi) = \left(\frac{1}{r(x, \xi)} + \frac{1}{r'(x, \xi)} \right).$$

3. Numerical Resolution

The most commonly used grounding grids consist of a number of interconnected bare conductors (cylindrical bars with circular cross sections). Hence, in practical cases, direct solution by BEM of the above 2D integral equation should imply a very high computational cost that could not be accepted in practice. Therefore, our scope is to reduce the problem complexity.

Let y be a generic point at the boundary of a bar, let $\hat{y}$ be its orthogonal projection over the middle line, let $a(\hat{y})$ be the radius of the cross section of the bar at this point (considered much smaller than the bar length), and let L be the set of middle lines of all the buried conductors. If one assumes that the current leakage density σ is constant over the cross sections of the boundary, one can write

$$V(\hat{x}) \simeq \frac{1}{2\gamma} \int_L a(\hat{\xi}) \sigma(\hat{\xi}) \hat{K}(\hat{x}, \hat{\xi}) \, dL \qquad \forall \hat{x} \in L$$

where

$$\hat{K}(\hat{x}, \hat{\xi}) = \frac{1}{\hat{r}(\hat{x}, \hat{\xi})} + \frac{1}{\hat{r}'(\hat{x}, \hat{\xi})},$$

$$\hat{r}(\hat{x}, \hat{\xi}) = \sqrt{|\hat{x} - \hat{\xi}|^2 + a^2(\hat{\xi})}, \qquad \hat{r}'(\hat{x}, \hat{\xi}) = \sqrt{|\hat{x} - \hat{\xi}'|^2 + a^2(\hat{\xi})}.$$

If the boundary condition $V(x) = 1$ is replaced by its approximation $V(\hat{x}) = 1$, and using the previous expression, one gets a 1D integral equation, which solution may be obtained by BEM at a lower computational cost. As it is usual in this type of techniques, a geometrical interpolation is performed on the 1D domain L and a functional interpolation is carried out on the leakage current density σ. Let t and s be the local coordinates on L of the points $\hat{\xi}$ and $\hat{x}$, respectively. Then, the potential $V(\hat{x}(s))$ at every point of L could be approximated by

$$\frac{1}{2\gamma} \int_I a(t) \sigma(t) \hat{K} \left(\hat{x}(s), \hat{\xi}(t) \right) J(t) \, dt$$

where I is the definition interval of t, and $J(t)$ is the jacobian determinant of the transformation from global to local coordinates.

Formulations corresponding to Point Collocation, Galerkin and Mean Squares have been developed for the numerical solution of the approximated 1D integral equation [Moreno, 1989]. Point Collocation and Galerkin schemes have been implemented considering different types of elements (constant, linear and quadratic) in functional and geometric interpolations. Several numerical tests have been performed, and some of the results are given later on. Mean Squares schemes have not yet been implemented because of their greater computational cost.

It has not been possible to use Gauss quadratures on the discretized equations (as it is usual in FEM) with a moderate cost and satisfactory accuracy, due to numerical ill-conditioning induced by the kernel $\hat{K}(\hat{x}, \hat{\xi})$. Instead, a compound adaptative Simpson quadrature (using Richardson extrapolation to obtain error estimates) has been used and has proved to be accurate enough, although the computational time still required has been important.

It is easy to verify that some particular cases of the proposed BEM formulation (Point Collocation or Galerkin using constant density elements and simplified integration techniques) can de identified with classical intuitive methods, such as those of Dawalibi-Mukhedkar and Heppe [Moreno, 1989].

4. Application Examples

Two examples are presented: an academic problem (a single cylindrical bar) in order to compare the results with those obtained by one of the Average Potential Methods [Heppe, 1979], and a real case (a rectangular grid with ground rods).

4.1. SINGLE BAR PROBLEM

Problem characteristics are: Bar Length = 2 m; Radius of the Bar Cross Section = 0.01 m; Installation Depth = 0.25 m; Earth Resistivity = 1000 Ohm·m. The problem has been solved by Heppe's Method and by BEM. The following results are given: a) Equivalent resistances obtained for different levels of discretization (Fig. 1), b) Leakage current density distribution (Figs. 2.a and 2.c), and c) Error in boundary condition (Figs. 2.b and 2.d). One can notice the good agreement between BEM and APM results, although BEM discretization has been coarser. Although some of the problems reported for APM [Garrett et al., 1985] are still present, it seems from Fig. 2 that high accuracy results could be obtained by BEM with only a few high order elements. However, this seems to be prevented by instabilities near the bar edges.

4.2. GRID WITH GROUND RODS PROBLEM

The geometry is shown in Fig. 3. Bars characteristics and earth resistivity are the same of the previous example. The 8 horizontally buried bars (6 m. long) and the 4 ground rods at the corners (2 m. long) were discretized, respectively, in 30 and 10 linear equal-length elements. The following results are given: a) Ground surface potential distribution (Fig. 4), b) Leakage current density distribution (Figs. 5.a, 5.c and 5.e) and c) Error in boundary condition (Figs. 5.b, 5.d and 5.f).

5. Conclusions

A Computer Aided Design system based upon the suggested approach is under development at the present time. Several techniques for earthing grids analysis have already been implemented in the system. A Fredholm Integral Equation of the first kind is solved by means of BEM. The solution of the integral equation describes the leakage current density that flows from the grounding grid surface. Then, the equivalent resistance of the system and the ground surface potential distribution can be efficiently computed.

The presented formulation includes a number of classical and widespread simplified methods (APM) as particular cases, and allows to get more accurate results with a more reduced number of degrees of freedom. Although computational costs are still important (in particular those derived from numerical integration), and some of the problems reported for APM are still present, this kind of approach will allow to perform a rigorous mathematical analysis to identify the errors origin. Whenever this will be succesfully achieved, high acuracy results could be obtained by BEM with an extraordinary reduction in the computational cost.

6. Acknowledgments

This work has been partially supported by the "Plan Nacional de Investigación y Desarrollo Eléctrico del Ministerio de Industria y Energía" of the Spanish Government, throughout the OCIDE Research Project "Diseño de Tomas de Tierra Asistido por Ordenador", Grant Number TC0772 UPC–FECSA.

7. References

Dautray, R. and J.L. Lions, (1988), *Analyse Mathématique et Calcul Numérique pour les Sciences et les Techniques*. Vol. 6, Masson, Paris.

Durand, E., (1966), *Électrostatique*. Masson, Paris.

Garrett, D.L. and J.G. Pruitt, (1985), "Problems Encountered with the Average Potential Method of Analyzing Substation Grounding Systems". *IEEE Trans. on Power App. and Systems*, Vol. 104, No. 12, pp 3586–3596.

Heppe, R.J., (1979), "Computation of Potential at Surface Above an Energized Grid or Other Electrode, Allowing for Non-Uniform Current Distribution". *IEEE Trans. on Power App. and Systems*, Vol. 98, No. 6, pp 1978–1989.

Moreno, Ll., (1989), *Disseny Assistit per Ordinador de Postes a Terra en Instal·lacions Elèctriques*. Tesina de Especialidad, Escuela Técnica Superior de Ingenieros de Caminos, Canales y Puertos, U.P.C., Barcelona.

Stakgold, I., (1970), *Boundary Value Problems of Mathematical Physics*, MacMillan Co., London

Sverak, J.G., W.K. Dick, T.H. Dodds, and R.H. Heppe, (1981), "Safe Substations Grounding. Part I". *IEEE Trans. on Power App. and Systems*, Vol. 100, No. 9, pp 4281–4290.

METHOD	— Discretization	Equiv. Resist. (Ω)
APM	— 1 Segmentation	502.155
APM	— 10 Segmentations	496.509
APM	— 50 Segmentations	494.028
APM	— 100 Segmentations	493.225
BEM	— 10 Linear Elements	494.897
BEM	— 50 Linear Elements	493.204
BEM	— 5 Quadratic Elements	494.756
BEM	— 25 Quadratic Elements	493.146

Figure 1.—Single bar problem: Equivalent resistances obtained by APM and BEM for different levels of discretization.

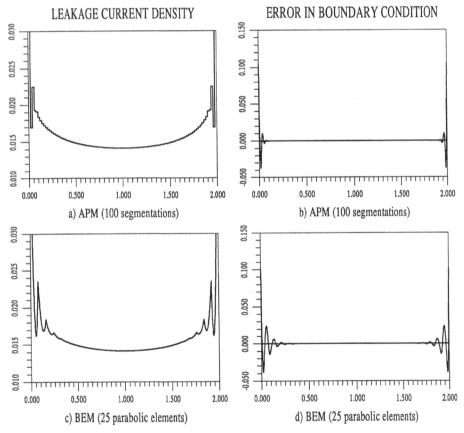

LEAKAGE CURRENT DENSITY

ERROR IN BOUNDARY CONDITION

a) APM (100 segmentations)

b) APM (100 segmentations)

c) BEM (25 parabolic elements)

d) BEM (25 parabolic elements)

Figure 2.—Single bar problem: Comparision between results obtained by APM (100 segmentations) and BEM (25 parabolic elements). Leakage Current Density and Error in Boundary Condition throughout the bar lenght.

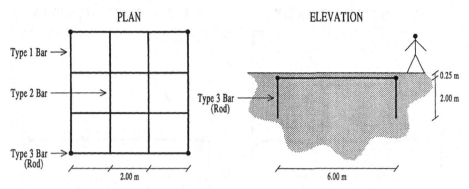

Figure 3.—Grid with ground rods problem: Plan and Elevation.

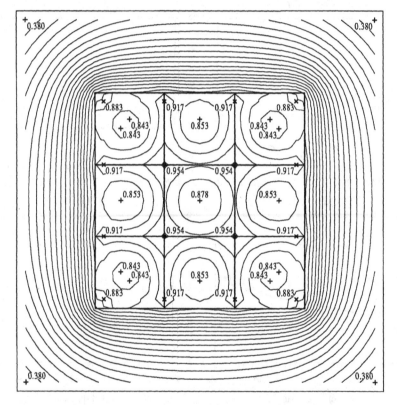

Figure 4.—Grid with ground rods problem: Results obtained by BEM (5 linear elements per meter). Ground surface potential distribution.

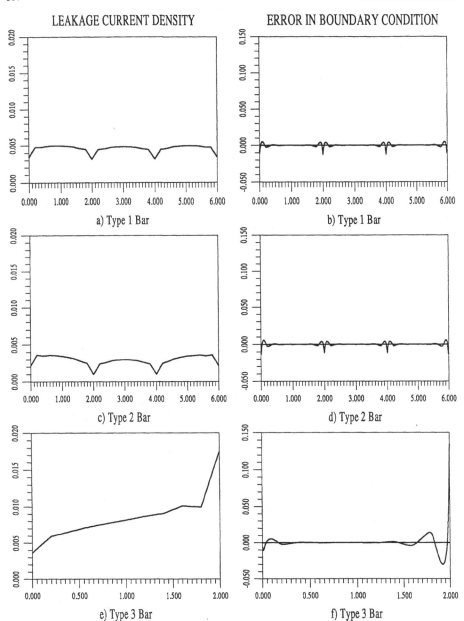

Figure 5.—Grid with ground rods problem: Results obtained by BEM (5 linear elements per meter). Leakage Current Density and Error in Boundary Condition throughout the bars lenght.

HEAT TRANSFER FROM THE COOLANT TUBES OF A POWER CONDENSER

A. A. NICOL
Department of Mechanical Engineering
University of Strathclyde
Glasgow, G1 1XH
Scotland

G. WILKS
Department of Mathematics
University of Keele
Keele, ST5 5BG
England

1. Introduction

The Nusselt model is commonly used as the basis for assessing the heat transfer characteristics of inundation and drainage flow under gravity over a horizontal tube. The model neglects inertia and as a result a universal film thickness over the tube is predicted for a given inundation flow rate. In [1] it is pointed out that accordingly there is no way of distinguishing between thin, high velocity sheet inundation as opposed to thick, low velocity sheet inundation when each may give rise to the same overall flow rate. Such variations however have been shown in [2] to have a significant influence on the associated heat transfer.

In the work that follows the alternative hydrodynamic model of [1], which remedies the above deficiency, is used to assess the heat transfer characteristics of flow over a single tube. In effect this leads to the possibility of assessing the effects of tube spacing on the flow and heat transfer characteristics within the large banks of horizontal tubes of a power condenser.

2. The Flow Model

The model of sheet inundation flow onto a single tube is illustrated in Figure 1. The sheet is represented by a two–dimensional liquid jet of semi–width h_0 and uniform downward velocity U_0. If $Q = U_0 h_0$ denotes a given volumetric flow rate over one side of the tube then varying U_0 can be associated with changes in relative tube spacing. The hydrodynamic film characteristics will vary accordingly.

Initially the jet at impingement experiences an inviscid deflection within which there is a deeply imbedded stagnation boundary layer of $O\left(\frac{\nu h_0}{U_0}\right)^{\frac{1}{2}}$. Away from the top of the tube a viscous boundary layer grows within the film, which is accelerating under gravity, until viscous effects penetrate the whole film at the end of Region I. For Prandtl numbers greater than unity the penetration of thermal effects to the free surface will be delayed over a span designated by Region IIa. In Region IIb viscous and thermal effects influence the whole film.

M. Heiliö (ed.), Proceedings of the Fifth European Conference on Mathematics in Industry, 315–318.
© *1991 B. G. Teubner Stuttgart and Kluwer Academic Publishers. Printed in the Netherlands.*

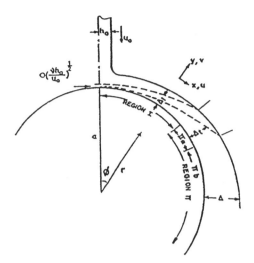

Figure 1. Jet spreading model of sheet inundation.

3. Governing Equations

In plane polar coordinates (r, θ) the velocity components are v_r and v_ϕ. On the assumption $h_0 \ll a$ the following transformations are invoked

$$y = \frac{r - a}{h_0}, \quad x = \phi; \quad u = \frac{v_\phi}{U_0}, \quad v = \frac{a}{h_0} \frac{v_r}{U_0}, \quad \theta = \frac{T_s - T}{T_s - T_w}.$$

Neglecting terms of $O\left(\frac{h_0}{a}\right)$ compared to unity the governing equations of conservation of mass, momentum and energy now read

$$\frac{\partial u}{\partial x} + \frac{\partial v}{\partial y} = 0$$

$$u\frac{\partial u}{\partial x} + v\frac{\partial u}{\partial y} = F(x) + \alpha\frac{\partial^2 u}{\partial y^2}$$

$$u\frac{\partial \theta}{\partial x} + v\frac{\partial \theta}{\partial y} = \frac{\alpha}{\Pr}\frac{\partial^2 \theta}{\partial y^2}$$

where $F(x) = \frac{\sin x}{F_r}$; Froude number, $F_r = \frac{U_0^2}{ag}$, $\alpha = \frac{\nu a}{h_0^2 U_0} = \frac{a^2}{h_0^2 \operatorname{Re}}$ and $\operatorname{Re} = \frac{U_0 a}{\nu}$ is the Reynolds number based on the tube radius.

The boundary conditions are

(i) $u = v = 0$, $\theta = 1$ on $y = 0$,

(ii) $\frac{\partial u}{\partial y} = 0$, $\theta = 0$ on $y = \delta(x)$ where $\delta(x)$ is the non–dimensional film thickness,

(iii) $\int_0^{\delta(x)} u \, dy = 1$.

The above equations have been solved by momentum integral techniques in the various Regions I, IIa, IIb. In particular, details of the hydrodynamics have been presented in [1].

4. Results

The results are illustrated in Figures 2—5. Figure 2 presents a selection of dimensionless film thickness profiles for several values of the volumetric flow rate Raynolds number $\text{Re}_f = \frac{2Q}{\nu}$. As the flow rate decreases the profiles beyond the initial deflection region appear more and more to mirror the Nusselt profile. At high flow rates the departures from the Nusselt profile are quite marked. In Figure 3 the significance of specific inundation condition for a fixed flowrate is illustrated and in Figure 4 the underlying structure of the regional pattern of solution is demonstrated. Viscous and thermal penetration of the free surface is illustrated for various Re_f. Finally in Figure 5 the implications for local heat transfer for a typical fixed flow rate and varying conditions at impingement are presented. The Nusselt equivalent clearly represents an underestimate as compared with the present theory.

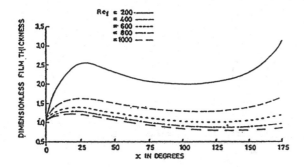

Figure 2. Dimensionless film thickness profiles for a range of flow rates of water at 20°C over a cylinder of radius 14mm with $U_0 = 0\cdot4$m/s.

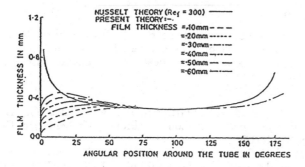

Figure 3. Comparison of predicted film thickness profiles with the Nusselt profile; $\text{Re}_f = 300$.

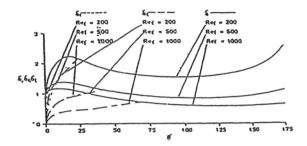

Figure 4. Dimensionless film, viscous and thermal layer thickness as a function of angle for Re = 200, 500 and 1000. $U_0 = 0.25\text{m/s}$, $T_s = 60°\text{C}$, $T_w = 50°\text{C}$.

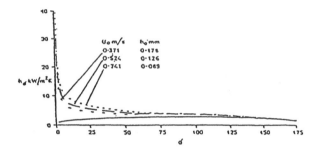

Figure 5. The influence of tube separation on local heat transfer coefficient for $\text{Re}_f = 500$, $T_s = 60°\text{C}$, $T_w = 50°\text{C}$.

5. Concluding Remarks

The theory presented in this paper remedies certain limitations of the Nusselt model. It provides the basis for a variety of more detailed studies which may lead to improved condenser design and assessment of heat transfer characteristics.

6. References

Abdelghaffer, M.A., Nicol, A.A., Gribben, R.J. and Wilks, G. (1989) 'Thin film inundation flow over a horizontal cylinder', Mathematical Engineering in Industry 2, 143-155.

Mitrovic, J. (1986) 'Influence of tube spacing and flow rate on heat transfer from a horizontal tube to a falling liquid film', Proc. 8th Intl. Heat Transfer Conference, San Francisco, Hemisphere Publishing Corporation.

NUMERICAL PREDICTIONS OF CIRCULATING FLUIDISED BEDS WITH MONO- AND MULTI-SIZED PARTICLES

M K Patel, K Pericleous and M Cross

Centre for Numerical Modelling and Process Analysis

Thames Polytechnic

London SE18 6PF.

ABSTRACT   The development of the multi-phase flow algorithm MIPSA (Multi-Interphase Slip- Algorithm) within a computational fluid dynamics (CFD) code called 'CASCADE' is described together with its application to the modelling of lean phase circulating fluidised beds.

INTRODUCTION : Circulating Fluidized Beds (CFB)

It is usual to talk of fluidized beds, as devices in which solids reside in a "bed", with a distinct free surface. The bed is agitated by gas blown from below, but the fluidizing velocity is such that the particles once free of the bed travel a short distance and then fall back into it (eg. Figure 1a). In recent years, a new fluidization regime has assumed significant importance and has grown in application; namely, the "fast" fluidization regime. The corresponding fluidized beds are known as "fast" or "circulating" fluidized beds (CFB). Consider Figure 1b, in which the gas velocity is so fast that any particles entering the reaction chamber are entrained by it. The residence time of solids in the vessel would be very short, generally too short for appreciable reaction, heating or combustion to take place in a single pass. Thus, the separated particles are fed back into the bed. This fluidization regime lies in the transitional range between fluidization and pneumatic transport.[1]   Several variations of CFB's exist as described in [2,3] and others.

The CASCADE code

Calculations performed on the CASCADE code [4].   A modification of the well-known IPSA [5] technique extended to multiple phases, called MIPSA, [4] is used.   CASCADE   uses the SIMPLE [6] iterative algorithm and has been used successfully in bubbling fluidized beds [7], blast furnaces [9] and moving bed in iron ore pellet induration process [8,10]. All examples given above are concerned with dense particle concentrations. In the present application, the particles are held in fairly dilute suspensions with particle motion governed usually by gas drag.   Recent CFB predictions that have appeared in literature include [11,12,13].

DESCRIPTION OF THE PROCESS TO BE MODELLED

The process to be modelled is depicted in Figure 2.   The following are the basic requirements for successful operation:
* particles do not drop below the throat of the nozzle,
* solids temperatures kept below the melting temperature (no segregation/accretion)
* pressure drop is kept to a minimum given the loading and heat transfer

319

M. Heiliö (ed.), Proceedings of the Fifth European Conference on Mathematics in Industry, 319–324.
© 1991 B. G. Teubner Stuttgart and Kluwer Academic Publishers. Printed in the Netherlands.

DESCRIPTION OF THE MODEL : Geometry and Grid

The geometry of the problem together with a typical grid, is shown in Figure 2. A distorted cylindrical polar grid, formed by expanding/contracting a regular polar grid in the radial direction, to fit the calculation domain.

Differential Equations

The flow is characterised by the time-averaged Navier-Stokes equations written in multi-phase form and by the equations of continuity and energy conservation. In adopting this procedure the dispersed phase is treated as a continuous fluid, interacting with the carrier gas through interphase source terms, representing momentum and energy exchange.

All dependent variables take the general form of the differential equations as follows:-

$$\frac{\partial}{\partial t} (\gamma \varphi) + \text{div} (\gamma_\varphi \underline{u} \varphi - \Gamma_\varphi \text{ grad } \varphi) = S\varphi \qquad (1)$$

where Γ_φ and S_φ are termed the exchange coefficients and source of φ

respectively, details of these can be found in [4,7].

The volume sharing principle is adopted, where

$$r_g = 1 - \sum_{j=1}^{N_p} r_j. \qquad (2)$$

Conservation equations can thus be solved for some, or all individual volume fractions.

Auxiliary equations

A number of auxiliary equations are required to effect closure of the problem. Gas density is evaluated from the ideal gas law. The temperature is computed directly from the enthalpy,

$$T_g = h/c_p \qquad (3a)$$

where c_p is the specific heat, h denotes the enthalpy. The particles are assumed to have infinite conductivity. Hence, surface temperature equally bulk temperature.

Also,

$$T_p = H/C_{pp} \qquad (3b)$$

For the case of circulating fluidized beds, a more appropriate expression is obtained if dp/dx is replace by

$$f_i = \frac{3}{4} c_D r_g \frac{|V_g - V_i|}{dp} \rho_g r_i \ f(r_g) \qquad (4)$$

and c_D the drag coefficient is related to the Reynolds number, ie.

$$c_D = \frac{24}{Re} (1 + 0.15 \, Re^{0.687}) \; : \; Re < 10^3$$

and

$$c_D = 0.44 \qquad\qquad : \; Re > 10^3 \tag{5}$$

where

$$Re = \rho_g \frac{(V_g - V_i)}{\mu_g} d_p \tag{6}$$

The function $f(r_g)$ in equation (4) accounts for the presence of other particles in the fluid and acts as a correction term to the usual Stokes law for free fall of a single particle. For the present study, this was taken as 1. Other forms have been adopted in [7]. For dense phase fluidization CASCADE adopts the Darcy relation [14].

Laminar viscosity is assumed to that of air at the average temperature of the precess. A constant effective viscosity was used for these calculations. However, CASCADE has other options, some of which take into account the presence of particles [7].

Interphase heat transfer

The gas–particle heat transfer coefficient, e_i, is defined in a form analogous to f_i,

$$e_i = \frac{6N_u \, r_i \, k_g}{d_i} \times vol \tag{7}$$

where, the Nusselt number, Nu, is given by

$$Nu = 2 + 0.6 \, Re^{0.16} \, Pr^{0.33} \tag{8}$$

Details can be found in [4,7].

Boundary Conditions

Since the software solves the elliptic form of the differential equations, information about the boundary conditions surrounding the domain of interest is required. Hence,

o Gas mass, momentum and thermal energy at the inlet (assuming plug inlet profile) are specified.

o Particle and gas mass, momentum and thermal energy at the downcommer inlet are specified. Here, for steady–state simulations, the particle temperature has been assumed. However, in transient runs this may be coupled to the average outlet temperature, assuming a prescribed heat loss through the cyclone and downcommer which are not modelled here.

o Wall friction is supplied to the gas only; this may be a serious approximation in regions where particle concentrations are high. Its net effect would be an underprediction of the pressure drop.

o Heat transfer to the wall is through a prescribed heat sink. Again, all wall heat transfer is via the gas. Empirical expressions exist for particle to wall heat transfer, eg. Botterill [15]; however, these appear to be more appropriate for dense beds.

o A constant pressure boundary is assumed finally for the exit.

SUMMARY OF CASES STUDIED AND RESULTS OBTAINED

Three geometrical configurations were studied, with three different particle loadings, low, medium and high. For the case with a single mixing chamber and as a consequence very long riser, it has been necessary to include time dependence to procure convergence. This is because particles appear to accumulate in "dunes" and then travel in masses until they reach the exit. This is a real effect, also observed in pneumatic conveying applications [16] and possibly undesirable due to consequent pressure oscillations. Due to space limitations, a small selection of results are presented here. The flowfields within the CFB are depicted in the form of vector plots, contour maps of particle volume fractions, temperature and pressure in Figures 3–5.

The flowfield in the mixing chamber is best described in terms of velocity vectors. In broad terms, the flow is determined by a strong axial jet, which entering from the nozzle expands until it occupies the width of the opening at the top of the mixer. A toroidal vortex surrounds the jet which results in the downflow on the mixer walls. The centre of the vortex appears to lie in the lower part of the mixing chamber.

The presence of particles has a strong influence on the jet behaviour. Momentum is transferred from the outer portion of the jet to the particles leading to a rather concave looking profile. The jet has sufficient momentum to burrow through the particles along the axis. The particles reduce the effective area through which the gas can flow, hence leading to a maximum velocity down-stream of the nozzle.

There is considerable slip between particles and gas. In the jet region particles only reach one third of the gas velocity. Towards the periphery, the particles tend to fall under the action of gravity and hence enhance the recirculation process. In the jet region, the particle velocity profile assumes a more conventional parabolic shape.

Figure 4 shows particle volume fraction contours in the bottom mixer, for all the three configurations. The medium loading case is depicted.

Particles enter the mixing chamber through the downcommer duct, on the periphery of the mixer (since the simulation is axisymmetric, this is in fact a ring in the model). On entry, particles fall along the wall and cone, until they reach the nozzle. At this position, the jet velocity is so high that no particles penetrate into the inlet throat; instead they are thrown back to form a "wall" surrounding the jet. The weight of this wall is responsible for the jet contraction observed in the previous section.

Particle concentration tends to increase roughly in proportion to the mass inflow of particles.

Figure 5a shows the normalised gas temperatures as a function of the temperature difference between gas and particles at inlet.

It is clear that there exists an inner core extending from the nozzle to about two thirds of the way in the mixing chamber which is quite hot, whereas the outer core, which contains most of the resident particles, is at the lowest temperature. In some cases, the temperature in the outer most core falls below the inlet solids temperature due to excessive cooling.

The pressure drop, Table 3, for case 1a/1b predicts an oscillatory behaviour due to the effect of "dunes" forming in the long riser. Case 2a/b/c pressure drops vs holdup gave a straight line which has the same gradient as for the static test cases.

CONCLUSIONS

The CASCADE code has been used to model a simplified CFB. The simulations have shown a strong dependence between particle holdup and pressure drop, as expected. However, the value obtained are not necessarily those that would result by simply equating the weight of particles to pressure drop. Other forces come to play, namely particle acceleration/deceleration and also in the case of a fast moving axial jet, the shear force of the jet on the particles that surround it.

The main simplification adopted are more for convenience rather than for limitations in the technique demonstrated. It is planned that in the next version of the model many of these assumptions will be removed or replaced where appropriate by reliable correlations to account for

o   particle-particle interaction;
o   wall to particle heat transfer;
o   wall to particle friction,

and other observed effects such as agglomeration of particles.

A question remains as to the influence of gas turbulence on the particle equations. For the large particles used here, this was ignored. However, for much smaller particles sizes, it may have to be considered.

REFERENCES

[1]   Reh L (1971), Fluidized bed processing, Chem Eng Prog 67, 58-63.
[2]   Basu P (1986), Circulating Fluidized bed, Technology (Toronto: Pergamon).
[3]   Yerushalmi J (1986), in Gas Fluidization Technology, ed D Geldart, (Chichester: Wiley-Interscience) Chapter 7.
[4]   Patel M K, Wade K and Cross M (1989), Multiphase approach to CFD algor and appl, in Numerical Methods in Laminar and Turbulent Flow, ed Taylor, Gresho, Sani and Hauser, (Pineridge Press).
[5]   Spalding D B (1980), Mathematical methods in nuclear-reactor thermal hydraulics, Report HTS/81/3, Imperial College, London.
[6]   Patankar S V and Spalding D B (1972), A calculation procedure for heat, mass and momentum transfer in 3D parabolic flows. IJHMT, 15, p.1787.
[7]   Patel M K and Cross M (1989), The modelling of fluidized beds for ore reduction, in Num Meths in Laminar and Turbulent Flow, ed Taylor, Gresho, Sani and Hauser, (Pineridge Press).
[8]   Cross M and Wade K C (1989), Computer simulation of iron ore pellet induration with additives, in Proc of 5th Intl Symp Agglomeration (ed M Cross and R Oliver), pub IChem E, pp.291-298.
[9]   Wade K, Cross M and Smith R (1989). Heat trans within the lead/tin blast furnace in Num Meths in Therm Prob, ed Lewis and Morgan (Pineridge Press).
[10] Cross M, Patel M K and Wade K (1990). Analysis of the Gas Flow and Heat Distribution in Iron Ore Pellet Induration Systems in Control 90 - Mineral and Metallurg Proc (Eds R K Rajamain and J A Herbst), pub SME-AIME, 99-108.
[11] Yang Wen-Ching, Revay D, Anderson R G, Chelen E J, Keairns D L and Cicero D C (1983). Fluidization Phenomena in large-scale, cold-flow model in Fluidization, ed Kunil D and Toei R (Engineering Foundation).
[12] Hartge E U, Li Y and Werther J (1986), Flow Structures in Fast Fluidized beds in Fludization, Ed Østergaard K and Sørensen A (Engineering Foundation).
[13] Militzer J (1985). Numerical prediction of the fully developed two-phase (Air-solids) flow in a pipe. The 1st Int Conf on Circulating fluidized beds. Nova Scotia, Canada.
[14] Gidaspow B (1986). Hydrodynamics of fluidization and heat transfer: Supercomputing modelling. App Mech, Rev Vol 39, No.1.
[15] Botterill J S M (1975). Fluidized bed heat transfer (London: Academic) also in Fluidized beds: Comb and appl, ed J R Howard (1983) (London: Applied Sci).
[16] Mason D J, Cross M, Patel M K and Reed A R (1989). Numerical modelling of Dense Phase Pneumatic Conveying of bulk particulate materials. Proc 14th Powder and Bulk Solids Handling and Processing Conf, Chicago, USA.

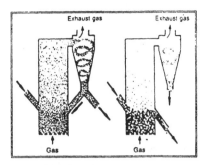

Figure 1 Fluidization regimes

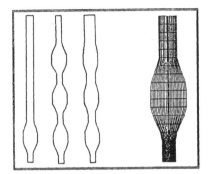

Figure 2 Geometry configurations and grid considered

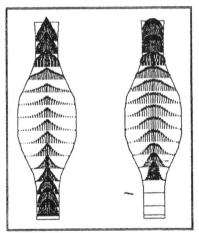

Figure 3 Typical gas and solids velocity vectors

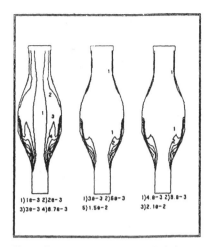

Figure 4 Typical particle volume fraction distributions

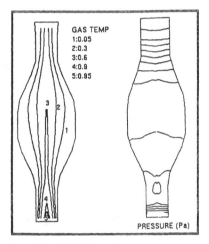

Figure 5 Typical distributions of
a) gas temperatures
b) pressure

COMPUTATION OF MAGNETIC FIELDS IN ELECTRIC MACHINERY USING A COUPLED FINITE ELEMENT-NETWORK ELEMENT MODEL.

D. Philips, L. Dupré and R. Van Keer.

1. INTRODUCTION

For the design and optimisation of an electric machine the magnetic field must be known in detail. To solve the field problem in both a time efficient and accurate way a new numerical technique is presented here. Currently, the method is embedded in a CAD-package [4].

In the magnetic network method, (see e.g. Ostović [2]), the magnetic circuit of the whole machine is built up from network elements with uniform field patterns. The time efficiency results from the low number of unknowns (one loop flux per mesh), needed for the field description. However the accuracy is not always optimal due to regions with irregular field patterns, for which a network description is rather artificial.

These regions can be calculated with a high precision using the finite element method (FEM), introduced by Sylvester and Chari [5]. However, a considerable amount of unknowns is wasted in regions with uniform field patterns.

In this paper we therefore present a hybrid method which couples the network element method with a FEM intrinsically. We thus avoid the drawbacks of both methods, while retaining their advantages.

2. STATEMENT OF THE PROBLEM

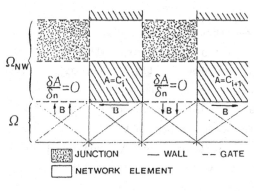

Fig. 1.

In fig. 1, Ω_{NW} represents that fraction of the entire machine region Ω' calculated with network elements. Essentially, such an element should have a 2-gate 2-wall structure [3]. Ω is the remaining part of the machine, sharing alternating walls and gates with Ω_{NW}. The union of the w walls is denoted by Γ_D.

M. Heiliö (ed.), Proceedings of the Fifth European Conference on Mathematics in Industry, 325–329.
© 1991 B. G. Teubner Stuttgart and Kluwer Academic Publishers. Printed in the Netherlands.

In the 2D magnetostatic field calculation the known current density is $\bar{J} = J(x,y).\bar{e}_z$, with the z-axis parallel to the shaft. In Ω the magnetic induction is written $\bar{B} = \text{rot } \bar{A}$, with the vector potential $\bar{A} = A(x,y).\bar{e}_z$. It is by now standard [1], that from Maxwell equations $A(x,y)$ obeys a 2nd order BVP in Ω, which in variational form reads :
Find $A \in H^1(\Omega)$ for which

$$\int_\Omega \nu(x,y,|\text{grad } A|^2).\text{grad } A.\text{grad } v.dx.dy = \int_\Omega J.v.dx.dy$$

$$\forall v \in V = \{v \in H^1(\Omega) \mid v = 0 \text{ on } \Gamma_D\}$$

(2.1)

$$A = c_i \qquad \text{on wall i,} \qquad 1 \leq i \leq w \qquad (2.2)$$

Here the reluctivity ν gives rise to the non-linearity. The homogeneous B.C. $\frac{\partial A}{\partial n} = 0$, forcing the flux lines normal to the gates, is incorporated in (2.1). The constants c_i in the B.C. (2.2) (prohibiting any flux leakage through the walls) are unknown. In fact c_i is identified with the mesh flux around wall i [4]. These loop fluxes appear as the unknowns in the network equations for the $\ell \geq w$ fundamental loops

$$\int_L \nu.\bar{B}.d\bar{\ell} = \int_S J.dx.dy \qquad (2.3)$$

(L being a contour corresponding to a mesh and passing through the network elements; S is the enclosed surface).

The coupling of the set of the network equations (2.3) with a finite element discretisation of (2.1)-(2.2) is thus realised in a natural way by identification of the common unknowns. In the next section the method is justified using a unified FE-like structure underlying the coupled problem.

3. UNIFIED APPROACH

To the f free nodes in the linear mesh in Ω we assign the classical pyramid-shaped basic functions ψ_j, $1 \leq j \leq f$. For the wall $i \in \Gamma_D$, $1 \leq i \leq w$, the regions K_i and $K_i' \supset K_i$ are shown in fig. 2. Let ψ_{f+i} be the continuous piecewise linear functions on Ω', taking the value 1 on K_i and vanishing outside K_i'. Similar 'platform' functions ψ_{f+w+k}, $1 \leq k \leq (\ell-w)$ are assigned to the $(\ell-w)$ fundamental loops not intersecting $\partial\Omega$.

The coupled set of field equations in Ω and Ω_{NW} mentioned above can be rewritten as a single discrete variational problem for

$$A^*(x,y) = \sum_{j=1}^{f+\ell} A_j.\psi_j(x,y) \qquad (x,y) \in \Omega' \qquad (3.1)$$

with $A_1, \ldots, A_f$ the nodal values of the FE-approximation of the potential A considered in (2.1)-(2.2) and with $A_{f+1}, \ldots, A_{f+\ell}$ the fundamental loop fluxes, obeying the network equations (2.3), recalling that A_{f+i}, $1 \le i \le w$, is identified with the constant potential on wall i.

More precisely we have :

THEOREM. Let for the fundamental loop around wall i, $1 \le i \le w$, the path L in (2.3) contain the midparallels of the triangles in the mesh neigbouring the wall (fig.3). Assume that $\nu = 0$ in the junctions of Ω_{NW}. Then the function A*, (3.1), obeys :

$$\int_{\Omega'} \nu(x,y, |grad\ A^*|^2).grad\ A^*.grad\ \psi_i\ dx\ dy = \int_{\Omega'} J.\psi_i.dx\ dy, \qquad 1 \le i \le f+\ell$$
$$(3.2)$$

Proof. For $1 \le i \le f$, (3.2) directly becomes the FE-discretisation of (2.1)-(2.2) by definition of A* and the underlying basic functions.

For $f+1 \le i \le f+w$, (3.2) first reads

$$\int_{K_i} J.dx.dy = \int_{K_i' \backslash K_i} \nu.grad\ A^*.grad\ \psi_i.dx.dy \qquad (3.3)$$

Let L_r be a line in the shaded domain D_r of fig. 3, $1 \le r \le 3$, then

$$\int_{D_r} \nu.grad\ A^*.grad\ \psi_i.dx.dy = \int_{L_r} \bar{H}.d\bar{\ell}$$

when equaling the flux through a network element $(= \frac{H}{\nu}$ .width) to the difference of the two neighbouring fundamental loop fluxes. A similar result follows for D_4 when L_4 consists of the midparallels of the triangles, introducing $\bar{B}$ = rot $\bar{A}$. Summing up, taking $\nu = 0$ in the junctions and combining with (3.3), we end with the network equations (2.3). Similarly for $f+w < i \le f+\ell$, (3.2) reduces to the set of the remaining fundamental loop equations.∎

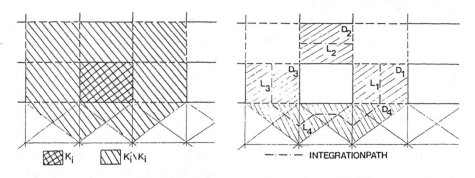

Fig. 2   Fig. 3

4. ALGORITHM

The discussion of the magnetic field equations in Ω and Ω_{NW} may now start from (3.1)-(3.2). For instance we may establish a standard Newton-Raphson linearization method for the variables A_j, $1 \leq j \leq f+\ell$, obeying

$$F_i(A_1,\ldots,A_{f+\ell}) \equiv \int_{\Omega'} \nu(x,y,|\text{grad } A^*|^2).\text{grad } A^*.\text{grad } \psi_i dx.dy - \int_{\Omega'} J.\psi_i dx.dy$$

$$= 0 \qquad\qquad 1 \leq i \leq f+\ell \qquad\qquad (4.1)$$

In particular, at each step of the Newton-Raphson iteration, the resulting linear set of equations has a unique solution :

THEOREM. The Jacobian matrix of (4.1), i.e. $(\dfrac{\partial F_i}{\partial A_j})_{1\leq i,j\leq f+\ell}$, is symmetric and positive definite (hence invertible) $\forall$ $(A_1,\ldots, A_{f+\ell}) \in \mathbb{R}^{f+\ell}$.

Proof. Putting, for brevity, $\alpha = |\text{grad } A^*|$, we directly have

$$\frac{\partial F_i}{\partial A_j} = \int_{\Omega'} \nu.\text{grad } \psi_j^*.\text{grad } \psi_i^* \, dx \, dy$$

$$+ 2 \int_{\Omega'} \frac{\partial \nu}{\partial(\alpha^2)} .(\text{grad } A^*.\text{grad } \psi_j^*).(\text{grad } A^*.\text{grad } \psi_i^*) dx \, dy$$

$$\equiv G_{ij} + H_{ij}$$

Clearly, this matrix is symmetric. Moreover, for any real column matrix $\eta = (\eta_1,\ldots,\eta_{f+\ell})^T$, we have $\eta^T.G.\eta \geq 0$ and $\eta^T.H.\eta \geq 0$, with $G = (G_{ij})$, etc. Indeed $\nu(x,y,\alpha^2)$ is a positive and increasing function of α for all (x,y). Finally $\eta^T.G.\eta = 0$ implies that grad $(\sum_{i=1}^{f+\ell} \eta_i.\psi_i) = 0$ (almost everywhere) outside the junctions (where $\nu > 0$), from which it successively follows that $(\sum_{i=1}^{f+\ell} \eta_i.\psi_i) = 0$ outside the junctions and $\eta = 0$, due to the definition of the functions ψ_i, $1 \leq i \leq f+\ell$. ∎

5. NUMERICAL RESULT AND CONCLUSION

Both the accuracy and time efficiency of the coupled method are clearly demonstrated by the calculation of the 4-pole P.M. machine in fig. 4.

The stator is built up from classical network elements for the back iron, while the air gap is modelled by position dependent network elements [3], totaling 54 fundamental loops. The linear FE mesh on the rotor features 502 free nodes.

We calculated the flux in the stator teeth for a pole pitch. Fig.5 reveals a good agreement between the results obtained by the present method and those found by a more elaborate adaptive FEM with 2000 nodes

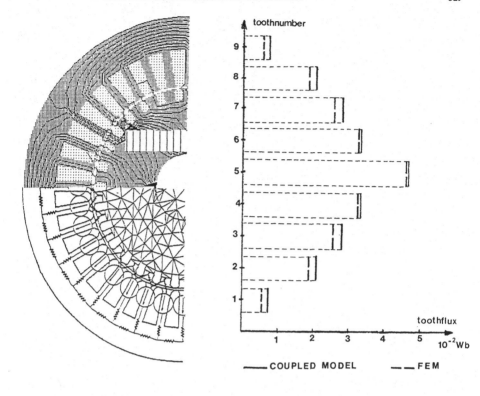

Fig. 4.   Fig. 5.

6. REFERENCES

[1] Glowinski R., Marrocco A.: "Analyse numérique du champ magnétique
    d'un alternateur par éléments finis et sur-relaxation ponctuelle
    non linéaire", Comp. Meth. Appl. Mech. Eng., 3, pp. 55-85 (1974).

[2] Ostović V.: "Dynamics of Saturated Electric Machines", Springer
    Verlag, N.Y. (1989).

[3] Philips D.: "A hybrid network-finite element model for switched
    reluctance motors", Journ. Eng. Design, 1, pp. 55-64 (1990).

[4] Philips D., Loukipoudis E., Dupré L., Melkebeek J.: "A New Approach
    to CAD in Electric Machinery", Journ. Eng. Design, (submitted).

[5] Silvester P., Chari M.: "Finite Element Solution of Saturable
    Magnetic Field Problems", IEEE Trans. on PAS, vol. PAS-89, pp.1642-
    1651, (1970).

Laboratory of Industrial Electricity & Seminar of Mathematical Analysis
State University of Ghent, St.Pietersnieuwstraat 41, 9000 Gent, Belgium

ADAPTIVE SPEED SYNCHRONIZATION OF SEGMENTED CONVEYOR BELTS

D. Prätzel–Wolters and S. Schmid
Fachbereich Mathematik
Universität Kaiserslautern
P.O. Box 3049
D–6750 Kaiserslautern, FRG

ABSTRACT. We consider a number of conveyor belts transporting goods of uncertain weights. The goal is to keep the belt speeds synchronized even under varying loads. By analyzing the mathematical model of a conveyor belt adaptive controllers are designed that achieve this goal. This was shown rigorously and is demonstrated by simulations.

1. INTRODUCTION

Conveyor belts are frequently used in factories. If goods are to be transported around corners or if the path of transportation should be kept flexible, the use of several belts positioned in a suitable order will often serve this purpose. Clearly, the belts should have approximately the same speed in order to avoid jams or damages when a mass is passing from one belt to another. To this end controllers are sought that are capable of keeping the belts at an equal speed no matter what load they are carrying. The unpredictable load is introducing some uncertainty in the system parameters of the conveyor belts, and this makes the use of adaptive controllers advisable.

2. THE MATHEMATICAL MODEL OF A SIMPLE CONVEYOR BELT TYPE

We consider a simple conveyor belt driven by a direct current motor (Fig.1)

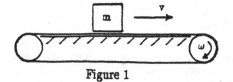

Figure 1

It can be modelled by the following equations (2.1a), (2.1b):

M. Heiliö (ed.), Proceedings of the Fifth European Conference on Mathematics in Industry, 331–334.
© 1991 B. G. Teubner Stuttgart and Kluwer Academic Publishers. Printed in the Netherlands.

$$\begin{bmatrix} \dot{I} \\ \dot{\omega} \end{bmatrix} = \begin{bmatrix} -\dfrac{R}{L} & -\dfrac{\psi}{L} \\[2mm] \dfrac{\psi}{J+mr^2} & 0 \end{bmatrix} \begin{bmatrix} I \\ \omega \end{bmatrix} + \begin{bmatrix} \dfrac{1}{L} \\[2mm] 0 \end{bmatrix} U + \begin{bmatrix} 0 \\[2mm] -\dfrac{\overline{M}(m,\omega)}{J+mr^2} \end{bmatrix}, \; v = \begin{bmatrix} 0 & r \end{bmatrix} \begin{bmatrix} I \\ \omega \end{bmatrix}$$

with U the input voltage, I the current, R the resistance, L the inductivity, ψ the intensity of the exciting field (constant), ω the angular velocity of the driving pulley, r the radius of the pulley, v the belt velocity, m the mass on the belt, J the total moment of inertia of the system (without mass on the belt), $\overline{M}(m,\omega)$ the angular momentum necessary to keep the belt at constant velocity, ωr when transporting a mass m.

Assuming the belt is sliding over a surface the properties of sliding friction lead us to the following form of $\overline{M}(m,\omega)$:

$$\overline{M}(m,\omega) = \begin{cases} \overline{M}(m,) & \omega > 0 \\[2mm] 0 & \omega = 0 \\[2mm] -\overline{M}(m) & \omega < 0 \end{cases} \quad \text{with } \overline{M}(m) = \alpha m + \beta, \; \alpha > 0, \; \beta > 0. \quad (2.1c)$$

Due to the unknown mass m the model equations have uncertain parameters.

3. DESIGNING THE CONTROLLER

The task of the controller is to keep the belt velocity constant, i.e. the output v should track a given constant reference velocity v^* independent of the specific value of m in (2.1). This requires an adaptive tracking strategy. In [1] we have shown that the adaptive tracking problem can be reduced to an adaptive stabilization problem by augmenting the original system by a suitable prefilter of order zero. In this special case of a constant reference signal the Laplace transformation of the transfer function of the prefilter takes the PI–form $K(1+\frac{1}{Ts})$ with $K > 0$ the proportional constant and $T > 0$ the integral time constant. The augmented system can then be written in the form:

$$\begin{bmatrix} \dot{I} \\ \omega \\ x \end{bmatrix} = \begin{bmatrix} -\dfrac{R}{L} & -\dfrac{\psi}{L} & \dfrac{K}{L} \\[2mm] \dfrac{\psi}{J+mr^2} & 0 & 0 \\[2mm] 0 & 0 & 0 \end{bmatrix} \begin{bmatrix} I \\ \omega \\ x \end{bmatrix} + \begin{bmatrix} \dfrac{K}{L} \\[2mm] 0 \\[2mm] \dfrac{I}{T} \end{bmatrix} w + \begin{bmatrix} 0 \\[2mm] -\dfrac{\overline{M}(m,\omega)}{J+mr^2} \\[2mm] 0 \end{bmatrix}, \; v = \begin{bmatrix} 0 & r & 0 \end{bmatrix} \begin{bmatrix} I \\ \omega \\ x \end{bmatrix}$$

(x the state of the prefilter), short:

$\dot{z} = A(m)z + bw + h(m)$, $v = cz$, with input w and output v.
The friction term h(m) (uncertain but time–independent as long as m is constant) can be taken care of as follows: clearly, $h(m) \in \text{im } A(m)$, hence for some $g(m): A(m)g(m) = h(m)$, and we have:

$$(\dot{z} + g(m)) = A(m)(z+g(m)) + bw$$
$$v + cg(m) = c(z+g(m))$$

Following the ideas in [1] and taking $v^*+cg(m)$ as the modified reference velocity for the above system we can write:

$$(\dot{z}+g(m)-\bar{z}(m)) = A(m)(z + g(m) - \bar{z}(m)) + bw$$
$$v - v^* = c\,(z + g(m) - \bar{z}(m))$$

where the existence of such a time dependent function $\bar{z}(m)$ is guaranteed by the performed augmentation. All that remains to be done is to design an adaptive stabilizing controller for this system. It can easily be checked that the system is of minimum phase and has relative degree two. The damping coefficient of the system as defined in [2] is $a = \dfrac{R}{L} - \dfrac{I}{T}$. Since we assume knowledge of the parameters R and L – only m is unknow – we can choose $T > \dfrac{L}{R}$, so that $a > 0$. Applying a result in [2] we obtain the following

THEOREM: Given the system (2.1) with unknown mass parameter m and a constant reference velocity v^*, the adaptive controller

$$u(t) = K\left(w(t) + \frac{1}{T} \int_{t_0}^{t} w(\tau)d\tau\right) + \gamma,\ T > \frac{L}{R},\ \gamma \text{ any constant}$$

$$w(t) = -k(t)\,(v(t) - v^*), \qquad \dot{k}(t) = (v(t) - v^*)^2$$

stabilizes the closed loop system in the sense:

$$\lim_{t\to\infty} v(t) = v^*,\quad \lim_{t\to\infty} k(t) = k_\infty < \infty$$

for arbitrary initial values $I(t_0) = I_0$, $w(t_0) = w_0$.

Using the concept of high gains k is plausible in this case: in order to keep the eigenvalues of $A-kbc$ in the half plane $\{s: \mathrm{Res} \leq -\epsilon\}$, $\epsilon > 0$ fixed, as m tends to infinity, k has to tend to infinity, too. Of course, m can also decrease here, that is why one can intoduce an additional pole to let the gain decrease, too (cf. [3]):

$$\dot{k} = (v - v^*)^2 - \alpha\,k, \qquad \alpha > 0.$$

4. SIMULATIONS

Coupling between two conveyer belts occurs when a mass is passing from one belt to another. When modelling the coupling one has to distinguish between the case when the mass is resting on one belt and sliding on the other and the case when the mass is sliding on both belts. Additional uncertain parameters play a role then, like the lengths of the goods transported and friction coefficients. Instead of writing down the equations here and applying theoretical synchronization results of [4] we are demonstrating the synchronized behaviour of the conveyor belts equipped with the derived controllers by a typical simulation example. The diagrams in Fig. 2 depict the behaviour of a conveyor belt positioned between two others when different masses are transported along them. At certain points of time masses are entering or leaving the belt.

$\alpha = 0$ $\qquad\qquad\qquad\qquad\qquad$ $\alpha > 0$

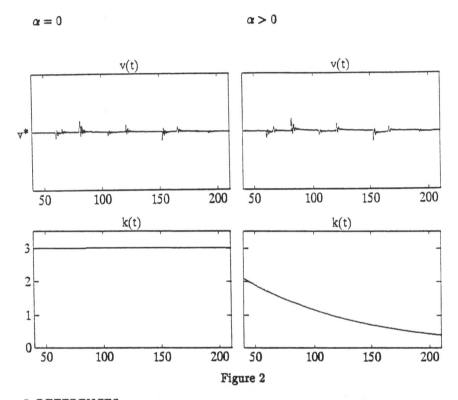

Figure 2

5. REFERENCES

[1]   U. Helmke, D. Prätzel–Wolters, S. Schmid: Adaptive tracking of scalar minimum phase systems, to appear in the series "Progress in Systems and Control Theory", Birkhäuser–Verlag

[2]   A.S. Morse: Simple algorithms for adaptive stabilization, Proceedings of the IIASA Conference on Modelling and Adaptive Control, Spinger–Verlag, 1986

[3]   P. Joannou: Robust direct adaptive control, Proc. 23rd Conf. on Decision and Control, 1984

[4]   U. Helmke, D. Prätzel–Wolters, S. Schmid: Adaptive synchronization of interconnected linear systems, submitted to Intl. J. Control.

Eddy Current Computation

M.Reißel, J.Wick

A big source for loss of energy in a generator are eddy currents. In the past it was possible only to measure eddy currents on a real generator and then to transfer this knowledge into the construction of the next generator. This is a very slow iteration procedure. Hence with the first large computers in the 70^{ies}, several companies created research groups to get computational information about eddy currents before manufactoring the generator. This sounds quite simple, since the mathematical model are obviously Maxwell's equations. But they don't want a solution theory, they want a numerical solution for a concrete generator. This is a huge machine with complicated geometry and nonhomogeneous material. The companies invested a lot of money and man-power. But at the end there was no running code for the full problem. Now, they went to the mathematicians for help and keeping their investment. This means, we have to deal with a problem, where the model and its discretization is given but the strategy for the numerical solution is open.

Of course, we must start our investigation with the model again. The basis are Maxwell's equations, where the engineers have added two important assumptions:

1) the displacement current can be neglected: $\dot{D} = 0$

2) every field has the form: $\mathcal{F}(t, x, y, z) = F(x, y, z)e^{i\omega t}$

For a first investigation we have defined a test problem with the engineers. This contains all the structural difficulties of the real problem, but the geometry is simpler and some reference results are known. In a rectangular box of air we have imprinted currents J_e and two rectangular boxes of laminated and solid iron. The air is considered as a perfect insulator, the laminated iron is a perfect insulator in one direction, a conductor in the two others.

The magnetic field in the air H_a will be splitted into two parts

$$H_a = H_i + H_c$$

335

M. Heiliö (ed.), Proceedings of the Fifth European Conference on Mathematics in Industry, 335–338.

with

$$\mathrm{rot} H_i \; = \; J_e$$

$$\mathrm{div} H_c = -\mathrm{div} H_i \qquad ; \qquad \mathrm{rot} H_c = 0. \tag{1}$$

If the air region is simply connected, we can compute H_c from a potential Φ

$$\Delta\Phi = \mathrm{div} H_i. \tag{2}$$

With arbitrary boundary conditions (1) is easy to solve. In (2) we must add boundary conditions. On the outer surface they are given using H_i. On the inner surface we have transition conditions, which couple the different regions. The magnetic properties of laminated iron are modeled by homogenization, this means we have a constant permeability in the plates

$$\mu_\| = 0.9\mu_{Fe} + 0.1\mu_a$$

and for the perpendicular direction

$$\mu_\perp = \left(\tfrac{0.9}{\mu_{Fe}} + \tfrac{0.1}{\mu_a}\right)^{-1}.$$

Since laminated iron is assumed to be a perfect insulator in one direction, one component of the magnetic field depends on the two others. Hence we can choose a representation

$$H_{li} = (0,0,H_z)^t - \mathrm{grad}\Phi. \tag{3}$$

This simplifies also the transition condition between air and laminated iron. Solid iron is assumed to be homogeneous, so the conductivity and the permeability can be expressed by constants σ and μ resp. Hence we get

$$\text{rot rot} H = -i\omega\sigma\mu H. \tag{4}$$

The engineers have chosen finite differences in the air and the laminated iron region and finite volumes for (4). Since the grid for the finite difference approximation differs from the grid for the finite volume approximation the adjustment of the transition condition becomes a bit intricately.

For this reason the engineers have introduced a potential Ψ in solid iron, too. Now, the transition conditions are simple. Furthermore the unknowns can be separated in the potentials - now we call them Φ generelly - and the H- components, which leads to a linear system

$$\begin{pmatrix} A_{\Phi\Phi} & A_{\Phi H} \\ A_{H\Phi} & A_{HH} \end{pmatrix} \begin{pmatrix} \Phi \\ H \end{pmatrix} = \begin{pmatrix} b_\Phi \\ b_H \end{pmatrix} \tag{5}$$

Starting with a guess H^0,

$$\begin{aligned} A_{\Phi\Phi}\Phi^i &= b_\Phi - A_{\Phi H}H^{i-1} \\ A_{HH}H^i &= b_H - A_{H\Phi}\Phi^i \end{aligned} \qquad i \in \mathcal{N} \tag{6}$$

is a quite natural iteration scheme.

But the additional equation, taken by the engineers to determine Ψ, depends linearly on (4). Therefore the matrix in (5) is singular. Unfortunately, the iteration (6) converges in some cases, where the subsystems are solved with SOR. Hence it was hard to convince the engineers that their singular approach must be corrected. In addition, for the correct system with (4) the iteration (6) failes, which requires another algorithm, solving (5). This leads to a large change in the program, hence the engineers are interested, why SOR failes. This investigation is not already finished, since it was of minor interest. The essential part was to find an algorithm solving the eddy current problem

Since A is a complex matrix, we tried the bi-conjugate gradient method [1] with an incomplete LU-decomposition as preconditioning. For the standard case

$$k := \mu_{Fe}/\mu_a = 100$$

we get after 50 iterations a residual norm less than 10^{-4}. With the bi-conjugate squared method [2] the convergence is about 30% faster.

If we decrease k for the solid block, the covergence rate also decreases and it diverges for k = 1. By rearranging the equations we get faster convergence without problems for k = 1.

$$
\begin{aligned}
k &= 1 & : & \quad 80 \text{ iteration steps} \\
k &= 10 & : & \quad 35 \text{ iteration steps} \\
k &= 100 & : & \quad 18 \text{ iteration steps}
\end{aligned}
$$

But until now we have no mathematical explanation for these effects.

References

[1]   D.A.H. Jacobs: The Exploitation of Sparsity by Iterative Methods in
                               I.S.Duff: Sparse Matrices and their Uses, Academic Press
                               1981, pp. 191 - 222

[2]   P.Sonneveld: CGS, a fast Lanczos-Type Solver for Nonsymmetric Linear
                               Systems, SIAM J. Sci.Stat. Comput. 10 (1989) 36 - 52

M. Reißel, J. Wick

Dept. of Mathematics

University of Kaiserslautern

Erwin-Schrödinger-Str.

D 6750 Kaiserslautern , Germany

The Coupling of
Acoustical Membrane and Cavity Vibrations

S.W. Rienstra

Abstract

To support the design of kettledrums a mathematical model is made, including a uniform membrane with airloading, a cylindrical cavity, and an infinite field. The model is simple enough to be implemented on a small computer, but yet describes some of the most important effects and acoustical mechanisms in detail, like the generation of a modal spectrum, the free field radiation damping, and the coupling of membrane modes and cavity modes.

Since the solution method breaks down when the system mode becomes equal to a cavity mode (the membrane amplitude vanishes), this phenomenon is further analysed by a one dimensional model.

Introduction

Although the basic principles of most musical instruments are known for more than a century, a rational design based on a complete understanding is still impossible in any practical situation. On the one hand this is due to the inherently rather complex configurations allowing all sorts of coupling and types of vibrational modes, in addition to the naturally limited amount of measurable information. On the other hand the sound is not meant for measuring instruments, but for the extremely sensitive and critical human ears, while at the same time the appreciation of beauty is too subtle and subjective to be represented by a formula. So any experienced designer will primarily rely on his ears, and will use measurements and model calculations only as a guide.

These observations couldn't be more true for the kettledrum, to the design of which a kettledrum factory invited us to give theoretical support by means of a mathematical model. This model had to include some of the most fundamental mechanisms (of the ones that are *supposed* to play a rôle), to allow the designer to order and interpret his experimental data.

The model and solution method we adopted is basically the one introduced by Christian, Davis, Tubis, Anderson, Mills and Rossing [1]. The good agreement they obtained with experimental results apparently justified the simplifications made. We extended the model to introduce the presence of a small vent hole in the bottom (which in the end appeared to have only a little effect). By analysing the necessary numerical calculations carefully, we were able to implement the model on a small computer ([2]).

The method is, however, inherently poor when a mode is close to a cavity mode. Then the coupling between the air vibrations in the cavity and the membrane vibrations disappears,

M. Heiliö (ed.), Proceedings of the Fifth European Conference on Mathematics in Industry, 339–344.
© 1991 *B. G. Teubner Stuttgart and Kluwer Academic Publishers. Printed in the Netherlands.*

and the radition damping vanishes. Therefore, this phenomenon is further analyzed by a very simple 1-D model, allowing a fully analytical solution.

The model

The cavity is modelled as a flanged finite cylinder, in cylindrical coordinates (r, θ, z) described by $r = a$, $0 \leq z \leq L$. The bottom $z = 0$ has a small hole $r \leq d$ $(d << a)$, and at the top side $z = L$ the cavity is closed by a uniform membrane of tension T and density σ. The air, with density ρ_a and sound speed c_a, vibrates with (acoustic) excess pressure p and velocity $\vec{v} = \nabla \phi$, while the membrane displacement is given by $z = L + \eta$. As we are interested in the acoustic regime, the equations may be linearized, and we may utilize Fourier analysis and look for a solution per frequency ω. In complex notation we write p, $\vec{v}$, $\eta \sim e^{-i\omega t}$, and then futher ignore the exponential. The prevailing equations are then:

$$\nabla^2 p + k^2 p = 0 \ , \quad p = i\omega \rho_a \phi \ , \qquad \text{(in air)} \tag{1}$$

$$T \nabla_0^2 \eta + \omega^2 \sigma \eta = [p] \ , \qquad \text{(at the membrane)} \tag{2}$$

where $k = \omega/c_a$ is the acoustic wave number, ∇_0 denotes ∇ restricted to the membrane surface, and $[p]$ denotes the pressure difference across the membrane. Note that the propagation speed of (transversal) waves in the membrane is $c_M = (T/\sigma)^{\frac{1}{2}}$. Boundary conditions are a vanishing normal velocity at the hard-walled surfaces: $(\nabla \phi \cdot \vec{n}) = 0$, and a matching membrane surface and air velocity at the membrane: $-i\omega\eta = \phi_z$. The bottom opening is modelled as a small orifice in an infinitely thin wall, so that the diffraction effects are acoustically equivalent to a dipole source, of which the strength is determined by the incompressible flow through the orifice (the inner region in a matched asymptotic expansion formulation). This flow is, on its turn, driven by the pressure of the incident acoustic wave. The resulting relation between the rate of change of volume velocity through and the pressure at the hole is [2]:

$$p(\vec{0}) = -i\omega \frac{\rho_a}{2d} \int\limits_0^{2\pi} \int\limits_0^d \phi_z(r, \theta, 0) \, r \, dr d\theta \quad (d \to 0) \ . \tag{3}$$

As there are no sources at infinity we have for $r \to \infty$, $z \to \infty$ Sommerfeld's radiation condition for only outward radiating waves.

Ideal modes

In the absence of air $(\rho_a = 0)$ the acoustic field vanishes, and the solution reduces to vibrations of the membrane alone:

$$\eta = \eta_{mn}^{(0)}(r, \theta) = J_m(x_{mn} r/a) \, e^{im\theta} / a\sqrt{\pi} \, J_m'(x_{mn}) \quad (m = 0, \, 1, ...; \, n = 1, \, 2, ...) \tag{4}$$

$$\omega = \omega_{mn}^{(0)} = x_{mn} \, c_M/a$$

where J_m is the m-th order Besselfunction of the first kind, and $J_m(x_{mn}) = 0$. Note that $\{\eta_{mn}^{(0)}\}$ forms an orthogonal basis.

When the density σ of the membrane tends to infinity, the deflection η vanishes, and the cavity field is decoupled from the (vanishing) outer field, so the solution reduces to cavity resonances. These are given by (for $d = 0$)

$$\phi = \phi_{mnl}^{(c)}(r, \theta, z) = J_m(y_{mn}r/a) \cos(l\pi z/L) \, e^{im\theta}$$

$$\omega = \omega_{mnl}^{(c)} = (y_{mn}^2 + l^2\pi^2a^2/L^2)^{\frac{1}{2}} \, c_a/a \tag{5}$$

where $m = 0, 1, 2, ...$, $n = 1, 2, 3, ...$, $l = 0, 1, 2, ...$, and $J_m'(y_{mn}) = 0$.

General solution

The approach is as follows. Both inside and outside the cavity the acoustic field is described, via a Green's function representation, as if it were driven by the membrane displacements. At the same time the field just below and above the membrane can be considered as a driving force to the membrane. So we can formulate an integral equation for the membrane displacement η. This equation is then written in matrix form by a suitable expansion in vacuum modes $\eta_{mn}^{(0)}$. Finally, this matrix equation is solved numerically. Note that this approach is obviously inadequate if a mode has a relatively small membrane deflection.

By standard techniques ([1, 2]) we obtain outside

$$\phi(r, \theta, z) = \frac{i\omega}{4\pi} \int_0^{2\pi} \int_0^a \eta(r', \theta') \, G_{out}(r, \theta, z; r', \theta', L) \, r'dr'd\theta' \ , \tag{6}$$

and inside

$$\phi(r, \theta, z) = -\frac{i\omega}{4\pi} \int_0^{2\pi} \int_0^a \eta(r', \theta') \, G_{in}(r, \theta, z; r', \theta', L) \, r'dr'd\theta' + \frac{d}{2\pi} \, G_{in}(r, \theta, z; \vec{0}) \, \phi(\vec{0}) \ . \tag{7}$$

It may be noted that $G_{in}(\vec{r}; \vec{0})$ is symmetric (only $m = 0$ modes), and singular in $\vec{r} = \vec{0}$. Indeed, this expression (7) is, with respect to the orifice effect, approximate and only valid for $d << a$ and $|r| >> d$. Furthermore, since the second term is a correction (as long as η is not small), the value $\phi(\vec{0})$ to be substituted is effectively the one obtained for $d = 0$. So we end up with

$$\phi(r, \phi, z) = -\frac{i\omega}{4\pi} \int_0^{2\pi} \int_0^a \eta(r', \theta') \, [G_{in}(r, \theta, z; r', \theta', L)$$

$$+ \frac{d}{2\pi} \, G_{in}(r, \theta, z; \vec{0}) \, G_{in}(\vec{0}; r', \theta', L)] \, r'dr'd\theta' \ . \tag{8}$$

After substitution of these (formal) solutions into eq. (2), we introduce the expansion

$$\eta(r,\,\theta) = \sum_{m=-\infty}^{\infty} \sum_{n=1}^{\infty} a_{mn}\ \eta_{mn}^{(0)}(r,\theta)\ ,$$

multiply left- and righthand side by $\eta_{m'n'}(r,\theta)^{*}r$ and integrate over the membrane surface. After utilizing the orthogonality, and some other details, we arrive at the infinite set of equations

$$(\omega^2 - \omega_{mn}^{(0)2})\sigma a_{mn} = 4\omega^2 \rho_a a \sum_{n'=0}^{\infty} a_{mn'}\ x_{mn}\ x_{mn'}\ [C_{mnn'} - \tfrac{1}{2}i I_{mnn'} - \frac{2d}{\pi a}\ \delta_{0,m}\ S_n\ S_{n'}]$$

where, per m, the set of ω's allowing a solution $(a_{m1},\ a_{m2},...)$ is to be found. This set is the sought spectrum. The notation used is

$$C_{mnn'} = \sum_{n''=1}^{\infty} \cotg(\gamma_{mn''}L/a)\ /\gamma_{mn''}(x_{mn}^2 - y_{mn''}^2)(x_{mn'}^2 - y_{mn''}^2)(1 - m^2/y_{mn''}^2)\ ,$$

$$I_{mnn'} = \int_0^{\infty} \lambda J_m(\lambda)^2/\gamma(\lambda)(\lambda^2 - x_{mn}^2)(\lambda^2 - x_{mn'}^2)\ d\lambda\ ,$$

$$\gamma(\lambda) = i\ \sqrt{i(\lambda - ka)}\ \sqrt{-i(\lambda + ka)},\qquad \gamma_{mn} = \gamma(y_{mn})\ ,$$

$$S_n = \sum_{n''=1}^{\infty} [(x_{0n}^2 - y_{0n''}^2)\gamma_{0n''}\ \sin(\gamma_{0n'}L/a)J_0(y_{0n''})]^{-1}\ .$$

By a suitable contour deformation the integral $I_{mnn'}$ is further prepared for efficient numerical evaluation. The set of equations is solved via iterations over ka after writing the equations, in matrix form, as a quasi-eigenvalue problem. Further details may be found in [2].

Numerical example

To illustrate the present theory, we plotted in fig.1 the $m = 1$ spectrum for a typical kettledrum as a function of tension: $a = 0.328$ m, $\sigma = 0.2653$ kg/m^2, $L = 0.4142$ m, $\rho_a = 1.21$ kg/m^3, $c_a = 344.0$ m/s. (The $m = 1$ modes are musically most important. The bottom hole affects only the $m = 0$ modes; we found for the present example only a small effect). The dotted lines denote the level of the cavity modes. We see that in general the frequencies increase steadily with tension. Only a mode near a cavity mode is, however, somewhat reluctant to increase. Since the presented frequencies are really complex (the imaginary part is related to the decay time) we plotted one mode in the complex plane (fig.2,3). We see that for certain values of T the mode indeed becomes very close to a cavity mode, as the imaginary part also vanishes. A purely real frequency, however, does not decay, and since this is only possible with a stagnant membrane, we may expect problems with our solution method (which indeed occur) in this regime, and we would like to see additional confirmation of this phenomenon. Furthermore, it would be interesting to see whether it is accidental (for example, because σ was relatively large), or whether it would occur always at some tension.

Therefore we analysed a model, complex enough to include the basic effects, but simple enough to allow a fully analytical approach.

1-D model

A first proposal was a low frequency limit ($ka \to 0$) of the present model, with $kL = O(1)$, since this reduces the cavity field into a simple plane wave, and the radiated field into that of a point source. This is not a convenient limit, however, since the impedance at the cavity opening, even without membrane, becomes infinitely large, implying total reflection and a vanishing outside field. A finite impedance is obtained if the outside space is also reduced to a (semi-infinite) cylinder:

$$\phi_{zz} - c_a^{-2}\phi_{tt} = 0 \ \text{ in } \ 0 < z < L, \ L < z < \infty \ ,$$

$$-Ta^{-2}\eta - \sigma\eta_{tt} = p(L+, \ t) - p(L-, \ t) \ \text{ at } \ z = L \ ,$$

with $\phi_z = 0$ at $z = 0$, $\phi_z = \eta_t$ at $z = L \pm 0$, and outgoing wave for $z \to \infty$.

For ϕ, η, $p \sim e^{-i\omega t}$, finding the solution is standard, and we end up with the eigenvalue equation for $kL = \omega L/c_a$

$$kL + i\frac{\rho_a L}{\sigma} - \left(\frac{c_M L}{c_a a}\right)^2 \frac{1}{kL} = \frac{\rho_a L}{\sigma} \ \text{cotg}(kL) \ ,$$

which can be solved easily numerically for varying $s = \sqrt{\rho_a L/\sigma}$ and $b = c_M L/c_a a$ (fig.4). Note that $\sin(kL) = 0$, i.e. $kL = n\pi$, correspond to cavity modes. It is seen that indeed for any s these cavity modes are passed for increasing b, while at the same time the contour of fig.3 is very similar to a corresponding one of fig.4. So the $1 - D$ model provides qualitative insight into the coupling phenomenon, and indeed it does not seem to be accidental that the modes of the full 3-D model passed through the cavity modes.

References

1. R.S. Christian, R.E. Davis, A. Tubis, C.A. Anderson, R.I. Mills, T.D. Rossing, "Effects of air loading on timpani membrane vibrations", Journal of the Acoustical Society of America, **76** (5), 1984, 1336-1345.

2. S.W. Rienstra, "The acoustics of a kettledrum", IWDE Report WD 89.10, 1989, Eindhoven University of Technology.

Department of Mathematics and Computing Science
Eindhoven University of Technology
P.O. Box 513, 5600 MB Eindhoven, The Netherlands

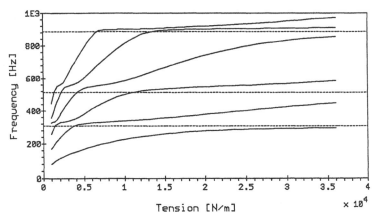

Figure 1. m=1 modes.

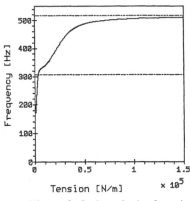

Figure 2. 2nd mode (real part)

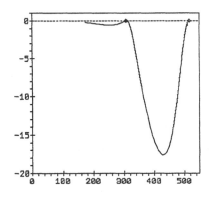

Figure 3. 2nd mode in complex k-plane

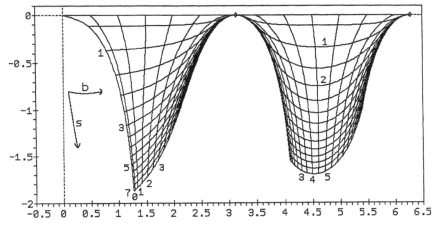

Figure 4. 1-D modes in complex kL-plane; contours for varying s and b.

IMPLEMENTATION ASPECTS OF USING NONLINEAR FILTERING FOR THE FAILURE DETECTION PROBLEM

TUULA RUOKONEN
Imatran Voima Oy, Research and Development Division
P.O.Box 112
SF-01601 Vantaa
Finland

ABSTRACT. This paper deals with Nonlinear Filtering as a method for failure detection in dynamic processes. The Failure Detection Problem is presented here as an Identification Problem, where no jumps between the possible operating modes are assumed, i.e. there is only uncertainty with regard to which is the present mode. This paper focuses on some implementation aspects of nonlinear filtering for failure detection.

1. Introduction

A fault detection and diagnosis system is an essential part of a safe, reliable and economical control system. Developing on-line methods for detecting and locating malfunctions in the process is an important goal towards full automation of systems. Model-based methods, which check the consistency of the measurements using the known functional relationships among the process variables, seem to have the potential of early detection of slowly developing faults in complex processes.

A popular nonlinear filter developed for failure detection in dynamic systems is the Identification Filter. The filter consists of parallel Kalman filters, each tuned to a different system mode of operation, whose estimates parametrize a finite-dimensional system os equations for the conditional probabilities of being at each mode [1]. The filter generates a sequence of decisions in finite time.

The goal of this paper is to focus on some implementation aspects of the Identification Filter. Theoretical proofs on the filter performance may not be valid in practice, due to approximations, modeling errors and wrong assumptions about the noise conditions [2]. Aspects, such as the global decision making interpreting a sequence of local decisions, the choice of the decision thresholds and filter reinitialization are of great practical value. As an example, changing the decision threshold and scaling the likelihood functions are described as methods to change the decision time [2].

2. Identification Problem

The Nonlinear Filtering Problem is the estimation of a stochastic process x_t from a related process y_t (the noisy observation process), when both the state equation and the measurement equation are subject to random jumps in the structure and/or parameters. The nonlinear filter generates, in addition to the best estimation of the state vector, a set of probabilities, conditioned on the measurements, of being at each mode [3].

A special case of the Nonlinear Filtering Problem is the Identification Problem. By "identification" it is meant here that there are N possible known modes of operation, but no transitions from one mode to another are assumed. There is only uncertainty with regard to which fixed mode the system is in. The problem is decomposed into N scalar filtering problems. A treatment of the linear Gaussian problem can be found in [1]. The filter is finite dimensional and consists of N Kalman filters in parallel (each tuned to a different mode of operation) whose means parametrize N nonlinear stochastic differential equations for p_j,

345

M. Heiliö (ed.), Proceedings of the Fifth European Conference on Mathematics in Industry, 345–348.
© 1991 *B. G. Teubner Stuttgart and Kluwer Academic Publishers. Printed in the Netherlands.*

$j=1,...,N$, the conditional probabilities of being in each mode. The general structure of the optimal nonlinear filter for the Identification Problem is as follows [1].

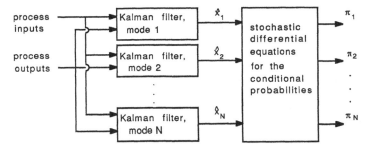

Figure 1. The structure of the Nonlinear Filter for the Identification Problem.

Process equations (v and w standard Brownian motions):

$$\begin{cases} dx = (a_i x + b_i u)dt + g_{vi} dv ; \ x(0) = x_i^o \\ dy = c_i x \, dt + g_w dw ; \ y(0) = 0 \end{cases}$$

(1)

Conditional probability equations:

$$d\pi_j = \frac{1}{g_w^2} \pi_j \sum_{k=1}^{N} \pi_k \left(c_j \hat{x}_j - c_k \hat{x}_k \right) \left(dy - \sum_{k=1}^{N} \pi_k c_k \hat{x}_k \, dt \right) ; \ j=1,...,N ; \ \pi_j(0) = \pi_j^0$$

(2)

The respective likelihood function $l_j(t)$ has the following exponential form:

$$l_j(t) = \exp\left\{ \frac{1}{g_w^2} \left(\int_0^t c_j \hat{x}_j(s) \, dy(s) - \frac{1}{2} \int_0^t c_j^2 \hat{x}_j^2(s) ds \right) \right\}$$

(3)

Conditional probabilities p_j in Eq. (2) may also be expressed as follows [1]:

$$\pi_j(t) = \frac{l_j \, \pi_j^0}{\sum_{k=1}^{N} l_k \, \pi_k^0}$$

(4)

3. Implementation Aspects

The performance of the nonlinear filter for the Identification Problem is studied in [2,4], and the conclusion is given, that a correct decision is expected on the average after large enough time. For practical implementation, the problems of interpreting a sequence of decisions, setting a decision threshold and reinitializing the filter have to be understood and solved.

The decision time is defined here as the time when the conditional probability p_j for one of the modes exceeds a designated threshold value L. The average decision time can be altered by changing the threshold value L or by "scaling" the likelihood functions by a factor K, i.e. by raising the likelihood functions l_j to the power of K [2]. Increasing K is equivalent to decreasing the threshold value L. This speeds up the decision making but increases the probability of faulty decisions, i.e. false alarms or misses. The optimal choice of the threshold value or equivalently the scaling factor is always a compromise between the speed of the decision and reliability. The relationship between the scaled likelihood functions is the following (p_{k0} are a-priori mode probabilities):

$$l_j^K = \frac{L}{1-L} \sum_{k=1, k \neq j}^{N} \frac{\pi_{k0}}{\pi_{j0}} l_k^K$$

(5)

The simulation results provided here are for the example process consisting of two tanks and a valve (described in detail in [2]). The normal mode <1> and the faulty mode <2> differ by the measurement gain for the second tank level. The time scale in the figures is the number of samples. In all simulations the same measurement noise path was taken. The conditional probabilities for the case of K=0.1 and L=90% are shown in Fig.2. In Fig.3 the threshold value is changed to L=75%. The same effect on the decision time can be obtained by changing the scaling factor to K=0.2 instead, i.e. multiplying it by two (Fig.4). This relationship could also be solved analytically using Eq. (5).

The decision time is normally not constant, but it depends on the operating point, control inputs and noise conditions. By using an adaptive choice of the scaling factor, it is possible to adjust the range of the decision time. If the decision making is too fast, the scaling factor is decreased, if the decision making is too slow the scaling factor is increased (compare Fig.2 and Fig.4). The scaling factor can be tuned based on a test run when starting the algorithm, or it can be adjusted on-line. Figures 5 and 6 show the effect of very simple adaptive scaling in order to change the decision time to 25 samples. The initial value for the scaling factor $K_0=1.0$ in Fig.5 and $K_0=0.2$ in Fig.6. A more advanced algorithm which uses information from all of the previous decisions may converge faster.

4. Conclusions

Model-based methods are promising in detecting failures which change the process parameters. In this paper nonlinear filtering is studied as a failure detection method. Theoretical proofs on the filter performance may not be valid in practice, due to approximations, modeling errors and wrong assumptions about the noise conditions. Thus, implementation aspects should be studied in more detail. This paper focuses on methods for changing of the decision time as an example of practical problems.

References

[1] Hijab, O. B., "The Adaptive LQG Problem - Part I", IEEE Transactions on Automatic Control, Vol. AC-28, No. 2, February 1983.
[2] Ruokonen, T., "Nonlinear Filtering Techniques for Failure Detection in Dynamic Systems". Ph.D. Dissertation, Florida Atlantic University, August 1989.
[3] Kushner, H. J., "Dynamical Equations for Nonlinear Filtering", Journal of Differential Equations 3, 1967, p.179-190.
[4] Ruokonen, T., Roth, Z.S., "Failure Detection Performance Analysis of the Optimal Nonlinear Filter for the Identification Problem". Proceedings of ACC, Vol. 1, p.876-878, June 1989.

Figure 2.
Filter performance with
the scaling factor K=0.1.

Figure 3.
Filter performance with
K=0.1 and L=75%.

Figure 4.
Filter performance with
K=0.2 and L=90%.

Figure 5.
Adaptive scaling with
initial value K_0=1.

Figure 6.
Adaptive scaling with
initial value K_0=0.2.

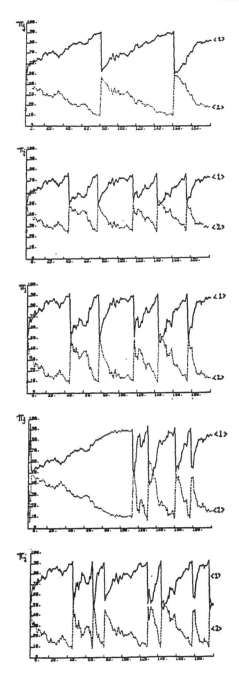

EXPECTED TIME TO FAILURE OF SOME GENERAL RELIABILITY SYSTEMS

Lennart Råde

Chalmers University of Technology, Gothenburg, Sweden

1. Introduction

The aim of this paper is to treat some applications of two general theorems about the expected time to failure of certain reliability systems. Explicit expressions for such expectations will be derived for some specific general systems. Furthermore, expressions for warm and cold redundancy coefficients of components $I_{WR}(i)$ and $I_{CR}(i)$ of such systems will be given.

Reliability systems with structures shown in figure 1 will be studied. In systems B, D and E one of the components is duplicated with a cold redundant component, which is activated upon failure of the component, to which it is attached.

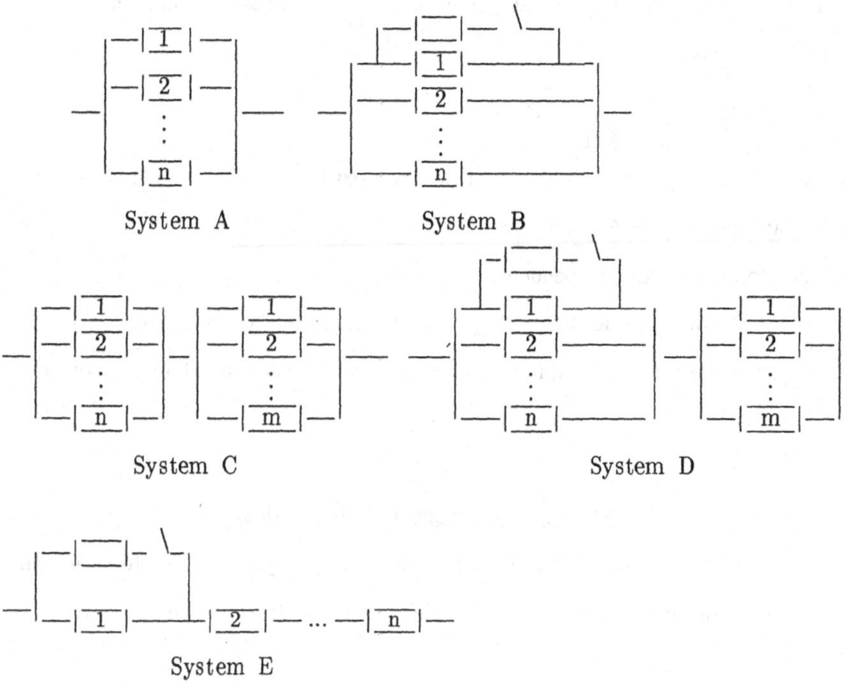

Figure 1

In [1] the following two theorems are proved.

M. Heiliö (ed.), Proceedings of the Fifth European Conference on Mathematics in Industry, 349–353.
© 1991 B. G. Teubner Stuttgart and Kluwer Academic Publishers. Printed in the Netherlands.

Theorem 1. Let T be the time to failure of a parallel reliability system of n

components (system A above) and let $\mu_{i_1 i_2 \ldots i_k}$ be the expected time to failure of the

series system of components $i_1, i_2, \ldots, i_k$ assuming that the joint distribution of the

components times to failure is the same as in the parallel system. Let S_k be the sum of

all such expectations for given k. Then

$$E[T] = \sum_{k=1}^{n} (-1)^{k+1} S_k .$$ (1)

Theorem 2. Let T be the time to failure of a series reliability system of two parallel

systems with n and m components, respectively (system C above). Let

$\mu_{i_1,\ldots,i_k,j_1,\ldots,j_\ell}$ be the expected time to failure of the series system of components

$i_1,\ldots,i_k$ from the first parallel system and components $j_1,\ldots,i_\ell$ from the second parallel

system, assuming that the joint distribution of the components times to failure is the

same as in the original series-parallel system. Let $S_{k,\ell}$ be the sum of all such

expectations for given k and ℓ. Then

$$E[T] = \sum_{k=1}^{n} \sum_{\ell=1}^{m} (-1)^{k+\ell} S_{k,\ell}$$ (2)

Note that it is usually much easier to find the expected time to failure for series systems

than for parallel systems. Also note that the theorems do not require that the lengths of

life of the components are independent.

In the following we denote by T_{WRi} and T_{CRi} the time to failure of the

system when component i is duplicated by a warm and a cold redundant component

respectively.

2. Systems with exponential distributions.

We will first consider system A with independent component times to failure all

of which are exponentially distributed with the same intensity λ. In this case we obtain

E[T] from theorem 1 with

$$S_k = \binom{n}{k} \frac{1}{k\lambda} \quad \text{and} \quad E[T] = \sum_{k=1}^{n} (-1)^{k+1} \binom{n}{k} \frac{1}{k\lambda} = \frac{1}{\lambda} \sum_{k=1}^{n} \frac{1}{k} = \frac{1}{\lambda} H_n$$ (3).

It follows immediately that

$$E[T_{WRi}] = \frac{1}{\lambda} H_{n+1} \quad \text{and} \quad I_{WR}(i) = (E[T_{WRi}] - E[T])/E[T] = \frac{1}{(n+1)H_n}$$ (4)

Thus for large n (γ is Euler's constant)

$$I_{WR}(i) \approx \frac{1}{(n+1)(\gamma + \ell nn)} \tag{5}$$

It is easily shown that under the same assumptions for system E

$$E[T] = \frac{1}{\lambda} \sum_{k=1}^{n} (\frac{1}{k} + \frac{1}{k^2}) \tag{6}$$

To find $E[T_{CRi}]$ for system A we consider system B. With use of theorem 1 and (6) we obtain

$$S_k = \binom{n-1}{k} \frac{1}{k\lambda} + \binom{n-1}{k-1}(\frac{1}{k} + \frac{1}{k^2})\frac{1}{\lambda} = \frac{1}{\lambda}(\binom{n}{k}\frac{1}{k} + \binom{n-1}{k-1}\frac{1}{k^2})$$

Thus

$$E[T_{CRi}] = \frac{1}{\lambda} \sum_{k=1}^{n} (-1)^{k+1} \frac{\binom{n}{k}}{k} + \frac{1}{\lambda} \sum_{k=1}^{n} (-1)^{k+1} \frac{\binom{n-1}{k-1}}{k^2}.$$

By integrating $(1-x)^n = \sum_{k=0}^{n} (-1)^k \binom{n}{k} x^k$ two times we can show that

$$E[T_{CR_i}] = (1 + \frac{1}{n}) H_n \text{ and } I_{CR}(i) = \frac{1}{n}. \tag{7}$$

Next we consider system C under the same assumptions. We now obtain from theorem 2

$$E[T] = \frac{1}{\lambda} \sum_{k=1}^{n} \sum_{\ell=1}^{m} (-1)^{k+\ell} \frac{\binom{n}{k}\binom{m}{\ell}}{k+\ell}.$$

But

$$\sum_{k=1}^{n} \sum_{\ell=1}^{m} (-1)^{k+\ell} \frac{\binom{n}{k}\binom{m}{\ell}}{k+\ell} = \sum_{i=2}^{n+m} (-1)^i \frac{\sum_{k=1}^{i} \binom{n}{k}\binom{m}{i-k}}{i} = \sum_{i=2}^{n+m} (-1)^i \frac{\binom{n+m}{i} - \binom{n}{i} - \binom{m}{i}}{i}$$

$$= \sum_{i=1}^{n+m} (-1)^{i+1} \frac{\binom{n}{i} + \binom{m}{i} - \binom{n+m}{i}}{i} = H_n + H_m - H_{n+m}.$$

Thus

$$E[T] = \frac{1}{\lambda} (H_n + H_m - H_{n+m}) \tag{8}$$

Thus for an arbitrary component of the parallel system with n components

$$I_{WR}(i) = \frac{m}{(n+1)(n+m+1)(H_n + H_m - H_{n+m})}. \tag{9}$$

Next we calculate $E[T_{CRi}]$, where i is one of the components of the left parallel system of system C. Thus we consider system D. We obtain this expectation from theorem 2

$$S_{k,\ell} = \frac{1}{\lambda} \{ \binom{n-1}{k}\binom{m}{\ell} \cdot \frac{1}{k+\ell} + \binom{n-1}{k-1}\binom{m}{\ell}(\frac{1}{k+\ell} + \frac{1}{(k+\ell)^2}) \} =$$

$$= \frac{1}{\lambda} \binom{m}{\ell} \cdot \{ \frac{\binom{n}{k}}{k+\ell} + \frac{\binom{n-1}{k-1}}{(k+\ell)^2} \} .$$

Thus

$$E[T_{CRi}] = \frac{1}{\lambda} \sum_{k=1}^{n} \sum_{\ell=1}^{m} (-1)^{k+\ell} \frac{\binom{n}{k}\binom{m}{\ell}}{k+\ell} + \frac{1}{\lambda} \sum_{k=1}^{n} \sum_{\ell=1}^{m} (-1)^{k+\ell} \frac{\binom{n-1}{k-1}\binom{m}{\ell}}{(k+\ell)^2} .$$

This can be simplified to

$$E[T_{CRi}] = \frac{1}{\lambda} \{ (1+\frac{1}{n})H_n + H_m - (1+\frac{1}{n+m})H_{n+m} \} \tag{10}$$

It follows that

$$I_{CR}(i) = \frac{\frac{1}{n} H_n - \frac{1}{n+m} H_{n+m}}{H_n + H_m - H_{n+m}} \tag{11}$$

3. Systems in random shock environment

We will in this section consider systems in random shock environment. Shocks are assumed to be generated by a renewal point process and we will measure times to failure of components and systems in terms of number of shocks to failure.

First we consider system A assuming that the components when hit by a shock independently of each other fail with probability $p=1-q$. We can then find $E[T]$ from (1) with $S_k = \binom{n}{k}/(1-q^k)$, because the expected time to failure of a series system of k such components is $1/(1-q^k)$. Thus

$$E[T] = \sum_{k=1}^{n} (-1)^{k+1} \frac{\binom{n}{k}}{1-q^k} = \sum_{i=0}^{\infty} (1 - (1-q^i)^n) \tag{12}$$

This result is derived with other methods in (2).

Next we consider system C under similar assumptions but with failure probability $p_1 = 1-q_1$ for the components of the left system and failure probability $p_2 = 1-q_2$ for the components of the right system. In this case we obtain $E[T]$ from (2) with

$$S_{k,\ell} = \binom{n}{k}\binom{m}{\ell}/(1-q_1^k q_2^\ell).$$

Thus

$$E[T] = \sum_{k=1}^{n} \sum_{\ell=1}^{m} (-1)^{k+\ell} \frac{\binom{n}{k}\binom{m}{\ell}}{1-q_1^k q_2^\ell} = \sum_{i=0}^{\infty} (1-(1-q_1^i)^n)(1-(1-q_2^i)^m). \tag{19}$$

Next we consider systems with 3-state components with three possible states: 0,1 and 2. These states can be described as follows.

0: good   1: degraded   2: failed.

We will assume that when a component in state 0 is hit by a shock it will pass to state 1 with probability $p_1 = 1-q_1$ and when in state 1 it will pass to state 2 with probability $p_2 = 1-q_2$, $p_1 \neq p_2$. The components pass or stay in states independently of each other. Let T be the number of shocks until failure of one such component. Then

$$P(T > i) = q_1^i + \sum_{j=0}^{i-1} q_1^j p_1 q_2^{k-1-j} = \frac{p_1 q_2^i - p_2 q_1^i}{q_2 - q_1}. \tag{13}$$

It follows that the expected number of shocks until failure of a series system of k such components is given by

$$\sum_{i=0}^{\infty} P(T>i)^k = \frac{1}{(q_2-q_1)^k} \sum_{i=0}^{k} (-1)^i \binom{k}{i} \frac{p_1^{k-i} p_2^i}{1-q_1^i q_2^{k-i}} \tag{14}$$

Thus for system A made up of 3-state components of this type we obtain from theorem 1

$$E[T] = \sum_{k=1}^{n} (-1)^{k+1} \binom{n}{k} \frac{1}{(q_2-q_1)^k} \sum_{i=0}^{k} (-1)^i \binom{k}{i} \frac{p_1^{k-i} p_2^i}{1-q_1^i q_2^{k-i}} = \tag{15}$$

$$\sum_{i=0}^{\infty} \{1 - (\frac{p_1(1-q_2^i)-p_2(1-q_1^i)}{q_2- q_1})^n\} \tag{16}$$

Most of the results in this paper can be generalized to k-out of-n-systems.

References
(1)   Råde, L., (1989) Expected time to failure of reliability systems. Math. Scientist 14, pp. 24-37.
(2)   Råde, L., (1976) Reliability systems in random environment. J. Appl. Prob. 13, pp. 407-410.

Lennart Råde
Mathematics Department
Chalmers University of Technology
S-412 96 Gothenburg

APPLICATION OF MODEL PREDICTIVE ADAPTIVE CONSISTENCY CONTROL IN TMP-MILL

Kari Saarinen
ABB Strömberg Process Automation

1. INTRODUCTION

In thermo-mechanical pulping, TMP, and chemical thermo-mechanical pulping, CTMP, processes, variations in chip size, density and moisture will all adversely affect pulp quality and refiner operation. Average refiner motor loads are typically reduced in order to prevent accidental overloads due to these process disturbances. Since neither the variations nor their effect on pulp quality can be measured fast enough for feedback or feedforward control, a novel advanced control approach is essential.

To solve this problem we have developed an on-line high consistency sensor to measure the pulp consistency just after the refiner and a control algorithm to compensate the mass flow variations.

The consistency measurement is based on infra-red light reflection directed in pulp stream. Advanced software for the calculation of consistency and sensor diagnostic is integrated in the control system.

The control system was not easy to put in practice because the process is non-linear and behaves in very different ways depending on, e.g., the refining gap, plate age and production rate. In order to run the process near its limits, the control system must react to the disturbances sufficiently fast. Because of the changes in process dynamics and static process gains, PI-controllers turn out not to fulfil these demants. That is why we decided to use adaptive controllers.

The basic element in our control system is the consistency control loop. In this loop we use model predictive controller where the process model is continuously identified.

2. PROCESS MODELL

Model predictive control requires a process model for predicting future process values. In literature there are models based on physics which describe the refining process, but they are usually stationary models and too much simplified to be used in this context.

In paper industry there are doubts about the ARMAX types of models, because the order of the process modell, the dead time of the process and almost anything may change depending on running conditions.

We have chosen an incremental type of process model

$$(1) \qquad \Delta y(t) = H(q^{-1})q^{-d-1}\Delta u(t) + \frac{C(q^{-1})\Delta\sigma(t)}{D(q^{-1})}$$

M. Heiliö (ed.), Proceedings of the Fifth European Conference on Mathematics in Industry, 355–359.

where d is the dead time of the process, q^{-1} is delay operator

$$q^{-1}f(t) = f(t-1)$$

and

$$H(q^{-1}) = h_1 + h_2 q^{-1} + , \ldots , + h_n q^{-n-1}$$
$$C(q^{-1}) = 1 + c_1 q^{-1} + , \ldots , + c_n q^{-n}$$
$$D(q^{-1}) = 1 + d_1 q^{-1} + , \ldots , + d_n q^{-n} \quad ; \quad n_h + n_c + n_d = m$$

y = consistency after refiner, σ = white noise
u = output to dilution water flow valve, $\Delta f(t) = f(t) - f(t-1)$

A major advantage of above the model (1) is that it is easy to
understand because the coefficients h_i are simply the process values
$y(t+d+1)$ $(i,t=1,2,\ldots)$ for a unit impulse at $t=0$. Another advantage is
that it can obtained without requiring knowledge of the system order.
The disadvantage is that the number of coefficients is larger than in
normal ARMAX model.

3. ON-LINE IDENTIFICATION

The parameters Θ are updated at every sampling interval, using the
following recursive identification method and stocastic Gauss-Newton
searching direction /1/:

$$(2) \qquad \Theta(t+1) = \Theta(t) + K(t)\varepsilon_l(t)$$
where

$$K(t) = P(t-1)G(t)[G(t)^T P(t-1)G(t) + \gamma(t)]^{-1}$$
$$P(t) = [I - K(t)G(t)]P(t-1)/\gamma(t)$$

In practice the following limitation seems to be quite important:

$$(3) \qquad \varepsilon_l(t) = \begin{cases} -\alpha & ; & \varepsilon(t) < -\alpha \\ \varepsilon(t) & ; & |\varepsilon(t)| < \alpha \\ \alpha & ; & \varepsilon(t) > \alpha \end{cases}$$

The $\varepsilon(t)$ is one step prediction error

$$\varepsilon(t,\Theta) = \Delta y(t) - \Theta^T(t)\Phi(t,\Theta)$$
where

$$\Theta^T(t) = [h_1, h_1, \ldots, h_n, d_1, d_2, \ldots, d_n, c_1, c_1, \ldots, c_n]$$
is the parameter vector and

$$\Phi^T(t) = [\Delta u(t-d-1), \Delta u(t-d-2), \ldots, \Delta u(t-d-n_h), \Delta v(t-1),$$
$$\ldots, \Delta v(t-1-n_d), \Delta \sigma(t-1), \Delta \sigma(t-2), \ldots, \Delta \sigma(t-n_c)]$$
is the data vector. $\gamma(t)$ is the weighting factor.

In this case the gradient G(t) is

(4) $\quad\quad\quad G(t) = [\; \Delta\tilde{u}(t),\ldots,\Delta\tilde{u}(t-n_h+1),\; -\Delta\tilde{v}(t),\ldots,-\Delta\tilde{v}(t-n_d-1),$
$\quad\quad\quad\quad\quad \Delta\tilde{\sigma}(t),\ldots,\Delta\tilde{\sigma}(t-n_c+1)]$
$\quad\quad \Delta\tilde{u}(t) = \Delta u(t-d-1) + d_1(t)\Delta u(t-d-2)+,\ldots,+d_{n_d}(t)\Delta u(t-d-1-n_d)$
$\quad\quad\quad\quad -c_1(t)\Delta\tilde{u}(t-1)-,\ldots,\; -\; c_{n_c}(t)\Delta\tilde{u}(t-n_c)$
$\quad\quad \Delta\tilde{v}(t) = \Delta v(t) - c_1(t)\Delta\tilde{v}(t-1)-,\ldots,-c_{n_c}(t)\Delta\tilde{v}(t-n_c)$
$\quad\quad \Delta\tilde{\sigma}(t) = \Delta\sigma(t) - c_1(t)\Delta\tilde{\sigma}(t-1)-,\ldots,-c_{n_c}(t)\Delta\tilde{\sigma}(t-n_c)$

The setpoint changes either in production rate, in housing pressure or in refining gap will produce a stepwise load disturbance in pulp consistency. The difficulties associated with using parametric adaptive control in the presence of load disturbance are well known. Various methods have evolved to cope with this problem. To cater for a load disturbance it has been common to include an estimation of a DC value in addition to the estimation of dynamic parameters. To cater for time varying dynamics an exponential forgetting factor is commonly used.

Because we use an incremental model there is no need to estimate the DC value. We calculate the exponential forgetting factor γ using the Fortescue /2/ method.

4. CONTROL ALGORITHM (a Model Predictive Controller /3/)

Let Y be the vector of the predicted process values, ΔU the vector of the future control outputs and S the vector of future setpoints $(Y, \Delta U, S\ R^N)$. Let us define vector Z

$$Z^T = [\; z_1, z_2, \ldots, z_N],$$

where the components are calculated in the same way as in /3/, using only the present measured value $y(t)$ and past data $\Delta y(t-1), \Delta y(t-2), \ldots, \Delta u(t-1), \Delta u(t-2), \ldots$ In addition to the calculations in /3/ we use also the noise model.

Let us also define matrix G in the following way:

$$G_{ij} = r_{i+1-j} \quad \text{for } i<j; \quad G_{ij} = 0 \text{ for } i>j; \quad i,j = 1,\ldots, N$$
$$r_0 = 0; \quad\quad\quad r_i = r_{i-1} + h_i$$

Now the predicted process values can be written in convenient vector-matrix notations as follow:

$$Y = G\Delta U + Z$$

The optimum control ΔU that minimizes the objective function

$$L = (S-Y)^T W(S-Y) + \Delta U^T Q\Delta U$$

is

$$\Delta U = (G^T WG + Q)^{-1} G^T W(S-Y).$$

W and Q are diagonal non-negative weighting matrices:

W=diag(w_1, w_2, w_3,..., w_N); Q=diag(q_1, q_2, q_3,..., q_N)

Because the matrix $(G^T WG + Q)^{-1}$ has a special structure, the inversion can be calculated recursively as in /3/.

5. TEST RESULTS

The consistency control has been tested on several TMP and CTMP mills in Finland and in Norway. In Kyröskoski, Finland (mill I), the system has been working over two years. Other tests were carried out between 1989 - 1990 (mills II - IV).

At first we had problems in mill I. Controller had to be re-initialized once per month. After adding the limitation (3), we have not had that kind of problems any more.

Off-line idenfication studies were carried out at different operation points in mill I. The results (figure 1) showed that a change in production rate or in refining gap lead into significant changes in gain.

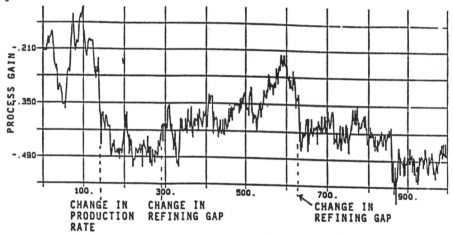

Figure 1. Static process gain in an experiment in mill I.

In figure 2 it can be seen that a carefully tuned PID-controller is some times almost as good as an adaptive controller, but when the process is changed the results are quite bad.

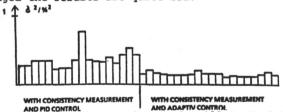

Figure 2. Consistency variances in mill I. Each bar corresponds 8 hour variance.

In table 1 shows a comparision of different control methods in different mills.

Table 1. Consistency and refining power variances in different mills with different control methods.

Mill/Test	Variance		Control method
	refining power [MW^2]	consistency [%^{2}]	
II/1-3	0.0083-0.0105	1.75-2.76	Specific energy
II/4-5	0.0046-0.0072	0.95-2.31	Manual
II/6-7	0.0026-0.0037	0.49-0.64	Adapt.
III/1	0.036	0.50	PID
III/2	0.053	1.00	PID
III/3	0.019	0.25	Adapt.
III/4	0.008	0.25	Adapt.
IV/1-3	0.0150-0.0360	2.13-2.34	Manual
IV/4	0.90	9.82	Manual
IV/5-8	0.0029-0.0140	0.27-0.57	Adapt.

Table I shows that there are consitency variations in TMP and CTMP pulp after the refiner. Also one can see that variances of refining power and consistency are at least twice larger with PID control than with adaptive control.

In mill III we made also lot of pulp quality tests. They showed that the consistency control reduced the freeness variance from 784 ml^2 to 289 ml^2 and long fiber variance from 81 %^{2} to 16 %^{2} compared to conventional methods.

REFERENCES

/1/ Ljung, L. and Söderström, T. (1983) Theory and practice of Recursive Identification; MIT press, Cambridge, Mass.

/2/ Fortescue, T.R., Kershenbaum, L.S. and Ydstie, B.E. (1981) Implementation of self tuning regulators with variable forgetting factors; Automatica 17 p 831.

/3/ Tung, L. (1983) American Control Conference, American Control; Green Valley, Ariz., p 349.

K.Saarinen
ABB Strömberg Drives OY
Process Automation
P.O. Box 606
SF-65101 Vaasa Finland

SHAPE DESIGN FOR MAGNETIC SHIELDING

Jukka Sarvas and Gösta Ehnholm

1. INTRODUCTION

In this work a magnetic shield was designed for a Magnetic Resonance Imaging device which is a hospital instrument for making cross-sectional planar images of patients. The mathematical problem involves a magnetostatic field computation for a magnetizing thin obstacle and a shape design task.

For the direct field problem a fast algorithm was developed based on a thin sheet approximation with infinite magnetic permeability. This approximation makes the problem an exterior Dirichlet problem for the magnetic potential.

The shape design problem is the following one. In an MRI-device a very strong and, in the imaging area, extremely homogeneous magnetic field is needed. This is obtained by a proper coil design. However, the device must be shielded (from outer noise and other reasons) by an iron shield. The shield reflects the primary field and disturbs the homogeneity of the total field. Therefore, the shape of the shield must be designed so that the reflected magnetic field is also homogeneous in the imaging area. This is a non-linear inverse problem. It was solved by developing an appropriate optimizing algorithm using the above fast field computation and a spherical harmonics expansion of the reflected field. Using this algorithm, an interactive computer program was constructed, with helpful graphics. It enables the user to design a shielding of practical size and shape, and satisfying the above reflection condition.

2. MAGNETOSTATIC FIELD PROBLEM

The direct field problem consists of computing the magnetostatic field for a given primary coil system and a given iron shielding, both being axially symmetric about the z-axis, see Figure 1. The major difficulty is the presence of magnetizable iron, which makes the field problem non-linear and numerically unstable. Furthermore, a fast field computing algorithm was needed in order to make the shape design task feasible for a 386-processor microcomputer. These problems were overcome by assuming that the shielding is thin (thickness less than a few centimeters) and with

M. Heiliö (ed.), Proceedings of the Fifth European Conference on Mathematics in Industry, 361–365.
© 1991 *B. G. Teubner Stuttgart and Kluwer Academic Publishers. Printed in the Netherlands.*

infinite magnetic permeability (in practice, the relative permeability more than 5000 is sufficient). This leads to the following formulation of the field problem.

The total magnetic field $\vec{H}$ consists of the primary field $\vec{H}_{pr}$, due to the primary coils, and of the secondary magnetic field $\vec{H}_m$ due to the magnetization in the shielding. Accordingly, $\vec{H} = \vec{H}_{pr} + \vec{H}_m$. Because the shielding is thin and can be considered as a surface S, $\vec{H}_m$ arises from a surface density q of (equivalent) magnetic charge. In other words, $\vec{H}_m = -\operatorname{grad} v_m$ with

$$(1) \qquad v_m(\vec{r}) = \frac{1}{4\pi} \int_S \frac{q(\vec{r})}{|\vec{r} - \vec{r}'|}\, da(\vec{r}),$$

which is the scalar potential of $\vec{H}_m$. Outside the coil system, also the primary field $\vec{H}_{pr}$ arises from a scalar potential v_{pr}, which we assume to be known. In fact, for a given coil system v_{pr} can be computed in a fast way using elliptic integrals, see [1]. We assume that both the coil system and the shielding are also symmetric about the xy-plane. Due to this symmetry, we can assume that v_{pr} and v_m vanish in the xy-plane. Because $\vec{H} = \vec{H}_{pr} + \vec{H}_m$, the total field $\vec{H}$, outside the coil system, has a scalar potential $v = v_{pr} + v_m$ with $\vec{H} = -\operatorname{grad} v$.

Because the magnetic permeability $\mu = \infty$ in the shielding, $\vec{H}$ is perpendicular to S due to the magnetic boundary conditions. This with the equation $\vec{H} = -\operatorname{grad} v$, implies that v is constant on S, and therefore, $v = 0$ on S, because $v = v_{pr} + v_m = 0$ in the xy-plane. So $v_m = -v_{pr}$ on S, and with (1) this implies

$$(2) \qquad \frac{1}{4\pi} \int_S \frac{q(\vec{r})}{|\vec{r} - \vec{r}'|}\, da(\vec{r}') = -v_{pr}(\vec{r})$$

for all $\vec{r} \in S$. This is the wanted surface integral equation for the unknown function q. We solved it using a boundary element method.

3. NUMERICAL SOLUTION OF THE SURFACE INTEGRAL EQUATION

Due to the axial symmetry the problem essentially is 1-dimensional. Let the curve C be the upper part of the intersection of S and the xz-plane. Then $q = q(t)$ essentially is a function of one variable t with t being the curve length on C. We approximate q by a piecewise linear approximation and discretize it so that at the nodal points $p_1, ..., p_{2n+1} \in C$ we have: $q_i = q(p_i)$, $i = 1, ..., 2n+1$, and between the nodal points $q(t)$ is linear. Because of the symmetry about the xy-plane, we may assume that $q_{2n+2-i} = -q_i$, $i = 1, ..., n$, and $q_{n+1} = 0$. We substitute this piecewise linear approximation of q into the equation (2) and demand that the resulting equation is satisfied at the matching points $u_1, u_2, ..., u_{2m+1} \in C$. This is a point collocation method, and it yields a matrix equation for the unknowns $q_1, ..., q_n$, of the form

$$(3) \qquad\qquad Fq = -v_{pr}$$

where $q = [q_1, ..., q_n]^T$ and $v_{pr} = [v_{pr}(u_1), ..., v_{pr}(u_m)]^T$. We overdetermined this equation by taking $m > n$. In particular, we distributed the matching points u_i more densely near the ends of C than near its center.

To obtain the elements of the matrix F, we for each u_i have to compute the integrals

$$(4) \qquad\qquad I_{ij} = \frac{1}{4\pi} \int_{S_j} \frac{q(\vec{r}')}{|u_i - \vec{r}'|} da(\vec{r}'), \qquad j = 1, 2, ..., 2n,$$

where S_j is the conical surface obtained by rotating J_j around the z-axis where J_j is the segment of line from p_j to p_{j+1}. On J_j the function $q(t)$ varies linearly from q_j to q_{j+1}. In the integral I_{ij} the circular part of the integration is easily recognized to yield an elliptic integral of the first kind and, accordingly, it can be computed by a very fast algorithm, see [2, sec.17.6]. The remaining integral along J_j was computed numerically by Romberg's rule. If the point u_i lies in the segment J_j, a logarithmic singularity occurs in the integral (4). This can numerically be treated in a fast way by the asymtotic behavior of the elliptic integrals, see [2, eq. 17.3.26].

Eventually, we solve the equation (3) for q. Altogether, the above procedure yields a fast algorithm by which we can solve q, for example, in a 386-microcomputer in less than 10 seconds for $n = 30$ and $m = 50$. After obtaining q, we get v_m from (1) and $\vec{H}_m$ from the equation: $\vec{H}_m = -\text{grad}\, v_m$. However, for the shape design problem, as it turns out, we only need q.

4. THE SHAPE DESIGN OF THE SHIELDING

For a given primary coil system we want to find the shape of the shielding S, i.e., the shape of its profile curve C, so that $\vec{H}_m$ is homogeneous, that means nearly constant, in the neighborhood of the origin. It is not difficult to show that it suffices to require that the z-component H_{mz} of $\vec{H}_m$ is homogeneous there. In fact, the requirement in our case was that

(5) $|H_{mz}(\vec{r}) - H_{mz}(0)|/|H_{mz}(0)| \leq 10^{-5}.$

for $|\vec{r}| \leq 0.25m$. To find an appropriate method for obtaining the wanted shape, we
expand H_{mz} in spherical harmonics at the origin:

(6) $H_{mz}(r, \theta) = \sum_{n=0}^{\infty} a_{2n} r^{2n} P_{2n}(\cos \theta),$

where $P_m(x)$, $-1 \leq x \leq 1$, $m = 0, 1, 2, ...$, are the Legendre polynomials. Note that
due to the symmetries of the coil system and the shielding, only even terms are
present in the above expansion. Using the generating function $(1 - 2xt + t^2)^{-1/2} =$
$\sum_{n=0}^{\infty} P_n(x)t^n$, $|x| < 1$, $|t| < 1$, for the Legendre polynomials, it is not difficult to
derive the following expression for the cofficients a_n in (6) directly in terms of the
surface density q:

(7) $a_m = -\frac{m+1}{2} \int_C q(\vec{r}) \frac{\sin(\theta(\vec{r}))}{|\vec{r}|^{m+1}} P_{m+1}(\cos \theta(\vec{r})) \, dl(\vec{r}),$

$m = 0, 1, 2, ...$, where $\theta(\vec{r})$ is the angle between the positive z-axis and the vector $\vec{r}$.

The complete homogeneity of H_{mz} is equivalent to the requirement that $a_{2n} = 0$
for all $n = 1, 2, 3, ...$, in (6). However, due to the tolerance given by (5) the sufficient
homogeneity demand is as follows:

(8) $|a_{2n}| \leq 4^{2n} \cdot 0.72, \quad n = 1, 2, 3, 4.$

In order to force the above conditions, we described the shape of the shielding in
terms of 8 shape parameters $(x_1, x_2, ..., x_8) = x$. The profile curve C is taken to be a
broken line with segments parallel to the coordinate axes. The parameters $x_1, ..., x_8$
describe the lengths and locations of these segments. Accordingly, the density q as
a function of the shape is also a function of x, and this dependence is numerically
found by the algorithm which computes q for a given shielding. Furthermore, due to
(7), the coefficients a_n depend on q, and so eventually, on x; the integration in (7)
was numerically performed by Romberg's rule. Now, we can force the conditions (8)
by minimizing the cost function $f(x) = \sum_{n=1}^{4} |a_{2n}(x)|^2$, where $x = (x_1, ..., x_8)$ runs
through reasonable shapes of the shielding. The required minimizing was carried
out by Marquardt's method [3].

For the practical shape design an interactive computer program was constructed
with the following features: the user control on the number of steps and the size of

the reqularization parameter in Marquardt's method, the possibility to freeze some of the current shape parameters for subsequent minimizing steps, and illustrative graphics to show the current shape with information on the cost function and other relevant facts. Figure 1 shows one of the shieldings designed by this program.

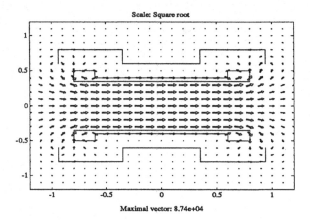

Figure 1: *The total magnetic field $\vec{H}$ due to the primary coil system (the inner cylinder) and the shielding (the outer cylinder). The shielding satisfies the homogeneity conditions (8).*

References:

[1] **Garrett, M. W.**: Calculation of Fields, Forces, and Mutual Inductances of Current Systems by Elliptic Integrals. – J.of Appl.Phys. Vol.34, No.9, 1963.

[2] **Abramovitz, M.** and **Stegun, I.**: Handbook of Mathematical functions. – Dover Pub. Inc., New York, 1973.

[3] **Press, Flannery, Teukolsky** and **Vetterling**: Numerical Recipes. – Cambridge University Press, Cambridge, 1986.

Jukka Sarvas
Rolf Nevanlinna -institute
Teollisuuskatu 23, Helsinki, Finland

Gösta Ehnholm
Instrumentarium Imaging Co
Teollisuuskatu 27, Helsinki, Finland

MATHEMATICAL CONTINUING EDUCATION
– A NEW TASK FOR UNIVERSITY

Marion Schulz–Reese

It is an extremely important task for mathematics to find ways to meet the requirements of an industrial and technological society.

There are different possibilities to fill up the gap, which undoubtedly is between university mathematics and the demand of industry.

First, mathematics has to realize that real–world problems are interesting subjects of mathematical research and that research cooperation between university and industry can show academic and practical profit.

Second, industry needs highly qualified mathematicians. University has to take into account that most of the graduates in mathematics will work their way up in industry and that they need an application–oriented education.

Third, a few years of full–time university education can no longer provide an adequate foundation for a lifetime of professional work. The progress in science and the rapidly changing requirements of industry call for a continuous updating of knowledge, as well in mathematics. Therefore mathematics must add continuing education to its charters. Mathematical continuing education is a very new field within the spectrum of university actvities. There are not many experiences in developing and organizing training courses for industrial staff. University has to answer a lot of questions to be successful in this field. Is there really a demand for mathematical continuing education? Which mathematical topics should be offered? What are the methodological and organizational aspects that should be taken into account?

Within the last years the Laboratory of Technomathematics at the University

M. Heiliö (ed.), Proceedings of the Fifth European Conference on Mathematics in Industry, 367–370.
© 1991 B. G. Teubner Stuttgart and Kluwer Academic Publishers. Printed in the Netherlands.

of Kaiserslautern has gained some experiences in the field of mathematical continuing education which can give some hints for further activities.

There is no doubt that nowadays mathematics plays an important role in industrial areas like research, development or production. Numerous interviews especially with engineers have confirmed the thesis that there is a demand for further mathematical training. It is true that there exist today a great variety of continuing education programmes, spanning a broad spectrum of disciplines, levels, intensities, and objectives. They are offered by industrial organizations, by professional societies, and by private firms specializing in adult educations. And there is a lot of money in the system (f.e. German industry spends 26 Billion DM per year only for intramural continuing education programmes). But there are only very few organizations offering courses in mathematics. Could this be an opening for university activities? Yes, but only if it will be done in a "professional" way. This means, first of all, to do an intensive demand assessment. Such an inquiry can be done in different ways, f.e. in form of interviews or in form of questionnaires. What matters is to pose the "right" questions. It is useless to ask for the need of special mathematical topics like hyperbolic differential equations or stochastic processes. It is better to obtain a general view by asking for present—day R & D activities, the use of software packages and mathematical methods, the problems in applying mathematics. And then university has to find out which kind of mathematics is useful for which problems and has to deduce which topics should be offered in continuing education. Our experiences have shown that there is a demand for continuing education on different levels and with different objectives. According to the objectives three types of mathematical continuing education can be defined (cp. [1], [2]).

1. **Basis—oriented** continuing education: Here the aim is to compensate deficits of former educational processes. This means to renew, extend or deepen mathematical knowledge and competences, to refresh or to add mathematical contents of the former university education or to fit vocational and professional knowledge into a scientific framework. Engineers in industry often use software packages without knowing the limits or the advantages of the implemented mathematical methods. Topics for this kind of continuing education are f.e. Mathematical methods of standard software packages, Numerical Methods for ODE's, Fourier analysis or Time series analysis.

2. **Demand oriented** continuing education: This type has to react on special, well formulated needs of different industrial fields. Industry has mathematical problems and wants to solve them. The task of university is to make available and transparent the right mathematical methods. Topics are f.e. Boundary elements, Nonlinear control, Numerical methods for vector/parallel computing.

3. **Supply—oriented** continuing education: Maybe this is the most exciting and challenging type of mathematical continuing education. Topics for it cannot be found at lists of wishes from industry. University is challenged to make an offer, to look for mathematical methods and theories (f.e. Wavelet analysis or Multiscale analysis) that promise successful applications but which cannot be seen or shown completely. Maybe in this way university can stimulate new ways for industrial problem solving or at least can give some new ideas.

There is a broad spectrum of possible organizational forms for mathematical continuing education: Lectures, workshops, short— or long—term courses, part—time courses, distance learning programmes, sequences of studies with modular structure etc. A very point is that continuing education will require the development of new educational tools and methodologies to fit the more mature and experienced audience. Mathematical continuing education will require breaking new educational grounds. Mathematicians are now being called upon to extend still further the scope of their activities to meet new needs of profession and of society arising from the galloping pace of technological innovation and the emergence of the knowledge intensive industry as a critical segment of the economy..

References:

[1]   Neunzert, H.: Continuing education in Mathematics — a way to discover and use "raw—material" mathematics. ECMI—Newsletter No.6, 1989

[2]   Schulz—Reese, M.: Mathematische Weiterbildung, Handlungsstrategien und Konzepte für eine neue Aufgabe der Mathematik. Erich Schmidt Verlag, Berlin 1990

Marion Schulz—Reese

Laboratory of Technomathematics

Department of Mathematics

University of Kaiserslautern

D—6750 Kaiserslautern

COMPUTER-AIDED QUALITY CONTROL OF THE TAPE RECORDER RECORDING-REPRODUCING CHANNEL

V. Šimonytė, V. Slivinskas

Abstract. The problem of digital measurement of the main parameters of the recording-reproducing channel (RRC) of a tape recorder which occurs in computer-aided quality control (QC) is considered. The difficulties one encounters in developing a QC system of the tape recorder RRC are discussed. The main methods used in software of the QC system developed are briefly described and compared with some other methods.

Keywords. Tape recorders; parameters of the recording-reproducing channel; quality control system; digital measurement methods.

INTRODUCTION

Estimation of the parameters of tape recorders is a complex and hard process which requires various special devices and much time. Measurement of the characteristics of tape recorders in industry is still connected with manual labour lessening the efficiency of QC of the equipment. With the aim of improving the measurement accuracy and reliability, different devices and instruments (voltmeters, generators, oscilloscopes, gauges of nonlinear distortions, etc) are used to measure the same characteristics. The process of generation of test signals also takes plenty of time and energy. Thus, the necessity of automation of QC of tape recorders arises.

This paper deals with methods of digital measurement of the magnitude response (MR) of the tape recorder RRC only, although the system developed also measures some other parameters as, for example, the coefficients of harmonics, detonation, and parasitic amplitude modulation.

DIGITAL MEASUREMENT OF THE MAGNITUDE RESPONSE

Assume that the initial data for the measurement of the MR is a finite sequence of the equidistant samples of the impulse response of the RRC. In practice

371

M. Heiliö (ed.), Proceedings of the Fifth European Conference on Mathematics in Industry, 371–374.
© *1991 B. G. Teubner Stuttgart and Kluwer Academic Publishers. Printed in the Netherlands.*

this sequence is obtained by passing through the analog- digital converter the output of the RRC which is the response of the channel to a sufficiently narrow rectangular pulse. Denote this sequence by

$$g_0, g_1, \ldots, g_{N-1} \tag{1}$$

where $g_i = g(i\Delta t)$, $i = \overline{0, N-1}$, g be the impulse response of the RRC, Δt the sampling interval.

The simplest way to get the estimate of the MR is to calculate the discrete Fourier transform of the sequence (1)

$$\widehat{H}(e^{j\omega\Delta t}) = \Delta t \sum_{i=0}^{N-1} g_i e^{-j\omega i \Delta t} \tag{2}$$

where $\omega \in \{m\Delta f \mid \Delta f = 1/(N\Delta t), \ m = 0, 1, \ldots, N-1\}$ by the FFT algorithm [1]. The well-known property of the FFT is its computational efficiency which makes it to be a widely-used method in spectral analysis. In spite of its efficiency, however, this method has some shortcomings. One of them occurs as a result of extrapolation of data by zeroes which is used by the FFT. It is obvious that in the case when the data does not comprise all the transient part of the impulse response such an extrapolation distorts the real information of the data. As a consequence of this, we get the distorted MR.

To avoid this phenomenon, we use a method based on another extrapolation principle. It can be shown that the sequence of the impulse response of a discrete-time linear stationary system satisfies the difference equation with constant coefficients

$$g_{n+k} = \alpha_n g_{n+k-1} + \alpha_{n-1} g_{n+k-2} + \cdots + \alpha_1 g_k, \quad k = 1, 2, \ldots. \tag{3}$$

The number n is called the order of the equation. The so-called problem of minimal partial realization of a finite sequence of scalars $\{m_1, \ldots, m_N\}$ searches the minimal order difference equation of the type (3) which this sequence satisfies. There exist a number of algorithms to solve this problem [2]. Thus, if we have a finite

sequence of the equidistant samples of the impulse response of the tape recorder
RRC we can extrapolate it by the minimal partial realization method. So we get
the extrapolated sequence $g_0, g_1, \ldots, g_{N-1}, \tilde{g}_N, \tilde{g}_{N+1}, \ldots$. Then the estimate of the
MR can be found from the estimate of the frequency response:

$$\tilde{H}(e^{j\omega\Delta t}) = \Delta t(g_0 + g_1 e^{-j\omega\Delta t} + \cdots + g_{N-1}e^{-j\omega(N-1)\Delta t} +$$
$$+ \tilde{g}_N e^{-j\omega N\Delta t} + \tilde{g}_{N+1}e^{-j\omega(N+1)\Delta t} + \cdots). \tag{4}$$

The method based on minimal partial realization avoids the distortion caused
by the extrapolation by zeroes, and thus improves the accuracy of estimation in
compare with the FFT. The errors of measurement of the impulse response samples
can, however, lead to the unstable equation of extrapolation. An effective way
to prevent this shortcoming is the so-called formant analysis of the extrapolated
sequence. By a formant we mean a function of discrete or continuous time which
equals the product of an exponential function and a polynomial whose coefficients
vary in time by a sinusoidal law of the same frequency but of distinct amplitudes
and phases, i.e., a formant is a function of the following type:

$$e^{\lambda t}(A_1 \sin(2\pi ft + \varphi_1) + A_2 t\sin(2\pi ft + \varphi_2) + \cdots + A_k t^{k-1}\sin(2\pi ft + \varphi_k)) \tag{5}$$

where $k \in \mathbf{N}$, $\lambda \in \mathbf{R}$, $A_j \in \mathbf{R}$, $\varphi_j \in [-\pi, \pi]$. In the case when $f = 0$ and $\varphi_j = \pi/2$,
a formant becomes an ordinary polynomial.

It is not difficult to see that the impulse response of a linear stationary system
equals the sum of formants. Thus, in order to prevent instability of the extrapolation equation, one can replace it by another equation which is obtained by
approximating the data by the sum of the stable formants. Practice shows that
these unstable formants are of small amplitudes compared with those of stable
of stable ones. Therefore, the error of approximation is sufficiently small in most
cases.

Above we considered the problem of estimation of the MR when the initial
data was the equidistant samples of the impulse response of the RRC. However,
some other data can also be used, for example, samples of the reproduced white

noise. In that case the estimate of the MR is obtained from the estimate of the spectral density of the reproduced noise. This approach has a shortcoming since it lost the information about the phase response. Thus, it can be used only for audio tape recorders because human ear is relatively insensitive to the phase distortions.

CONCLUSIONS

In this paper we present the techniques of digital measurement of the MR of the tape recorder RRC based on the so-called formant analysis of signals. These techniques compound a part of the methods used in the software of the QC system of tape recorders. The main advantage of the approach suggested lies in avoidance of the distortions of the MR caused by aliasing and by extrapoliation of data by zeroes inherent to the FFT.

REFERENCES

1. Oppenheim, A.V., Willsky, A.S. and I.T. Young. Signals and Systems. *Prentice Hall*, 1983.

2. Slivinskas, V. Methods of minimal realization of time-invariant linear dynamic systems. *Statističeskyje problemy upravlenija*. Issue **33**, 1977. P. 31–74 (in Russian).

Institute of Mathematics and Informatics
of the Lithuanian Academy of Sciences
Akademijos 4
232600 VILNIUS
LITHUANIA

APPLICATION OF NEIMARK'S CRITERION TO A STABILITY PROBLEM FOR THE FLOW THROUGH A COLLAPSIBLE TUBE.

Krystyna M. Sobczyk, University of Mining and Metallurgy,
                     Mathematical Institute, Cracow, Poland

    John W. Reyn, Delft University of Technology, Department
                  of Technical Mathematics and Informatics,
                  Delft, The Netherlands.

1. Introduction.

A collapsible tube is an elastic tube characterized by the fact that it changes drastically in cross-sectional shape (and area) owing to small changes in transmural pressure (the difference between the internal and external pressure) close to zero. Flows through such tubes are of physiological interest, whereas also technical applications are known, such as in flexible cold-water pipes used in ocean-thermo-energy installations. There have been a large number of laboratory experiments to investigate the properties of collapsible tube with segments of rubber tubes, often spanned between rigid inlet and outlet tubes and contained in a pressurized chamber. Incompressible fluid flows along the tube from a constant-head reservoir, and the flow rate can be controlled by adjusting the resistances of the rigid parts of the system upstream and downstream of the collapsible segment. Since a common result in these experiments has been the appearance of self-excited and self-sustained oscillations in the tube cross-sectional area and the outflow velocity for a range of values of the governing parameters it may be asked whether a steady flow is possible under these conditions and what are its stability properties. In this paper, using very simplifying assumptions concerning the type of flow, the collapsible tube and its connections to the rigid upstream and downstream tubes is investigated whether such a flow is stable at subcritical speeds, that is at flow speeds smaller than the speed of waves on the fluid filled tube.

Consider therefore a segment of collapsible tube. cyclindrical in shape in unloaded condition and also cylindrical but not necessarily circular if an incompressible inviscid fluid flows through it at constant speed. The relation between cross-sectional area of the tube A and transmural pressure P is given by the "tube law":

$$P = P(A) \tag{1}$$

The motion of the fluid is governed by the momentum equation

$$\frac{\partial V}{\partial t} + V \frac{\partial V}{\partial x} + \frac{1}{\rho} \frac{\partial P}{\partial x} = 0 \tag{2}$$

and the continuity equation

$$\frac{\partial}{\partial x} (VA) + \frac{\partial A}{\partial t} = 0, \tag{3}$$

where V is the flow speed, ρ the fluid density, x distance along the tube and t is time. Let the flow through the tube be a little different from that through a cylindrical tube with area A_* and let V_* be the speed, ρ_* the density and P_* the pressure of the fluid in the cyclindrical tube, respectively. Put, moreover, $V = v + V_*$, $A = a + A_*$, $P = p + P_*$, then (1), (2), (3) yields upon linearization

M. Heiliö (ed.), Proceedings of the Fifth European Conference on Mathematics in Industry, 375–379.
© 1991 B. G. Teubner Stuttgart and Kluwer Academic Publishers. Printed in the Netherlands.

$$p = \frac{\rho_* C_*^2}{A_*} \, a, \tag{4}$$

$$\frac{\partial v}{\partial t} + V_* \frac{\partial v}{\partial x} + \frac{1}{\rho_*} \frac{\partial p}{\partial x} = 0, \tag{5}$$

$$V_* \frac{\partial a}{\partial x} + A_* \frac{\partial v}{\partial x} + \frac{\partial a}{\partial t} = 0, \tag{6}$$

where $C_* = \sqrt{\frac{A_*}{\rho_*} \frac{dP}{dA}(A_*)}$ is the wave speed of small amplitude pressure waves.

The collapsible tube is upstream and downstream connected to rigid pipes with area A_* and mounting forces are considered to be low enough so that they can be neglected, also if there are disturbances to the cylindrical flow. The pressure at the upstream point of the tube ($x = -\ell$) is determined by the pressure far upstream and the resistance and inertia of the fluid in the upstream rigid pipe; the pressure at the downstream point ($x = \ell$) is determined similarly.
This leads to the boundary conditions

$$p(-\ell) = \beta_1 \, v(\ell) + \gamma_1 \frac{\partial v}{\partial t} \, (-\ell), \tag{7}$$

$$p(\, \ell) = \beta_2 \, v(\ell) + \gamma_2 \frac{\partial v}{\partial t} \, (\, \ell), \tag{8}$$

where $\beta_1 < 0$, $\beta_2 > 0$ represent resistance and $\gamma_1 < 0$, $\gamma_2 > 0$ represent inertia effects in the rigid pipes. Applying these boundary conditions to solutions of (4)-(6) only makes sense for $V_* < C_*$ (subcritical flow). Solutions of (4)-(6) satisfying (7),(8) of the form

$$C_1 \exp \omega \, (\frac{x}{V_*-C_*} - t) + C_2 \exp \omega(\frac{x}{V_*+C_*} - t),$$

where C_1 and C_2 are constants lead to the characteristic equation

$$e^{\lambda\omega} = \frac{(\gamma_1\omega-\beta_1-\rho_*C_*)(\gamma_2\omega-\beta_2+\rho_*C_*)}{(\gamma_1\omega-\beta_1+\rho_*C_*)(\gamma_2\omega-\beta_2-\rho_*C_*)} \tag{9}$$

where $\lambda = 4\ell C_*(V_*^2-C_*^2)^{-1}$ and $\omega \in \mathbb{C}$. These solutions are damping out if re $\omega > 0$ and if all solutions of (9) satisfy re $\omega > 0$ the flow is stable. In this paper (9) is analyzed, using Neimark's criterion [2], which is checked by a method indicated in [3]. Another approach is presented in [1].

2. Neimark's criterion for a quasipolynomial.
For the quasipolynomial

$$Q(s,e^{\tau s}) = q_0(s) + q_1(s)e^{\tau s} \tag{10}$$

where $q_0(s) = q_{00} + q_{01}s + q_{02}s^2$

$q_1(s) = q_{10} + q_{11}s + q_{12}s^2 \qquad q_{ij} \in \mathbb{R}, \; \tau > 0$

Neimark's criterion states that if:
1. the polynomials $q_0(s)$, $q_1(s)$ have no common roots   (11)

2. $Q(0,1) \neq 0$ (12)

3. $\left|\frac{q_{02}}{q_{12}}\right| \leq 1$ (13)

then the roots of $Q(s,e^{rs}) = 0$ are located in the left half of the complex plane (re s<0) if and only if

$$N(0) + 2 \sum_{j=1}^{L} (-1)^j \left[\frac{\beta_j - ry_j}{2\pi} \right] = 0 \qquad (14)$$

Here $N(0)$ is the number of roots of the equation $Q(s,1)=0$ in the right half of the complex plane (re s>0). The pairs (y_j, β_j) are the positive solutions of the equation $Q(iy, e^{i\beta}) =0$ ordered decreasingly $y_1 \geq y_2 \geq \ldots \ldots y_L > 0$ and $\frac{1}{2\pi}(\beta_j - ry_j) \notin Z$, $_{j=1...L}$.

If $\left| \frac{q_{02}}{q_{12}} \right| = 1$ then it should be additionally assumed that $|\rho(y)| < 1$ for $y >$ max $\{y_j\}$. (It was shown in [4] that this condition is not sufficient unless it can be shown that the roots of the characteristic equation are simple).

The difficulty in using this criterion is in calculating the pairs (y_j, β_j). These pairs can be interpreted as the points of intersection of the testing path $\rho = \rho(y)$, defined by $Q(iy, \rho(y)) = 0$ for $y \in [0,\infty)$, with the unit circle in the complex plane.

In [3] it has been shown that the roots of the equations re $Q(iy,e^{i\beta})=0 \wedge$ im $Q(iy,e^{i\beta})=0$ can be obtained from:

$$\left.\begin{array}{l} x \text{ re } \{q_0(iy) - q_1(iy)\} - \text{im } \{q_0(iy) + q_1(iy)\} = 0 \\ x \text{ im } \{q_0(iy) - q_1(iy)\} + \text{re } \{q_0(iy) + q_1(iy)\} = 0 \end{array}\right\} \qquad (15)$$

$$\text{where } \beta = \begin{cases} 2 \text{ arc tan } x, & x \geq 0 \\ 2\pi - 2 \text{ arc tan } x, & x < 0 \end{cases} \qquad (16)$$

Eliminating x from (15) the condition

$$\begin{vmatrix} (q_{00}-q_{10}) + y^2(q_{12}-q_{02}) & -y(q_{01}+q_{11}) \\ y(q_{01}-q_{11}) & (q_{00}+q_{10}) - y^2(q_{12}+q_{02}) \end{vmatrix} = 0 \qquad (17)$$

is obtained what leads to a quadratic equation for y^2. Since also only y>0 is considered the testing path can intersect the unit circle at most in two points and for its location with respect to the unit circle also $\rho(0)$ and $\lim_{y \to \infty} \rho(y)$ should be determined.

3. Stability conditions for the flow through a collapsible tube mounted in a rigid hydraulic circuit.

In order to use Neimark's criterion and the remarks of the previous section we should put $\omega=-s$ in (9). Rewriting the equation (9) in the standard form (10) we then obtain for $V_* < C_*$

$$q_{02}=-\gamma_1\gamma_2, \quad q_{01}=-[\gamma_1\beta_2+\gamma_2\beta_1+\rho_*C_*(\gamma_2-\gamma_1)], \quad q_{00}=-[\beta_1\beta_2+C_*\rho_*(\beta_2-\beta_1)-(\rho_*C_*)^2]$$

$$q_{12}= \gamma_1\gamma_2, \quad q_{11}= [\gamma_1\beta_2+\gamma_2\beta_1-\rho_*C_*(\gamma_2-\gamma_1)], \quad q_{10}= [\beta_1\beta_2-C_*\rho_*(\beta_2-\beta_1)-(\rho_*C_*)^2] \qquad (18)$$

Now let us check the conditions in the Neimark's criterion.

In our case the polynomials $q_0(s)$ and $q_1(s)$ have no common roots (condition (11)) if

$$C_*\rho_*(\gamma_1-\gamma_2) \neq \pm (\gamma_1\beta_2-\gamma_2\beta_1) \tag{19}$$

Let us assume that (19) is satisfied.

Condition (12) is satisfied since $Q(0,1) = q_{00}+q_{10} = \pm 2\rho_*C_*(\beta_2-\beta_1)\neq 0$ according to (8).

In our case condition (13) is also satisfied since $\left|\dfrac{q_{02}}{q_{12}}\right| = 1$ what means that the characteristic equation (9) is of the neutral type.

In order to check condition (14) the testing path should be determined. Since $q_{02}=-q_{12}$, condition (17) gives

$$y^2 = - \frac{q_{00}^2 - q_{10}^2}{[q_{01}^2 - q_{11}^2 + 2q_{12}(q_{00}+q_{10})]} \tag{20}$$

and using (18) we obtain

$$y^2 = - \frac{[\beta_1\beta_2-(\rho_*C_*)^2](\beta_1-\beta_2)}{\gamma_1^2\beta_2-\gamma_2^2\beta_1} < 0 \tag{21}$$

what means that there are no real solutions y and the testing path has no intersection points with the unit circle.

The starting point of the testing path can be determined from the equation

$$0 = q_0(0) + q_1(0)\ \rho(0) \tag{22}$$

then we obtain

$$\rho(0) = \frac{1 - \dfrac{(\rho_*C_*)^2}{\beta_1\beta_2} - \dfrac{\beta_1-\beta_2}{\beta_1\beta_2}\ \rho_*C_*}{1 - \dfrac{(\rho_*C_*)^2}{\beta_1\beta_2} + \dfrac{\beta_1-\beta_2}{\beta_1\beta_2}\ \rho_*C_*} \tag{23}$$

what gives $|\rho(0)| < 1$ since $\beta_1\beta_2 < 0$ and $\beta_1-\beta_2 < 0$.

As a result the end of the testing path is situated on the unit circle since

$$\lim_{y\to\infty} - \frac{q_0(iy)}{q_1(iy)} = 1$$

So the testing path for the characteristic equation is situated inside the unit circle.

As was mentioned in section 2, for neutral type equations the condition that $|\rho(y)| < 1$ for y sufficiently large should be additionally checked.

Now we can check condition (14) which is reduced to $N(0) = 0$.

The equation $Q(s,1)=q_0(s)+q_1(s) = 0$ has only one root

$$s = - \frac{q_{00}+q_{10}}{q_{01}+q_{11}} = \frac{1}{V_*-C_*} \frac{\beta_1-\beta_2}{\gamma_1-\gamma_2} < 0 \qquad (24)$$

so $N(0) = 0$.

Since the characteristic equation is of the neutral type we should also check whether its roots are simple. For this purpose we determine the

polynomial $Q'(s,e^{\tau s}) = \frac{d}{ds} Q(s,e^{\tau s}) = q_0'(s) + [q_1'(s) + \tau q_1(s)] e^{\tau s}$ (25)
Introducing

$$Q^*(u) = \lim_{|s| \to \infty} \frac{1}{s^2} Q'(s,u) = \tau q_{12} u$$

from the solutions of the equation $Q^*(u) = 0$ we can find the limit of the sequence $\{re \, s_k\}$ (where s_k are the roots of the quasipolynomial (25) with $im \, s_k$ ordered increasingly), namely $\lim_{k=\infty} re \, s_k = \infty$. Hence the roots of (25) differ from the roots of $Q(s,e^{\tau s})$ what proves that the roots of the quasipolynomial $Q(s,e^{\tau s})$ are simple.

Now let us return to condition (19) and assume that $q_0(s)$ and $q_1(s)$ have a common root. Then we can write

$$Q(s,e^{\tau s}) = (s-s_c)[(q_{00}^1+q_{01}^1 s) + (q_{10}^1+q_{11}^1 s)e^{\tau s}] =$$

$$= (s-s_c) \, Q_1(s,e^{\tau s}) \qquad (26)$$

Then the stability condition follows from $re \, s_c < 0$ and the Neimark's criterion for the quasipolynomial $Q_1(s,e^{\tau s})$. The analysis of this case shows that also in this case there is stability.

4. Conclusions.

Under very simplyfying assumptions concerning the type of flow, the collapsible tube and its connections to the rigid upstream and downstream tubes it is shown by analyzing the characteristic equation using Neimark's criterion that the flow is stable for subcritical flow (flow speed smaller than speed of small amplitude pressure waves).

5. References.

[1].   Reyn, J.W. and Sobczyk, K.M., 1990, "A stability problem in the theory of flow through collapsible tubes", Delft Progress Report,14, pp. 3-20.

[2].   Neimark, Yu.I., 1949, "D-composition of the space of quasipolynomes", Prikl.Mat.i Mech. 13, no. 4, pp. 349-380.

[3].   Szymkat M. 1983, "Analytical stability criterion for a system of linear difference-differential equations of second order obtained by an elimination method". Proceedings of III International Conference on functional-differential systems and related topics, Blazejewko, Poland.

[4].   Datko, R., 1983, "An example of an unstable neutral differential equation", Int.Journ. of Control 38, no. 1, pp. 263-267.

MARANGONI CONVECTION IN A WELD POOL

M. Vynnycky

§1 Introduction

In the TIG (tungsten inert gas) welding process, an arc is struck between an inert electrode and the metal substrate, the thermal energy thus generated causing the base metal, usually steel or aluminium, to melt (fig. 1). The solidification of the molten metal in the weld pool forms the bond between the pieces of metal that are joined in the process. The heat input for the process is typically $O(10^3)$W and the current is $O(10^2)$A.

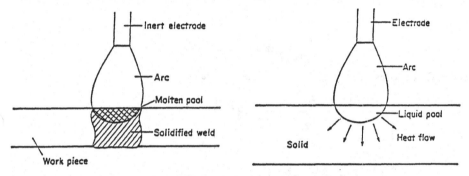

Fig. 1: Schematic sketch for the TIG welding process

For many years, the determination of weld pool shape rested solely on heat conduction models [1]. Later experimental work [2] demonstrated the existence of fluid flow in the pool, and it was shown that this could be due to buoyancy and the Lorentz force generated by the applied arc [3]. Most recently [4], attention has turned to the surface-tension (Marangoni) forces which arise at the pool surface as a result of the large temperature gradients induced by the arc source; in addition, the nature of these flows can be dramatically affected [5] by the presence of impurities, such as sulphur.

In this note, we shall attempt to provide a mathematical description of laminar Marangoni flows, assuming high Reynolds number ($Re \gg 1$) and neglecting thermal effects.

M. Heiliö (ed.), Proceedings of the Fifth European Conference on Mathematics in Industry, 381–385.
© 1991 B. G. Teubner Stuttgart and Kluwer Academic Publishers. Printed in the Netherlands.

§2 Formulation

We assume that the flow is driven in the direction OB (fig. 2) by a constant shear stress, τ, at OB and that BA is a no-slip boundary; in addition, OA is a symmetry plane at which there is zero shear stress, and there is no flow out of the region OAB. Since $Re \gg 1$, we expect a core region where the vorticity (ω_0) is constant (by the Prandtl-Batchelor theorem), boundary layers, all of thickness, $Re^{-\frac{1}{2}}$, on OA, OB and BA and corner layers at O, A and B. A balance between the driving force in the Marangoni shear layer at OB and the frictional force at BA determines the appropriate velocity scale, U, for the flow and the value of ω_0:

$$U = \left(\frac{\tau^2 a}{\mu\rho}\right)^{\frac{1}{3}}, \qquad (1)$$

where a is the pool radius, μ is the coefficient of viscosity and ρ is the liquid metal density;

$$\omega_0 = \left(\int_{BA}(\sin s - \cos s)\Omega_0\,ds\right)^{-\frac{2}{3}}, \qquad (2)$$

where Ω_0 denotes the vorticity (appropriately scaled) at BA.

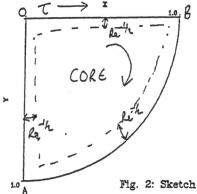

Fig. 2: Sketch of flow structure

The stream function, ψ, for the core flow is then given by

$$\nabla^2\psi = -\omega_0, \qquad (3)$$

subject to

$$\psi = 0 \quad \text{on} \quad OAB; \qquad (4)$$

ω_0 is found by solving the boundary-layer equations numerically by means of the Crank-Nicholson method, using the core solution for the boundary condition at the outer edge of the boundary layer.

§3 Modification

On integrating, the boundary-layer equations exhibit separation. This is not too surprising: the development of the externally imposed adverse pressure gradient as the flow proceeds along BA, resulting in the slowing down of the fluid motion, means that the fluid can no longer remain attached to the wall.

We are forced to modify the model, by allowing for the possibility of more than one recirculating region. The resulting model (fig. 3), obtained after the appropriate asymptotic analysis [6], consists of a vigorous constant-vorticity flow in the upper part of the pool, where $\psi \sim O(1)$, and an approximately stagnant flow in the lower part, where $\psi \sim O(Re^{-\frac{1}{3}})$. The flows are separated by a dividing streamline (CD), across which the mass flux is zero and the normal stress is continuous; the latter condition reduces, at leading order, to Bernoulli's equation for a streamline, which then implies that the tangential velocity on CD is constant, and that separation (at D) and reattachment (at C) must be tangential.

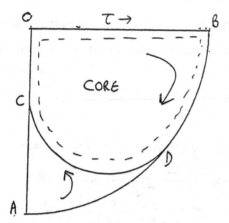

Fig. 3: Sketch of flow with dividing streamlines

This results in a free-boundary problem for CD; we require to solve (3), subject to (4), which now holds on $OCDB$, and the extra boundary condition on CD from Bernoulli's equation, which is necessary in order to determine the position of CD and the tangential velocity at CD. ω_0 may be found by using (2), except that the limits are now BD. Discussion of the position of the separation point D may be found in [6].

§4 Results

Fig. 4 shows the results obtained using a 32×32 mesh to solve (3) by means of SOR (successive overrelaxation) and a generalised *regula falsi* method [7]. The tangential velocity at CD was found to be -0.22 and the vorticity, ω_0, 2.11.

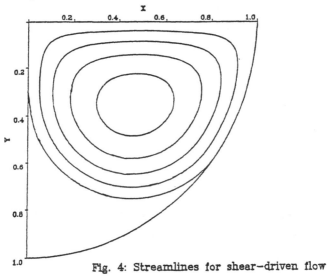

Fig. 4: Streamlines for shear-driven flow

In addition, a numerical solution of the full 2-d Navier Stokes equations with the boundary conditions discussed in §2 was obtained by means of the Galerkin formulation of the finite-element method. The plot shown (fig. 5) was obtained at $Re = 200$.

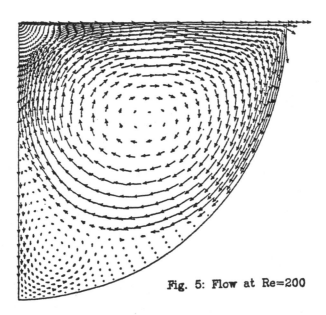

Fig. 5: Flow at Re=200

§5 Conclusions

In this account, two different methods have been used to determine the structure of laminar Marangoni flow in a weld pool for large values of Re. In both cases, boundary-layer separation is present and the recirculating flow in the upper part of the pool is more vigorous than the flow in the lower part.

Further extensions to the work will involve the solution of the fully-coupled problem, where the driving shear stress is no longer constant, but proportional to the streamwise temperature gradient. These extensions and a fuller account of the above work may be found in [6].

References

1. Rosenthal, D.: 1946 The theory of moving sources of heat and its applications to metal treatments. *Trans. ASME.* **68**, 849-866.

2. Kublanov, V. and Erokhin, A.: 1974 On metal motion in a stationary weld pool under the action of electromagnetic forces and gas flow velocity head in arc welding. Doc. No. 212-318-74, 1-12.

3. Atthey, D. R.: 1980 A mathematical model for fluid flow in a weld pool at high currents. *J. Fluid Mech.* **98**,787-901.

4. Oreper, G. M. and Szekely, J.: 1984 Heat- and fluid-flow phenomena in weld pools. *J. Fluid Mech.* **147**, 53-79.

5. Burgardt, P. and Heiple, C. R.: 1986 Interaction between impurities and GTA weld shape. *Weld J.* **65** *Res. Suppl.* 150-s to 155-s.

6. Vynnycky, M.: 1990 D. Phil Thesis, University of Oxford (to appear).

7. Fox, L. and Sankar, R.: 1973 The Regula-Falsi method for free-boundary problems. *J. Inst. Math. Appl.* **12**, 49-54.

M. Vynnycky
Mathematical Institute
24-29 St. Giles'
Oxford
OX1 3LB
United Kingdom.

SOLUTION OF FREE BOUNDARY PROBLEMS IN MOULD FILLING

W. WEINELT, K.-H. HARTWIG,
H. LIEBERMANN, J. STEINBACH
Sektion Mathematik
Technische Universität
PSF 964
9010 Chemnitz
BRD

ABSTRACT. For the production of plastic articles by injection moulding the computation of solutions of corresponding sufficiently efficient mathematical models plays an increasing role.

The process is, generally speaking, characterized by a "complex" physical situation in a geometrically "difficult" domain. Starting from a simplificated version of the mathematical model, it is desirable to improve both the physical and the geometric description of the process (fig. 1). By means of accepted versions of the model it is possible to control the construction of the injection moulds.

A survey concerning the theoretical foundation of plastics processing is given in [1]. A mathematical formulation of the mould filling processes is contained, for instance, in [2] and [3]. The basis of this formulation consists in the so-called Hele-Shaw approximation of viscous flows in flat moulds, and the mathematical problems arising from this approximation are free boundary problems in two dimensions.

This paper is devoted to a short summary of our studies in mathematical modelling of this injection moulding process (we refer especially to [4], [5] and [6]).

1. Hele-Shaw flow within a flat cavity

We consider flow motions in three-dimensional domains characterized by a thickness function $2d$ defined on the midsurface Ω (see fig. 2). The side view of this cavity is schematically shown in the upper part of figure 2. The places A, where the plastic melt streams from the gate into the cavity, are called the fill orifices (Ω_o in the ground-plan of fig. 2). In practice we can assume that the air pressure in the cavity is always the same and equal to the atmospheric pressure $p_o = 0$, because of the air porosity of the cavity at the side Z.

At every time $t \in (0,T)$ (T - the filling time) the filling level is characterized by the domain $\Omega_t \subset \Omega$ with the (free) boundary $\Gamma_o(t) \subset \Omega$ (the flow front).

M. Heiliö (ed.), Proceedings of the Fifth European Conference on Mathematics in Industry, 387–392.
© 1991 B. G. Teubner Stuttgart and Kluwer Academic Publishers. Printed in the Netherlands.

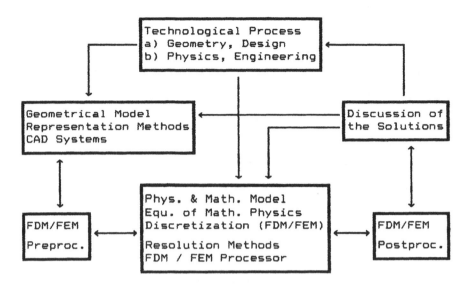

Figure 1. Mathematical Modelling of a Technological Process.

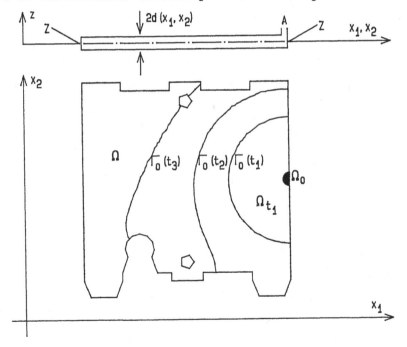

Figure 2. Side view and ground-plan of a cavity. There are shown some flow fronts.

The model is based on neglecting external and inertial forces, assuming the pressure to be $p=p(x_1,x_2,t)$, and $v_3=0$, $v_i=v_i(z,t)$, $i=1,2$. Moreover the fluid is supposed to be Newtonian, with viscosity η, incompressible, and the cavity is symmetric with respect to $z=0$. Then, starting from the impulse equation of the Navier-Stokes equations, one obtains the following relation between then mean velocity

$$\vec{v}=(\overline{v}_1,\overline{v}_2), \quad \overline{v}_i = d^{-1}\int_0^d v_i(x_1,x_2,z,t)\,dz, \quad i = 1,2$$

and the pressure gradient

$$\vec{v} = -\frac{k}{2d}\,\text{grad}\,p, \quad k= 2\int_0^d \frac{z^2}{\eta}\,dz.$$

From the equation of continuity then the elliptic differential equation

(1) $$- \text{div}\,(k\,\text{grad}\,p) = 2\,d\,v\,\chi(\Omega_o) \quad \text{in}\;\Omega_t$$

is obtained, where the given source term at the fill orifice Ω_o is denoted by v.

The boundary conditions are

$$p = 0 \quad \text{and} \quad \frac{\partial p}{\partial n} = -\frac{k}{2d}\,\overline{v}_n \quad \text{on}\;\Gamma_o(t),$$

(2) $$\frac{\partial p}{\partial n} + \alpha\,p = 0 \quad \text{on}\;\partial\Omega_t \cap \Gamma,$$

where $0 < \alpha \ll 1$ is a regularization parameter in the real boundary condition $\partial p/\partial n = 0$ describing the zero flux $\overline{v}_n = 0$ on $\partial\Omega_t \cap \Gamma$.

2. Variational inequality

Setting $p \equiv p_o = 0$ in $\Omega \setminus \Omega_t$ (in accordance with the physical behaviour discussed above) and using the Baiocchi transformation

$$u(x_1,x_2,t) = \int_0^t p(x_1,x_2,t')\,dt'$$

we get from the boundary problem (1), (2) the variational inequality for

$$u \in K = \{ \ v \in \overset{\circ}{W}_2^1(\Omega), \ v \geq 0 \text{ in } \Omega \ \} :$$

$$a(u, v - u) \geq \ < \Phi, \ v - u > \ \forall \ v \in K,$$

(3) $$a(v, w) = \int_\Omega (k \text{ grad } v, \text{ grad } w) \ dx + \alpha \int_\Gamma k \ v \ w \ ds,$$

$$<\Phi, \ v> = \int_\Omega f \ v \ dx,$$

$$f = - 2d \ (1 - \chi(\Omega_0)) + 2d \ \chi(\Omega_0) \int_0^t v \ dt'.$$

This problem (3) is uniquely solvable under convenient conditions on the data.

3. F D M — approximation and resolution methods

We use triangular grids (see [7] or [8]) and approximate the differential operator $L u = -$ div (k grad u) and the boundary conditions by a difference operator Λy using the integral balance method ([9], y denotes the grid function). This together with an approximation Φ_h of Φ yields a discrete variational inequality that can be formulated as

(4) $$\Lambda y \geq \Phi_h, \quad y \geq 0, \quad y (\Lambda y - \Phi_h) = 0 \ .$$

Under convenient conditions the order of convergence is of $O(h)$.
   For solving problems (4) we use the penalty method

(5) $$\Lambda y + \varepsilon^{-1} y^- = \Phi_h \ , \ \varepsilon > 0, \ y^- = \min \ \{0, \ y\}$$

that gives rise to an additional error of the order $O(\varepsilon)$. The nonlinear equations (5) then are solved iteratively and finally the M A F for large linear systems ([10]) is used.
   For a special moulded article figures 2 – 4 show single flow fronts, the field of flow directions together with some isobares and a pressure representation, respectively.

4. Generalizations

Generalizations concern the treatment of
   (i)   mouldings with curved midsurfaces ([11]),
   (ii)   composed mouldings ([12]) and
   (iii) plastic melts with a time – dependent viscosity
          $\eta = \eta(\vartheta(x, t))$, ϑ – temperature ([13], [14]).
Figure 5 demonstrates the filling process for a composed moulding (a "box" with 6 pigeon-holes). In this numerical experiment the orifices are located at 4 corners of the box.

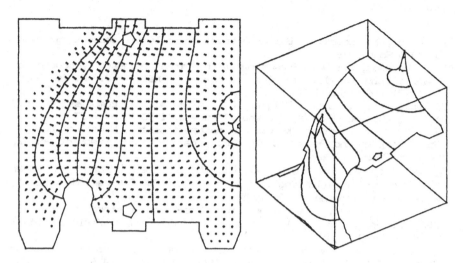

Figure 3. Field of flow directions and some isobares of the pressure field at a given time.

Figure 4. Pressure representation at a given time.

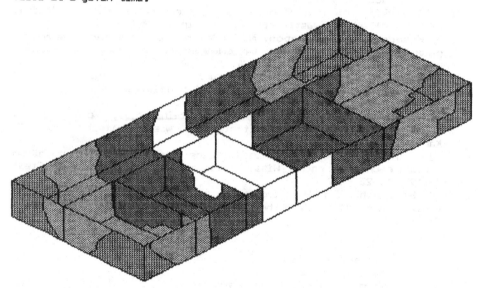

Figure 5. Filling process for a composed moulding: filling zones and flow fronts, 4 time steps. The orifices are located at the 4 lower corners of the box.

References

[1] Tadmor, Z., Gogos, C. G.: Principles of polymer processing. John
    Wiley & Sons 1979
[2] Richardson, S.: Hele-Shaw flows with a free boundary produced by
    the injection of fluid into a narrow chanel. J. Fluid Mech.
    56(1972)4, 609 - 618
[3] Tucker (III), C.L., Folgar, F.: A model of compression mold
    filling. Polymer Engineering and Science 23(1983)2, 69 - 73
[4] Weinelt, W., Hartwig, K.-H.: Zur mathematischen Modellierung des
    Spritzgießens und Pressens. TH Karl - Marx - Stadt, Preprint Nr. 2,
    Februar 1986
[5] Weinelt, W., Hartwig, K.-H.: Numerische Lösung linearer
    elliptischer Variationsungleichungen mittels FDM. TH Karl-Marx-Stadt,
    Wiss. Schriftenreihe Nr. 8, 1985, 1 - 77
[6] Hartwig, K.-H., Liebermann, H.: Pre- und Postprocessing für
    dünnwandige Bauteile. Wiss. Zeitschr. der TU Karl-Marx-Stadt
    31(1989)1, 60 - 65
[7] Lo, S. H.: A new mesh generation scheme for arbitrary planar
    domains. Int. J. Numer. Meth. Eng. 21(1985), 1403 - 1426
[8] Cavendish, J.C., Field, D.A., Frey, W.H.: An approach to automatic
    three-dimensional finite element mesh generation. Int. J. Num. Meth.
    Eng. 21(1985), 329 - 347
[9] Samarskij, A. A.: Theorie der Differenzenverfahren. Akadem.
    Verl.-Gesellschaft Geest&Portig K.G., Leipzig 1984
[10] Кучеров, А.Б., Макаров, М.М.: Метод приближенной фактори-
    зации для решения разностных смешанных эллиптических краевых
    задач. Изд. Моск. Ун - та 1984, 55 - 64
[11] Weinelt, W.: Numerisches Experiment zum Spritzgießen.
    II. Berücksichtigung der Krümmung der Mittelflächen. Wiss. Zeitschr.
    der TU Karl-Marx-Stadt 31(1989)2, 270 - 275
[12] Hartwig, K.-H., Liebermann, H., Steinbach,J., Weinelt, W.:
    Numerisches Experiment zum Spritzgießen I. Wiss. Zeitschr. der TU
    Karl-Marx-Stadt 29(1987)2, 204 - 208
[13] Steinbach, J.: Existenzsätze für eine Klasse von evolutionären
    Variationsungleichungen. Wiss. Zeitschr. der TU Karl-Marx-Stadt
    30(1988)1, 22 - 26
[14] Steinbach, J.: Untersuchungen zu evolutionären Variations-
    ungleichungen. Wiss. Zeitschr. der TU Karl-Marx-Stadt 31(1989)2,
    257 - 264

Wilfried Weinelt et al.
Sektion Mathematik
Technische Universität Chemnitz
PSF 964
9010 Chemnitz
BRD

MULTI-FIBRE EXTRUSION - A NONLINEAR INTEGRAL EQUATION

P. Wilmott

1) Introduction: We consider the extrusion of fibres from an infinite
bath of Newtonian viscous fluid (see fig.1). The fibres are under
tension with the speed of each
fibre being prescribed at a
fixed distance from the bath.
We present a model in which
the mass flow between fibres
is coupled. We investigate
the effect of pulling one
group of fibres faster than
its neighbours. The outlets
are assumed to be sufficiently

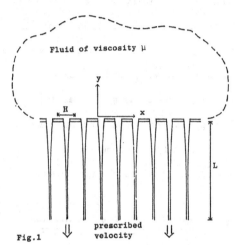

Fig.1

numerous to be considered as a continuous distribution of mass sinks
along the outlet face.

2). Coupled flow model: We separate the fluid flow into three distinct
regions as follows. Cartesian axes are shown in fig.1 with the x-axis
being the outlet face. Although this problem is three-dimensional, we
shall assume that variations only exist in the x and y directions (the
z direction is thus not shown in fig.1).

(i) Bath flow - the bath is assumed to be of infinite extent with
outlets along the plane y = 0 close together compared with the length
of the fibres. The flow in the bath can then be modelled by a
continuously distributed mass sink.

393

M. Heiliö (ed.), Proceedings of the Fifth European Conference on Mathematics in Industry, 393–396.
© 1991 B. G. Teubner Stuttgart and Kluwer Academic Publishers. Printed in the Netherlands.

(ii) Fibre flow - here the geometry is sufficiently slender for the extensional model, often used for glass and textile fibres, to be applied [1]. We take the outlets to be circular so that the fibre cross section remains circular also.

(iii) Jet flow- this is the region close to an outlet, that is, within a lengthscale of the order of the fibre radius. This region presents us with difficulties which we overcome by making two modelling assumptions. To treat this region accurately would detract from the simplicity of the model without adding to the insight gained. Let us consider each region in detail.

3) Bath flow: Within the bath (y>0) the equations of motion for the fluid velocity, q, and the pressure, p, are those of Stokes flow with boundary conditions $p \to p_0$ as $|x| \to \infty$ and $q = (0,-w(x))$ on $y = 0 +$. Here p_0 is the prescribed pressure at infinity and $w(x)$ the unknown normal speed on the outlet face. The normal stress on the $y = 0 +$ plane is related to the normal velocity by

$$s_b(x) = - p_0 + \frac{2\mu}{\pi} \int_{-\infty}^{\infty} \frac{w'(\xi)d\xi}{x-\xi} . \tag{1}$$

4). Fibre flow: We can exploit the slender geometry as in [1] to find that to leading order

$$3\mu \, Av_y = A_0{}^* s_f(x) \tag{2}$$

and $$\rho Av = - m(x). \tag{3}$$

In this extensional model, we have ignored surface tension so that the stresses at the fibre's free surface are zero. v is the negative extensional velocity of the fibre which is a function both of distance along the fibre (-y) and the fibre itself (x). A is the local cross-sectional area of the fibre, again a function of x and y, A_0^* is the area of the fibre as $y \to 0 -$, ρ is the fluid density and s_f the axial stress in the fibre, which the extensional theory shows is independent of distance along the fibre. m is the mass flux in the fibre, a function of x. Equation (2) is stress balance and equation (3) conservation of mass.

Equations (2) and (3) are trivially solved for v and we then impose the boundary condtion $v = - V(x)$ at $y = - L$ where $V(x)$ is the prescribed pulling speed.

Finally, local conservation of mass from bath to fibre gives $w(x) = m(x)/H^2 \rho$, where H is the distance between the centres of the outlets and we have taken the outlets to form a square array in the x-z plane.

5). <u>Jet flow</u>: Near the outlet the flow is fully three-dimensional and complicated by the presence of both fixed and free surfaces. In order to keep our model manageable we propose the following two assumptions to link together the flows in the bath and the fibres.

<u>Assumption (A)</u>: The fibre area for the extensional flow model can be taken to be the same as the area of the outlet nozzle, A_0 say. That is $A_0^* = A_0$. This, together with the solution of (2) and (3) leads to

$$s_f(x) = \frac{3\mu m}{\rho A_0 L} \ln \left[\max\left(\frac{A_0 V \rho}{m}, 1 \right) \right], \tag{4}$$

a relationship between the stress in the fibre and the mass flow. We include the maximum function since we expect that the fibre cannot go into compression.

Assumption (B): That the mass flow is proportional to the normal stress jump from the fibre to the bath i.e.

$$s_f(x) - s_b(x) = \frac{3\mu\alpha}{\rho LA_0} m(x),\qquad(5)$$

where α is a constant. This rule is suggested by Darcy's law.

6). The integral equation: Upon bringing together (1), (4) and (5) and nondimensionalizing with $x = 2LA_0\bar{x}/3\pi H^2$, $m(x) = \rho A_0 V_0 \bar{m}(\bar{x})$ and $V(x) = V_0\bar{V}(\bar{x})$ where V_0 is a typical value of V we find that

$$m \ln\left[\min\left(\frac{m}{V},1\right)\right] + \int_{-\infty}^{\infty} \frac{m'(\xi)d\xi}{x-\xi} = \beta - \alpha m,\qquad(6)$$

where $\beta = Lp_0/3V_0\mu$ and we have dropped overbars.

In general equation (6) must be solved numerically. However, we can solve a linearized version in the case when the pulling speed is a small perturbation to uniform i.e. $V = 1 + V_1(x)$ where $V_1(x)$ is prescribed and $V_1 \ll 1$. In this case $m = m^* + m_1(x)$ where $m^*\ln m^* = \beta - \alpha m^*$ and

$$m_1 = \frac{m^*}{D} \int_{-\infty}^{\infty} V_1(x-\xi)K(\pi\xi/D)d\xi,\qquad(7)$$

where $D = \ln m^* + 1 + \alpha$ and $K(x)$ has Fourier transform $(1+|s|)^{-1}$. Since the kernel in (7) is everywhere positive we have shown (at least in the linearized case) that an increase in the pulling speed of any group of fibres will lead to a global increase in mass flux.

Acknowledgement: This work was undertaken jointly by the author and E.L. Terrill.

Reference:

[1]   Dewynne, J., Ockendon, J.R. & Wilmott, P. 1989. On a mathematical model for fiber tapering. SIAM J. Appl. Math. 49, 989-990.

Mathematical Institute,
University of Oxford,
24-29 St Giles, Oxford OX1 3LB.

MODELLING THE THERMISTOR

A.S. Wood

Abstract. In this paper we outline a preliminary investigation into a problem of heat and current flow arising from the thermistor. An isotherm migration formulation is used and an efficient numerical simulation is proposed, based on finite differences. Numerical results are given which show that this approach is an effective tool for dealing with thermo-electric device simulation.

The thermistor is a thermo-electric device that may be used as a current-surge protector. As the device is heated, its thermally dependent electrical conductivity drops by several orders of magnitude over a relatively small temperature range. Cracks have been observed to appear on the surface of these devices during their working life (which may degrade their operational ability) and, using numerical simulation techniques, it is the purpose of this investigation to assertain whether this defect might be caused by high temperature gradients. The numerical experiments are based on a simple one-dimensional model in which the current flow (potential) is known analytically and that has an exact steady-state solution [*Westbrook, 1989*].

Typically the device has a flat cylindrical geometry with connections, at the top and bottom circular faces, to an external circuit. It is electrically insulated on its vertical sides. If these sides are also thermally insulated then a one-dimensional problem in temperature T and potential ϕ results which can be described by the following model, in which symmetry and scaling have been taken into account:

$$\frac{\partial T}{\partial t} = \frac{\partial^2 T}{\partial y^2} + \alpha \sigma(T)\left(\frac{\partial \phi}{\partial y}\right)^2, \quad 0 < y < 1; \quad \frac{\partial T}{\partial y} = 0, \quad y = 0; \quad \frac{\partial T}{\partial y} + \beta_1 T = 0, \quad y = 1,$$

$$(1)$$

and

$$\frac{\partial}{\partial y}\left(\sigma(T)\frac{\partial \phi}{\partial y}\right) = 0, \quad 0 < y < 1; \quad \phi = 0, \quad y = 0; \quad \phi = 1, \quad y = 1,$$ (2)

in which β_1 is a positive heat transfer co-efficient and α represents the ratio of internal heat generation to heat diffusion. The electrical conductivity is modelled by the step function

$$\sigma(T) = \begin{cases} 1, & T \leq 1, \\ \delta, & T > 1, \end{cases}$$

in which, typically, $\delta = 10^{-5}$. Consequently, there may exist a *cold* region of high electrical conductivity separated, by a moving interface $s(t)$, from a *hot* region of low electrical conductivity. The expected time sequence of phases is *cold*, *warm* (with *hot* and *cold* regions) and *hot* with the steady-state situation occuring in one of the phases. Diesselhorst [*1900*] and Cimatti [*1989*] give proofs for the existence and uniqueness of solutions. In the case of Dirichlet boundary conditions Cimatti [*1988*] obtains an upper bound for the temperature and Howison [*1988*] provides some order results on the size of the hot region. Howison [*1989*] shows that these results are essentially independent of the geometry of the device. The particular steady-state phase is determined by the values of α, β_1 and δ (see, for example, Westbrook [*1989*]). The exact solution to the potential problem

397

M. Heiliö (ed.), Proceedings of the Fifth European Conference on Mathematics in Industry, 397–400.

(2) is available, namely $\phi = y$, $0 \le y \le 1$, during the *cold* and *hot* phases and, during the *warm* phase,

$$\phi = \begin{cases} (\overline{\phi}y)/s, & 0 \le y \le s, \\ ((1-\overline{\phi})y + \overline{\phi} - s)/(1-s), & s \le y \le 1, \end{cases}$$

in which $\overline{\phi}$ is the potential at $y = s$. Our interest centres on simulating the movement of the interface and the development of the temperature distributions in the *hot* and *cold* regions (when they exist).

The steady-state solutions indicate a very small $O(\delta)$ temperature range in the *hot* region and an $O(1)$ range in the *cold* region, at least while both regions exist. Thus, numerical solutions based on finite differences are unlikely to be sufficiently sensitive to describe the detail in the *hot* region, unless very fine grid sizes are used (which is computationally expensive). For this reason the temperature problem (1) is rewritten using the isotherm migration (*IMM*) formulation [*Dix & Cizek, 1970*] in which the rôles of the dependent (temperature) and independent (spatial) variables are reversed. With little effort we obtain

$$\frac{\partial y}{\partial t} = \left(\frac{\partial y}{\partial T}\right)^{-2} \frac{\partial^2 y}{\partial T^2} - \alpha\sigma(T)\frac{\partial y}{\partial T}\left(\frac{\partial y}{\partial \phi}\right)^{-2} \tag{3}$$

subject to appropriate initial and boundary conditions. Using the exact solution for the potential ϕ, the second term on the right-hand side of (3) takes the form $-\alpha$ and $-\alpha\delta$ for the *cold* ($s = 0$) and *hot* ($s = 1$) phases, respectively. For the *warm* phase it has the form

$$-\alpha\delta\left(\frac{\overline{\phi}}{s}\right)^2, \quad 0 < y < s; \qquad -\alpha\left(\frac{1-\overline{\phi}}{1-s}\right)^2, \quad s < y < 1.$$

At $y = 0$ the term $\partial y/\partial T$ becomes unbounded due to the boundary condition shown in (1). Thus, in a neighbourhood of $y = 0$ we would not expect a numerical scheme to perform particularly well. Additionally, the *IMM* approach relies on the fact that the range of valid temperatures remains fixed, otherwise the location of certain isotherms that are being tracked may become undefined. If $T_0 = T(0,t)$ and $T_1 = T(1,t)$ then the change of variable

$$U = \frac{T - T_0}{T_1 - T_0} + \overline{\omega}y$$

fixes the boundary temperatures to $U(0,t) = 0$ and $U(1,t) = 1 + \overline{\omega}$ and introduces a non-zero flux $\overline{\omega}$ at $y = 0$. Equation (3) is now

$$\frac{\partial y}{\partial t} = \left(\frac{\partial y}{\partial U}\right)^{-2} \frac{\partial^2 y}{\partial U^2} - \frac{\alpha\sigma(U)}{T_1 - T_0}\frac{\partial y}{\partial U}\left(\frac{\partial y}{\partial \phi}\right)^{-2}. \tag{4}$$

We solve (4) using finite difference methods and the most straightforward approach uses the fully explicit algorithm

$$y_i^{m+1} = y_i^m + 4k\frac{y_{i+1}^m - 2y_i^m + y_{i-1}^m}{\left(y_{i+1}^m - y_{i-1}^m\right)^2} - k\frac{\alpha\sigma(U_i)}{T_1^m - T_0^m}\frac{y_{i+1}^m - y_{i-1}^m}{2\Delta U}\left(\frac{\partial y}{\partial \phi}\right)^{-2}\Bigg|_{t=t_m} \tag{5}$$

in which $y_i^m \sim y(U_i, t_m)$, ΔU is the temperature grid size and k is the time step. From stability considerations this scheme has identifiable restrictions on the range of valid time steps which makes the process computationally inefficient. This can be overcome by applying thecideas of Gourlay's [1970] hopscotch paper. At each time step, (5) is first applied at nodes (U_i, t_m) for which $i + m$ is **odd**. Then, at the same time level, the values of y_i^{m+1} at the intermeadiate nodes (for which $i + m$ is **even**) are predicted using the fully implicit replacement for (4) - the second and third terms on the righthand side of (5) have the superscript m replaced by $m + 1$. Since the values $y_{i\pm1}^{m+1}$ will be known from the application of (5) we can write the implicit algorithm in the *explicit* form

$$y_i^{m+1} = \frac{(y_{i+1}^{m+1} - y_{i-1}^{m+1})^2 y_i^m + 4k(y_{i+1}^{m+1} + y_{i-1}^{m+1}) - k\frac{\alpha\sigma(U_i)(y_{i+1}^{m+1} - y_{i-1}^{m+1})^3}{2(T_1^{m+1} - T_0^{m+1})\Delta U}}{(y_{i+1}^{m+1} - y_{i-1}^{m+1})^2 + 8k}. \qquad (6)$$

T_0 and T_1 are advanced by using a central difference form of equation (4) at $y = 0$ and $y = 1$, respectively, which incorporates a difference replacement for the appropriate boundary condition shown in (1).

Experiments were conducted with 10 specified isotherms and the parameter values $\alpha = 0.1$, $\beta_1 = 0.2$ and $\delta = 10^{-5}$ (indicating a *cold* steady-state solution). The process was initialised at $t = 0.1$ using the exact solution to the *cold* phase of (1), namely

$$T(y, t) = \alpha\left(\frac{1}{2} + \frac{1}{\beta_1}\right) - \frac{\alpha y^2}{2} - 2\alpha\beta_1 \sum_{n=1}^{\infty} \frac{\cos(\lambda_n y)}{\lambda_n^2(\lambda_n^2 + \beta_1^2 - \beta_1)\cos\lambda_n} e^{-\lambda_n^2 t}, \quad 0 \le y \le 1, \quad (7)$$

in which the λ_n are the *positive* roots of $\lambda\tan\lambda - \beta_1 = 0$.

The steady-state solution was well represented by the numerical process, a (numerically) steady-state situation being attained when the rate of change of heat content in $0 \le y \le 1$ fell below a certain tolerance (in this case 10^{-4}). At this time ($t = 27.9$) a weighted 1-norm measure of the relative error distribution of the isotherm locations was 4.8×10^{-5} indicating excellent agreement.

Moving to a value of $\alpha = 0.5$ puts the steady-state solution into the *warm* phase. A similar accuracy (as above) at the steady-state was obtained. The progress of the numerical solution was checked at the end of the *cold* phase, using (7), at which time the point $y = 0$ has just reached the critical temperature ($T = 1$) above which the electrical conductivity will drop. At this time ($t = 2.414$) the weighted 1-norm error measure (as above) was 5.5×10^{-5} and that of the actual isotherm values was 4.7×10^{-4}. Again, the numerical process is performing very accurately.

Figure 1 shows the movement (location and speed) of the interface s during the *warm* phase, indicating good agreement with the analytic value $s(\infty) = 1.9 \times 10^{-5}$. At no time were steep temperature gradients observed.

Finally, a word on the amount of computation required. In the above experiments a time step of $k = 0.005$ was used. Thus, for the second experiment, about 463 time steps were required to reach the end of the *cold* phase. The simulation was repeated using the fully explicit algorithm alone with a heuristic time step control designed to ensure the correct physical movement of the isotherms. This scheme produced results comparable to those indicated above, but required 840 steps to reach the end of the *cold* phase.

We conclude from this *simple* model that the observed cracking of operational thermistors does not appear to be due to large temperature gradients which agrees with the conclusions of Veen [1989]. Thus, the cracking is, perhaps, just a function of the quality control in force during the manufacture of the devices although Fowler & Howison [1990] discuss a circuit which, under certain conditions, can produce rapid temperature surges in the thermistor. However, we have demonstrated that the *IMM* approach can be used effectively in problems involving different temperature scales and that the hopscotch scheme is an efficient solver for the resulting model.

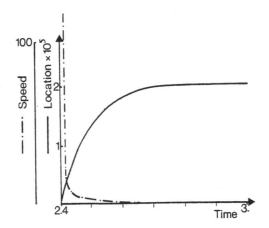

<u>Figure 1:</u>   Location and speed of the interface s(t)

References

Cimatti, G. (1988) A bound for the temperature in the thermistor problem, *IMA J. Appl. Math.*, **40**, 15-22.

Cimatti, G. (1989) Remark on existence and uniqueness for the thermistor problem under mixed boundary conditions, *Quart. Appl. Math.*, **47**, 117-121.

Diesselhorst, H. (1900) Über das Probleme eines elektrisch erwärmter Leiters, *Ann. Phy.*, **1**, 312-325.

Dix, R.C. & Cizek, J. (1970) The isotherm migration method for transient heat conduction analysis, In *Heat Transfer* **1** (*4th Int. Heat Transfer Conf., Paris*), Elsevier, Amsterdam.

Fowler, A.C. & Howison, S.D. (1990) Temperature surges in thermistors, In *Proc. 3rd Euro. Conf. Maths. in Industry* (Glasgow, 1988), J. Manley et al [Eds.], Kluwer Academic, Teubner.

Gourlay, A.R. (1970) Hopscotch: A fast second-order partial differential equation solver, *J. Inst. Maths. Applics.*, **6**, 375-390.

Howison, S.D. (1988) Complex variables in industrial mathematics, In *Proc. 2nd Euro. Symp. Maths. in Industry* (Oberwolfach, 1987), H. Neunzert [Ed.]. Kluwer Academic, Teubner.

Howison, S.D. (1989) A note on the thermistor problem in two dimensions, *Quart. Appl. Math.*, **47**, 509-512.

Venn, M. van der (1989) Thermistor cracking, Tech. Univ. Eindhoven, *OWE Report*, **89-02**.

Westbrook, D.R. (1989) The thermistor: A problem in heat and current flow, *Numer. Meths. PDEs*, **5**, 259-273.

A.S. Wood

University of Bradford, Dept. of Mathematics, Bradford, West Yorkshire BD7 1DP, U.K.

Printed in the United States
By Bookmasters